THE USES OF EXPERIMENT

STUDIES IN THE NATURAL SCIENCES

EDITED BY

DAVID GOODING
University of Bath

TREVOR PINCH
University of York

SIMON SCHAFFER
University of Cambridge

CAMBRIDGE
UNIVERSITY PRESS

Published by the Press Syndicate of the University of Cambridge
The Pitt Building, Trumpington Street, Cambridge CB2 1RP
40 West 20th Street, New York, NY 10011–4211, USA
10 Stamford Road, Oakleigh, Melbourne 3166, Australia

First published 1989
Reprinted 1993

British Library cataloguing in publication data

The uses of experiment : studies of
experimentation in the natural sciences
1. Science. Methodology – Philosophical
perspectives
I. Gooding, David, 1947– II. Pinch, J.
(Trevor J.) III. Schaffer, Simon, 1955–
501'.8

Library of Congress cataloguing in publication data

The uses of experiment : studies in the natural sciences / edited by
David Gooding, Trevor Pinch, Simon Schaffer.
p. cm.
Bibliography: p.
Includes indexes.
ISBN 0 521 33185 4 ISBN 0 521 33768 2 (paperback)
1. Science–Experiments–Philosophy. 2. Science–Methodology–
Case studies. I. Goding, David, 1947–   . II. Pinch, T.J. 1952–
III. Schaffer, Simon. 1955–
Q182.3.U74 1989
507 .2–dc1988-11630 CIP

Transferred to digital printing 1999

Van Helsing: 'He is experimenting, and doing it well.'
Harker: 'But how is he experimenting? The knowledge may help us to defeat him!'

Bram Stoker, *Dracula* (1897)

CONTENTS

CONTRIBUTORS

ALEXI ASSMUS is a Mellon Fellow at Harvard University, working towards a Ph.D in the history of modern physics. She received her B.A. in history at Stanford University where she began research with Peter Galison on the cloud chamber. She is interested in twentieth-century American physics, in particular the rise of theoretical physics early this century.
Address: Department of the History of Science, Harvard University, Science Center 235, Cambridge, Mass. 02138, USA.

JIM BENNETT is Curator of the Whipple Museum of the History of Science in Cambridge. He is author of *The Mathematical Science of Christopher Wren* (Cambridge, 1982) and *The Divided Circle* (Phaidon/Christies, 1987).
Address: Whipple Museum of the History of Science, Free School Lane, Cambridge, CB2 3RH, England.

GEOFFREY CANTOR lectures in history and philosophy of science at the University of Leeds. He is the author of *Optics after Newton* (Manchester, 1983) and co-editor of *Conceptions of Ether* (Cambridge, 1981), *The Figural and the Literal* (Croom Helm, 1987), and *The Companion to the History of Science* (Croom Helm, 1988). He was recently a Gifford Fellow at the University of Glasgow, preparing a book on Michael Faraday and the Sandemanians.
Address: Department of Philosophy, University of Leeds, Leeds LS2 9JT, England

ALLAN FRANKLIN is Professor of Physics at the University of Colorado. Although trained as an experimental high-energy physicist he has worked on the history and philosophy of science for the last dozen years. His book *The Neglect of Experiment* was published in 1986 by Cambridge University Press.

ix

Address: Department of Physics, University of Colorado at Boulder, Campus Box 390, Boulder, Colorado 803090, USA.

PETER GALISON is Associate Professor at Stanford University in the Department of Philosophy, Department of Physics, and Program in the History of Science. He is the author of *How Experiments End* (Chicago, 1987) and is working on two projects: a history of the transition between wartime and postwar physics in the United States and a book, *Image and Logic: Instrument and Arguments in 20th Century Physics*. Professor Galison is the recipient of the NSF Presidential Young Investigator Award.
Address: Program in the History of Science, Building 200-031, Stanford University, Stanford, California 94305, USA.

DAVID GOODING lectures in history and philosophy of science at the University of Bath and is a Visiting Research Fellow of the Royal Institution Centre for the History of Science and Technology. His study of the experimental rendering of nature, *The Making of Meaning*, is published by Nijhoff/Kluwer. He is co-editor of *Faraday Rediscovered* (Macmillan/Stockton, 1985) and is writing a biography of Faraday, *Nature's Apprentice*.
Address: Science Studies Centre, School of Humanities and Social Sciences, University of Bath, Claverton Down, Bath BA2 7AY, England.

WILLEM HACKMANN is Assistant Curator at the Museum of the History of Science in Oxford and Fellow of Linacre College, Oxford. He is the author of *Electricity from Glass: the history of the frictional electrical machine 1600-1850* (Sijthoff and Noordhoff, 1978) and *Seek and Strike: sonar, anti-submarine warfare and the Royal Navy 1914-1954* (HMSO, 1984).
Address: Museum of the History of Science, Old Ashmolean Building, Broad Street, Oxford OX1 3AZ, England.

JOHN KRIGE has doctorates in physical chemistry and in philosophy. He is the author of *Science, Revolution and Discontinuity* (Harvester, 1980) and is presently based in Geneva as a member of a team writing a history of the European Organization for Nuclear Research (CERN). The first volume of the history was published by North-Holland in 1987.
Address: Study Team for CERN History, Building 54, C.E.R.N., CR-1211, Genève 23, Switzerland.

DONALD MACKENZIE lectures in sociology at the University of Edinburgh. He is the author of *Statistics in Britain 1865-1930* (Edinburgh, 1981) and

co-edited *The Social Shaping of Technology* (Open University Press, 1985).
Address: Department of Sociology, University of Edinburgh, 18 Buccleuch Place,
Edinburgh EH8 9LN, Scotland.

RON NAYLOR is Principal Lecturer in the History and Philosophy of Science and Head of the Philosophy Division at Thames Polytechnic. He is also currently Course Director of the Division's Master's Degree in Modern European Thought. He is completing a book on Galileo's mechanics.
Address: School of Humanities, Thames Polytechnic, Wellington Street, London SE18 6PF, England.

THOMAS NICKLES has taught at the University of Illinois (Urbana) and at the University of Chicago. He is now Foundation Professor of Philosophy at the University of Nevada, Reno. He has published papers on explanation, intertheoretic reduction, scientific discovery and problem-solving. He is the editor of *Scientific Discovery: Logic and Rationality* and *Scientific Discovery: Case Studies*, both published by Reidel in 1980.
Address: Department of Philosophy, College of Arts and Sciences, University of Nevada at Reno, Reno, Nevada 89557-0056, USA.

ANDY PICKERING is Associate Professor in the Department of Sociology and in the Program in Science, Technology and Society at the University of Illinois (Urbana–Champaign). He is the author of *Constructing Quarks: a sociological history of particle physics* (Chicago and Edinburgh, 1984). He is joint editor of the annual series *Knowledge and Society* and is writing a book about pragmatism and the sociology of science entitled *Making Sense of Science*.
Address: Department of Sociology, College of Liberal Arts and Sciences, University of Illinois at Urbana–Champaign, 326 Lincoln Hall, 702 South Wright Street, Urbana, Illinois 61801, USA.

TREVOR PINCH lectures in sociology at the University of York. He is author of *Confronting Nature: the Sociology of Solar Neutrino Detection* (Reidel, 1986) co-author (with Harry Collins) of *Frames of Meaning* (Routledge and Kegan Paul, 1982) and co-editor (with W. Bijker and Thomas Hughes) of *The Social Construction of Technological Systems* (MIT Press, 1987).
Address: Department of Sociology, University of York, Heslington, York YO1 5DD, England.

SIMON SCHAFFER lectures in history and philosophy of science at the University of Cambridge. He is co-author (with Steven Shapin) of *Leviathan and the Air Pump: Hobbes, Boyle and the Experimental Life* (Princeton, 1985) and co-editor of the journal *Science in Context*, published by Cambridge University Press.
Address: Department of History and Philosophy of Science, Free School Lane, Cambridge, CB2 3RH, England.

JIM SECORD is Lecturer in History of Science at Imperial College of Science and Technology, London. He received his Ph. D. from Princeton in 1981, and subsequently held research fellowships at University College London and Churchill College Cambridge. His interests focus on the social history of sciences and its audiences and the development of the life and earth sciences since 1750. His book, *Controversy in Victorian Geology: the Cambrian-Silurian Dispute* (Princeton) was published in 1986. He is currently studying early Victorian controversies about evolution, creation and materialism.
Address: History of Science & Technology Group, Imperial College, Exhibition Road, London SW7, England.

JOHN WORRALL is Reader in Philosophy of Science at the London School of Economics. Author of many articles in philosophy and history of science, he also edited Imre Lakatos' *Proofs and Refutations* and *Philosophical Papers*, I and II (Cambridge University Press).
Address: Department of Logic, Philosophy and Scientific Method, London School of Economics and Political Science, Houghton Street, London WC2A 2AE, England.

PREFACE

Experiment is a respected but neglected activity. It is a commonplace that experiment is one of the hallmarks of science. Since this bit of conventional wisdom happens to be true, it is all the more surprising that students of science have paid so little attention to how and why this particular activity has become so significant. Of course, the *results* of experiment – observations and data – are universally acknowledged to be important. These outcomes have received much attention from philosophers and the influence of experimental results is largely taken for granted by historians of science. Thus it is common to treat science as nothing more than sets of statements about how nature is. In such a picture, there is no place for the practical activity which gives these statements their power. So the *process* of experimentation is taken to be either unproblematic or uninteresting. This account of science licenses a neglect of the process by which meaning is made. The neglect of experiment is symptomatic of a prejudice against practical activity and in favour of speech acts. Epitomised by the linguistic turn in philosophy, this vision of western science seeks to preserve scientists' power over the world while distancing their reasonings from practical engagement with that world. That such a literary and cerebral activity might have such authority over our imagination and even our experience is not surprising, since many other aspects of our culture do so; we think it *is* surprising that it could also have power over nature.

Many students of the accomplishments of western science will find these claims contentious. Experiment generates results which do not appear to depend on any of the features peculiar to experimenters' practices. After all, one of the most striking attributes of natural science is that its statements have such general, if not universal significance and practical application in settings very distant from those in which

xiii

they were first constructed. We contend that the practicalities and par-
ticularities of experimental work are central to the understanding of
the success and power of what scientists do. Far from denigrating intellect, we argue that its power is not really appreciated unless we recognise
that it deals with the practicalities of a natural, material and social
world and not only with disembodied, self-evident phenomena.
Recognising experiment's significance even implies a new view of the
uses of theory.

Experiment is more interesting and significant than the received
stories about science imply. Study of experimental work shows that
the ways in which scientists' endeavours give rise to their statements
about nature help explain the content and influence of those statements. There are features of experiment which illuminate the means
scientists use to transform specific events in the laboratory into general
world-views. A key aspect of this process – also a perennial problem
of philosophy – is the move from the particular to the general. According
to one received view of science, the solution to this problem lies in
similarities between the structure of nature and the processes of the
mind. Such an account pauses only briefly to contemplate this version
of the problem of generalisation. More considered and informative
studies of experimental work show two results of importance. First,
the production of universals implies a co-ordination of particular experiences. Experimental work sets out to achieve this. Similarities between
such structures are not ready-made universals disclosed in experience
– they are made by people. The second finding is that creation and
construction are integral to what experiments show. The experimenter's
task is therefore the persuasion of others. The experimenter is never
alone with nature: there is always an audience, real or implied, which
must be addressed and persuaded that what one experimenter makes
is meaningful and important even in their very different circumstances.
Experimental products must make sense for these others, and this sense
is made successfully only if the original experimenter can enable the
meaning of a trial to transcend the space in which that trial is performed.
Experimental reasoning will try to get rid of any feature of a trial which
ties it to particular settings.

Is science so different from the received view? If so, then it is extraordinary that much philosophical ink has been spilt on a view of science
bearing little relation to actual practice. Here, we must acknowledge
that scientists' own accounts often support the received view in two
ways. First, their published accounts of their use of experiment show
that experiment speaks directly to theoretically well-formulated prob-

lems. Second, they seek to show that decisions between alternative theories seem to rest wholly on the verdict of experiment. In scientists' accounts, experiment appears to invoke nature as an independent judge of the questions scientists put to her. This is because their public accounts focus on results and the arguments for them. Given that the purpose of science is to transcend the particular, there are good reasons why scientists may ignore the vicissitudes of construction and discovery. It does not follow that a similarly restricted view of science is appropriate for disciplines with other objectives and responsiblities, such as philosophy, history or sociology. Students of science need to examine actual practice, not merely reconstructed practice.

Like a growing number of historians and sociologists, and not a few scientists, contributors to this volume want to show why experiment is more interesting and more important than traditionally realised. This importance warrants careful attention rather than honorific neglect. We focus on different uses of experiment as a way of drawing attention to the range and diversity of experimental activities since the sixteenth century. A glimpse of this diversity may be gathered from the rare but welcome focus on experimental practice in the studies collected by Conant in *Harvard Case Histories in Experimental Science* (1948) and by Harré in *Great Scientific Experiments* (1981). Experiment has many uses apart from supporting or refuting knowledge claims: active observation, invention, the construction of models, imitation of natural phenomena, or the design of instruments to extend the senses. Not the least important use is the provision of evidence for rival philosophies of science, even if this support usually relies on a cursory account of experimental work. Our case studies include Galilean mechanics, Newtonian optics, early Victorian electromagnetism, experiments on insects, on clouds and thunderstorms, on quarks and on the accuracy of nuclear missiles. This sample hardly exhausts the fields in which experiment matters. Generalisations across such a range are likely to be provisional and we do not imply that there is some essential and unchanging activity called experiment. Yet there are some important lessons about the way this kind of human activity has developed and the uses it serves. These include human agency and skill, the role of persuasion and of rhetoric, and the significance of the site of experiment and of instrumentation both to learning and persuasion.

There is a surprising degree of unanimity both in the analytical resources which our authors use to study experiment and the image of experiment which they develop. Tools drawn from disciplines such as historical epistemology, constructivist sociology and literary criticism are used here to imply an account which sees experiment as an active process of argument and persuasion: human agency in the production of agreement about the contents of nature becomes the principal locus of enquiry in the studies which follow. Yet those seeking a new revisionist orthodoxy in this collection of studies and reflections on the experimental life and its implications will not find it here. Within the limits imposed by their admirable restraint, contributors have developed rival arguments and the editors have been keen to encourage this debate. To aid this process we have grouped the chapters in five parts, each representing a theme central to understanding experiment: instrumentation, the deployment of experiment in written argument, the realisation and representation of phenomena, the constituencies which use and perform experiments and relationships between experimentation and testing for reliability. In our introduction, we offer a survey of the contents of this book and indicate some ways in which studies of experiment might develop in the future. The selected bibliography at the end of this volume was assembled by inviting contributors to nominate titles from chapter references which they considered to be of general importance. It contains few studies which predate 1960; even fewer from classical philosophy of science. No doubt a citation analyst is already at work, charting this feature of our referencing patterns. We are convinced that the scarcity of older work which seems helpful is a mark of welcome reorientations in science studies, not a consequence of the discipline's modish ignorance and neophilia.

This book arises out of a meeting at Bath in September 1985, one of the first conferences devoted to experimentation in the natural sciences. The organisers were surprised and overwhelmed by the response to initial soundings which showed that a large audience was concerned with the process of experiment. The book is not the proceedings of this conference (only 14 of the 39 papers given at the conference are published here). The collected papers have been substantially revised and, in some cases, such as the chapters by Geoffrey Cantor and Jim Bennett, we should signal that these are contributions explicitly commissioned by the editors as comments upon the other papers in their respective sections. Allan Franklin has revised a chapter from his recent book *The Neglect of Experiment* for inclusion here.

Such a project incurs many debts. This is an appropriate place to thank those other contributors and discussants who made the conference a success, and Joyce Brown and Eve Gonty, who ran the show. We also thank the British Society for the History of Science, British Society for the Philosophy of Science and the Science Studies Group of the British Sociological Association, who co-sponsored the meeting, and the Royal Society and the International Union for the History and Philosophy of Science, who provided funds towards travel expenses of some of the overseas contributors. Frank James, at the Royal Institution Centre for the History of Science and Technology, has offered hospitality and encouragement on numerous occasions. We are also very grateful for help, whether deliberate or unwitting, from our friends and colleagues Harry Collins, Steve Shapin, Jan Golinski, Alan Simpson, Owen Hannaway, Mike Dennis, Rom Harré and Pat Burdett. The following have generously permitted us to use illustrations and other material: Cambridge University Library, the Masters, Fellows and Scholars of Churchill College in the University of Cambridge, the British Museum (Natural History), the British Library, the Institution of Electrical Engineers, the Public Record Office (Kew), the Science and Engineering Research Council, Edinburgh University Library, the Edinburgh Meteorological Office, the Royal Society, the Staatsbibliothek Preussischer Kulturbesitz at Berlin and the Museo di Storia della Scienza, Florence. We express our gratitude to Susan Sternberg and her colleagues at Cambridge University Press, as well as to an anonymous referee for very helpful comments and criticism. The editors thank all the contributors to this book for their unstinting patience, forbearance and goodwill. Trevor Pinch and Simon Schaffer take this opportunity to indicate their gratitude to David Gooding, who has carried the mass of administration, coordination, and sheer hard work upon his shoulders. As a result, of course, all the errors are his responsibility alone.

David Gooding
Trevor Pinch
Simon Schaffer

Bath–York–Cambridge   October 1987

In the 1993 impression we have added to the select bibliography on p 467, a number of monographs and collections of studies of experiment in the natural sciences which have appeared since 1987.

INTRODUCTION: SOME USES OF EXPERIMENT

PART I: INSTRUMENTS IN EXPERIMENT

The development of experimental science has been accompanied by a spectacular growth in the range of instrumentation which experimenters use. It has become a commonplace to refer to the significance of this growth in the interpretation of experiment. This is where our examination of the uses of experiment begins, with an examination of the ways in which instruments function in the experimental workplace. The riches of scholarship in the history of instruments provides an indispensable resource for the analyst of experiment. Many contributors to this book pursue this approach: Jim Secord and David Gooding both describe electrical devices of British experimenters in the 1820s and 1830s; Peter Galison and Alexi Assmus document the development of C.T.R. Wilson's cloud chamber; John Krige examines the British decision to help fund the particle accelerators at CERN; Allan Franklin reports on a wide range of devices employed in contemporary physics. It is clear from these cases that the uses of experimental instruments are many and various.

Yet it is hard to construct a taxonomy which adequately charts this variety and displays the forms in which instrumentation appears in experiment. This is partly because the relative priority of instruments in the experimental enterprise seems problematic. Some cases suggest that the experimenter's tactics are dictated by instrumental capacity, and reliance on instrumental testimony seems a precondition of experimental endeavour. In this sense, the experimenter is constrained by the instruments used. Other cases, however, indicate the constructive role which the experimenter plays in making any device count as an instrument precisely by using it in a trial and working to establish its

1

reliability. Furthermore, several chapters in this volume attend to the process through which phenomena are realised in and by instrumental manipulation. In Gooding's account of the electromagnetic explorations by Michael Faraday and Peter Barlow, or the examination of C.T.R. Wilson's condensation physics by Galison and Assmus, the emphasis is placed upon the activity of the experimenter in establishing the character of an instrument and of the effects produced with it. Barlow's magnetic compass or Wilson's cloud chamber had no self-evident character: their users had to work hard to fix what that meaning might be, and this fixity was revised by others. This activity is prior to, and enables, experimental efforts to model, imitate and measure phenomena.

Marketing and the experimental network

The three chapters in Part I examine the active role of the experimenter in interaction with instrumentation and display the range of such interactions available in the history of experimentation. Willem Hackmann classifies the instruments of the early modern period with respect to this kind of activity. He discriminates between 'passive' devices used for measurement and observation and 'active' instruments such as the 'philosophical' air pump and electrical machine, whose purpose was to produce novel phenomena for the experimenter. This classification suggests further considerations of importance for the relation between instrument and experimenter. More than one community was involved in the production and use of instruments: communities of makers and salesmen interacted with experimenters in complex ways, and this market-place was a site of key importance for the establishment of the authority and repute of the various devices of experimental science. In his commentary on Hackmann, for example, Jim Bennett argues for the categories of the early modern instrument market, where divisions between 'active' and 'passive' devices are less visible, and where contemporaries constructed their own taxonomies of the instruments on sale. As both Hackmann and Bennett demonstrate, the dissemination of devices such as the electrical machine, the barometer or the telescope to a wide audience of customers and clients involved the establishment of reliable means of assessing instrumental performance and definition. The public standards to which these instruments were subjected had important effects upon the ways experimenters behaved and, in particular, upon the demands they could expect to make of their tools. Furthermore, the creation of markets for instruments was accompanied by standardisation of techniques and artifacts, a development of great significance for the survival and growth of experimental communities capable of communicating among themselves.

Standards and models

A second feature of experimental instrumentation discussed here is also related to these issues of reliability and standardisation. In his examination of Newton's experiments with glass prisms, Simon Schaffer employs the term 'transparency' to describe the attribute an instrument possesses when it is treated as a reliable transmitter of nature's messages. Schaffer shows that devices do not appear in the laboratory as ready-made 'instruments'. Considerable work is necessary in order to establish the character of an object as an instrument, and this process is accompanied by efforts to win the assent of a community to the object's reliability and transparency. The example Schaffer chooses is that of the common glass prism, a mundane device whose use became emblematic of Newtonian optics. He argues that the Newtonian prism became 'transparent' only when agreement had been reached about the tenets of Newtonian optics. Closure in theoretical dispute was accompanied by the appearance of a new commodity: the marketable English clear glass prism. Bennett develops this thought by pointing out the similar processes which accompanied the establishment of the 'transparency' of the giant telescopes of William Herschel and the Earl of Rosse. By documenting the troubles of identity and replicability which bedevilled the dissemination and status of the prism and the telescope, these authors argue for an account of the use of instruments which recognises the role of the credibility and reputation of the makers of instruments and their users.

Replication and credibility are both key terms in the history of instrumentation and in the analysis of experiment. Many of the contributors to this volume examine the problems of making experiments reliable, and of establishing claims to reliability with fellow experimenters. Alongside the standardisation produced by instrument markets and the local achievement of reliability, an important aspect of this process is the production of phenomena before appropriate audiences. Hackmann charts the development of themes of analogy and modelling in the demonstrations of experimental natural philosophy, indicating the manifold ways in which experimenters devised instruments which purported to model nature. He describes models of electric fish and storm clouds, plant physiology and falling bodies. An enormous theoretical investment is involved in experimenters' claims that such devices are adequate simulacra of the real world. This seems to be a quite general feature of confidence in instrumental tactics. Measurement devices, optical instruments and philosophical machines all involve the more or less tacit claim that they successfully model and represent nature. Galison and Assmus demonstrate that in Wilson's late nineteenth-century programme, the cloud chamber was designed to make models of meteorological events in the laboratory. They describe this as the 'mimetic' approach in nineteenth-century science. Hackmann shows that mimesis was a common aim in eighteenth-century natural philosopy, and Gooding indicates that this imitative enterprise was also a striking feature of the research of the London electromagnetic network, especially when it sought to map geomagnetic phenomena in the 1820s. An implication of this strategy for ways of rendering phenomena and making them meaningful is explored later in this introduction.

The invisibility of instruments

The arguments developed by Gaston Bachelard and his colleagues in philosophy of science hint at ways in which such modelling might be analysed. Bachelard used the term 'phenomeno-technics' to pinpoint this aspect of instrumentation: effects are realised through active instrumental work, rather than recovered immediately by a passive observer from an all-powerful nature. In just this sense, instruments come to embody the theories they are used to support.[1] More recent sociology of experiment, such as that of H.M. Collins, Trevor Pinch and others, is equally forceful in its emphasis on the means experimenters use in order to save the standing of their instruments when in dispute. Devices can be calibrated, but only if the disputing parties accept that the well-known phenomenon used to standardise an instrument is an adequate surrogate for the unknown phenomenon whose character is questioned. Experimenters gain authority if their trials use instruments whose status is hard to challenge–and whose reliability is accepted. Instruments which are 'black-boxed' in this way are indispensable resources in controversy.[2]

The significance of this process is examined later in this volume by Tom Nickles and David Gooding. Here it is sufficient to note an impor-

1. Bachelard (1938), chapter 1.
2. Collins (1985), pp. 100–3; Pinch (1986), pp.212–14.

tant implication this has for our analysis of experimental instruments. An instrument's size and drama often aid the status which the device acquires: consider Van Marum's electrical jars, Rosse's telescopes or CERN's accelerators. But successful instrumentation is rendered invisible, through transparency and black-boxing. The invisibility of instruments is thus an important if paradoxical consequence of experimental achievement. Recovering the role of instruments in experiment represents an important advance in the understanding of how scientists achieve certainty.

PART II: EXPERIMENT AND ARGUMENT

Experiments are powerful resources for persuasion and conviction. Since at least the seventeenth century, arguments that appeal to experiment have often seemed more persuasive than those that do not. In order to fulfil this role, the work of the experimenter must be transcribed and disseminated. Transcriptions of experimental activity are arguments. They are efforts to establish a particular reading of nature and its behaviour. Experimental work leaves many written traces: graphical displays, laboratory notebooks, tables of data, brief reports, lengthier and more public articles and books. Traditionally, analysts of experiment have used only the most public experimental texts, in which the work of the experimenter is presented in forms fit for consumption by a readership of colleagues. The textual deposits left by experiment are such an obvious feature of experimental life that until recently very little attention seems to have been paid to the processes by which such traces are made. An approach due to Bruno Latour and Steve Woolgar concentrates on what they term the 'inscription devices' which are used to make written material in the laboratory.[3] Inscriptions are worked up by laboratory techniques into forms which circulate inside and beyond the original experimental setting. In their initial versions, inscriptions are harder to read and more open to attack by critics. The aim of the experimenter is to transform such inscriptions to a stage where they seem capable of but one reading and become powerful weapons in argument. On this showing, public texts which purport to describe or depend upon experimental work are unlikely to involve simple representations of some original trial. On the contrary, published transcripts will need very careful interpretation if they are to be used as means of access to the experimental setting.

3. Latour & Woolgar (1979), chapter 2.

This attitude to public stories about experiment implies that they are untrustworthy guides. If publication is designed to serve a polemical or persuasive purpose, then only after agreement is accomplished will such texts receive a consensual reading, and this reading will obscure the process by which the trials in question came to acquire a settled sense. However, this salutary suspicion must not license the conclusion that celebrated public statements by experimenters are epiphenomenal to the experimental life. In fact, such texts are constitutive of the course of experimental argument and its closure. They cannot be dismissed as exercises of no import just because they are rhetorical. In his contribution, Geoffrey Cantor draws attention to the considerable efforts of historians and sociologists of science to the recovery of the experimental setting from laboratory notebooks and transcripts. What are the implications of this enterprise? One is that analysis of the process of interpretation and argument is an indispensable technique for the analyst of experiment. The chapters presented in this Part show that experimenters' writing is rhetorical and interpretative in a large number of ways.

Reading the 'Book of Nature'

The ancient image of the experimenter as reader of the 'Book of Nature' already sustained the conviction that successful experiment gained authority from identification with the 'Book's' divine author. In this sense, experimental work looked like scriptural interpretation, bolstered by appropriate techniques of reading. A famous formulation of the image was that of Galileo, for whom Nature was a divinely authored book written in mathematical language. This Galilean theme was but one of many versions of the right way to interpret Nature. The chapters in this Part examine two key moments in the history of experimental writing and argument. Ron Naylor analyses the forms through which Galileo wrote about experiment, notably in his dialogues of the 1630s on the new astronomy and the new science of motion, while John Worrall examines the workings of the early nineteenth-century Laplacian régime in French physics, when experiments were described in the proceedings of committees of the Academy of Sciences appointed to award prizes to competing essays in experimental physics. The written dialogue and the transcript of the proceedings of a committee are accompanied by different rhetorics and different ways of treating experimental writing. Both depict the resolution of an experimental argument in deliberately dramatised form.

Naylor and Worrall reflect on these histrionics and their historiography. In the past, the Galilean dialogues have been scanned for evidence of Galileo's actual experimental activity. It has even been alleged that since the characters of his drama report trials in idealised or exaggeratedly accurate versions, no such experiments could have been performed. Naylor rejects such readings, and draws attention instead to Galileo's move from what he calls a 'pre-empirical episteme', towards one in which experiments could be offered as decisive confirmation of truth. The implication, discussed here both by Naylor and by Cantor, is that readers should attend to the different functions which invocation of an experiment might fulfil in different rhetorical settings. The dialogue is a particularly suggestive means of presenting trials in argument. Cantor argues that in some of the Galilean dialogues, Nature appears as an interlocutor. The process of conversion can be dramatised, the authority of Nature invoked. This is by no means always the pattern of Galileo's drama, and Naylor documents changes in the 1630s between the didactics of the dialogues on the system of the world and the demonstrations of those on the science of motion. Further, he points out the range of experimental determination which Galileo allows his actors to employ; some trials work as refutation of rival positions, others are written as the qualitative establishment of a matter of fact, and yet others seem to imply specific quantitative estimates of the behaviour of moving bodies. That there are exaggerations in Galileo's accounts of experiments is not denied: but, as Naylor shows, these exaggerations can only be understood in terms of the aims of the dialogues' author.

The experimental myth

Two sets of trials described in Galileo's books soon acquired exemplary status: the isochronicity of pendulums and the free fall of bodies from a great height. In both cases, Naylor shows, Galileo confronts the interlocutors and readers of *Two New Sciences* (1638) with reports of trials of extraordinary accuracy and scale. The size of the falling weights is increased from 10 to 200 pounds, the height enlarged to 360 feet. In the pendulum trials, the different sizes of arcs described by different bobs, and their different rates of oscillation, are ignored. The effect of these moves in the Galilean text was considerable: both experiments entered the mythology of experimental philosopy and its triumph. A very similar process of transformation of written transcript into mythic drama was accomplished in the case of the history of optics analysed by Worrall. The episode in question involves the alleged response of a Laplacian committee to Fresnel's prize essay on the wave theory of optical diffraction. In the received story, a critic of Fresnel pointed out a paradoxical consequence of the wave theory, but it was then revealed that this paradoxical consequence could indeed be produced in experiment. It is claimed that the effect of this surprise was devastating: the 'white spot' observed at the centre of the shadow of a circular disc was a self-evident mark of the undulationists' victory. Worrall shows that the proceedings of the prize committee did not follow this course, and that the story of the 'white spot' has the character of a salutary fable. His concern is to establish that in this case, actors in the experimental drama were not influenced by the novelty of an unexpected phenomenon in their reaction to a successful theory.

Worrall shares an important concern with other contributors to this volume, such as Pickering, Franklin and Nickles. What is the relationship between the stories which historians can tell about the course of experimental science and the more normative lessons which philosophers teach? Worrall notes that any account of what it is to be virtuous as a scientist must respect the behaviour of exemplary scientists. However suspicious of so-called 'rational reconstructions' of the course of science, all those who study science must be aware that transcriptions of experimental practice always carry normative content. The implication is that stories about experiment and the resolution of argument are often told to illustrate rival accounts of proper conduct in the sciences.

This point is borne out by Cantor's discovery that the 'white spot' story is a myth of recent origin. Both the Galilean and Fresnel episodes provide rich material for the examination of the writing of experiment. As Cantor argues, the texts in which experimental work is invoked play crucial roles in dramatising and reproducing images of the experimental life. Key problems of the history, philosophy and sociology of experiment can be investigated through these dialogues and reports. In Galilean dialogues, an answer is offered to the question of whether shared experience generates a shared belief about the world. In the romanticised *dénouement* of Fresnel's competition, an answer is offered to the question of whether novel predictions should give preferential support to theory. No doubt these answers rely on highly idealised, if not downright mendacious, accounts of experiment. This does not deprive them of power and significance.

PART III: REPRESENTING AND REALISING

In chapters 7–10 the significance of experiment is shown to lie in its use in construction and articulation of those aspects of the world that scientists take to be real. Gooding shows how electromagnetic phenomena, produced on a small scale in workshops, lecture theatres and laboratories, were made visible through imagery adapted from the mapping of global, geomagnetic and meteorological phenomena. He draws attention to a network of practices and images out of which rudimentary concepts emerged, some of which became important to electromagnetic field theory. He argues that new ways of doing things led to new meanings. These meanings were at first context-dependent. They described phenomena experienced and understood by a very small number of practitioners familiar with the immediate experimental situations. They reached public awareness through public demonstrations and displays and finally gained acceptance, independent of particular practices and places, through Faraday and Thomson's theory of lines of force. Here theory was used to give wider significance to experimental phenomena. Following Fleck, Gooding argues that the acceptance of the imagery of magnetic lines and the physical concepts it implied was due in part to the fact that it articulated an explanation of practices and phenomena widely disseminated by the mid-nineteenth century.[4]

Articulation and demonstration were central to C.T.R. Wilson's work, the subject of Peter Galison and Alexi Assmus's chapter. Wilson's concern to make phenomena visible in the laboratory without transforming the appearance they have in nature affected the selection and development of his experimental methods. For Wilson, it was legitimate to use experiment to reproduce the process of cloud formation within the laboratory. However, this move from uncontrolled natural phenomena to artificially controlled conditions could be relied upon only if the experimenter's art imitated nature closely. When considering the role of instruments in experiment we have already noted that experiments which imitate nature by images or models have a long history. Galison and Assmus argue that the mimetic approach had particular importance for Wilson's selection of methods of producing, controlling and enhancing natural processes and of making them visible away from the site of experiment, for example, through photographic techniques.

Galison and Assmus's chapter also includes an illuminating account of the transmutation of the cloud chamber into the bubble chamber.

4. Fleck (1979).

Divergent interests and methodologies led to different uses of ostensibly the 'same' instrument. Here is one of the ironies of experimentation. Wilson's meteorological and holistic interests focused his attention on real clouds in the sky *and* clouds he could at first take to be equally real in the cloud chamber. J.J. Thomson and his researchers at the Cavendish Laboratory, Cambridge, were in pursuit of sub-atomic entities. This drew their attention to laboratory clouds whose artificiality was unimportant: for them, Wilson's chamber soon became a means of detecting real particles rather than a tool for producing artificial clouds. An experimental anomaly undermined Wilson's confidence in the mimetic authenticity of his clouds. This meant that he resisted the assimilation of his methods into the Cavendish tradition of analytical, quantitative experimentation. Methods he developed to preserve the real, natural status of his clouds became important for other reasons. In particular, they gave control over parameters affecting cloud formation enabling quantitative predictions to get to grips with a quantified version of the phenomena. This control licensed inferences from cloud chamber phenomena to literally invisible but theoretically desirable particles.

Whereas Gooding shows that new concepts implicit in practices led to a new theory, Galison and Assmus show how changing epistemological and theoretical rationales for the use of instruments and methods change the significance of what they produce. The 'same' device moved from one context of assumptions, skills, interests and phenomena into another. In a chapter in which mimesis features so prominently, we may also ponder the striking resemblance of their chart of the conception and disintegration of condensation physics to a Feynman diagram of the interaction of short-lived particles and fields.

Practices, skills and beliefs

The first two chapters in Part III confirm that instruments and techniques remained as important in the nineteenth and early twentieth centuries as Schaffer and Hackmann showed them to be in the seventeenth and eighteenth centuries. This endorses what Hacking, Heilbron, Price and others have argued: the history (and philosophy) of theory is inseparable from the history of instrumental practices.[5] Most contributors to the present volume draw attention to the real man-made environment of skills and material artifacts with which

5. Heilbron (1979); Hacking (1983); Price (1984).

natural phenomena are first experienced, rendered visible and then, through more literary forms, made significant and compelling. Wilson learned how to use the cloud chamber and allied photographic techniques before he could claim to see any natural phenomenon with them, while the acquisition of skills in producing and rendering magnetic phenomena is central to Gooding's argument that new experience must be represented in ways that enable practical demonstration as well as verbal argument. Nickles' essay on the use of experiment in justification draws attention to the fact that a body of skills or 'taken for granted' practices is essential to the evidential significance of experiment generally.

Pickering's chapter shows the importance of the same instrumental and pragmatic concerns that Gooding and many other contributors identify. He examines the composite and collective skills of a team investigating the existence of fractional (non-integral) sub-electronic charges, or quarks. They managed to achieve a coherent view of their own experimental and theoretical activity by recovering the stability of the three interacting components of experimental practice (theory, instruments and phenomena) each time this stability was upset by the outcomes of both theoretical and experimental investigations. Pickering defines the problem typically facing Morpurgo's quark-hunting team – whether before, during or after an experimental run – in terms of the 'destabilisation' of a tenuous working balance between three items: (i) phenomenal expectations specified by theories which require that charges come either to integral wholes or as certain fractions, but not others; (ii) a working model of how charges of any value may be produced, attached to oil drops or other nuclei, detected and measured (drawing on instrumentation developed in a later part of the particle-tracking tradition described by Galison and Assmus); and, (iii) the phenomena and data actually produced. The overall objective was to maintain stability, a coherent working understanding, and to avoid violating certain precepts (for example, that charges may come in sizes other than those compatible with the available theories). This required frequent adjustments to one or more elements of the triad of material, conceptual and instrumental components.

Experiment and Epistemology

Adjustments and changes of practice and skill may be short-term, as in Pickering's case, or longer in duration, as are those discussed by Gooding and by Galison and Assmus. The duration and temporal order

of experiment are important variables, as Worrall shows in his examination of the relative significance of novel results. This problem of order
is often associated with that of discovery and justification. Nickles puts
the context of practices and skills to philosophical use in handling the
problem. He argues that this aspect of science requires that we change
our views about how experiment enables the empirical justification of
theory. Traditionally, philosophers have divorced the unruliness of discovery and discovery arguments from the logicality of justification.
Nickles first makes a case for generative justification, that is, the use
of empirical knowledge to support theories which include that knowledge. He doubts the extent to which the generative form of justification
was ever abandoned by scientists (in favour of consequentialist or
hypothetico–deductive methodologies). He argues that this form of
justification remains important in science (as a matter of fact) and that
therefore it cannot be neglected in philosophical accounts. This assertion
is underlined by Nickles' distinction – familiar to students of innovation
and discovery in science – between invention or construction, on the
one hand, and the giving of accounts of discovery, on the other. Nickles
uses the term 'construction' for the sort of work which historians such
as Schaffer, Gooding, Galison and Assmus describe, and 'reconstruction'
for the reporting of that work in scientists' published accounts. Most
philosophy of science is based upon reconstructions of experimental
and theoretical work, as if the only rationale that matters is the last
one to be given. We argue for the implausibility of this position below.

  Philosophers distrust generative justification because it is thought to
be viciously circular. Using what is known to construct new experimental trials is legitimate, but using that same knowledge to justify their
outcomes is not. This is one reason why so much importance is attached
to novel prediction (discussed in Worrall's chapter). Nickles claims that
the body of taken-for-granted knowledge, embodied in skills, instruments and so on, is necessary to the construction of evidential
arguments. This body of knowledge also meets the regress objection
traditionally made against inductive forms of justification. Philosophers
dislike inductive support because it invites a justificatory regress. A
sociological version of this regress argument has been developed by
H.M. Collins.[6] He claims that the epistemological warrant of experiment, or rather its persuasive force, resides outside the conduct of the
experiment itself. It resides instead (or perhaps, as well) in judgments
made by other scientists about the quality of the experiment. Potentially,
every aspect of an experiment can be challenged. These aspects range

6. Collins (1985), pp. 83–4.

from theoretical plausibility and the correctness of theoretical computations and interpretations, through theories of instrumentation, technical competence with instruments and data analysis, to the reputation and trustworthiness of the researchers involved. Acceptance or rejection of a result ultimately rests outside the experiment itself in evaluative judgments made by groups of experts. What they evaluate is the sort of work discussed by nearly every contributor to this volume, especially (and in detail) by Schaffer, Bennett, Pickering, Galison and Assmus, MacKenzie and Franklin. Judgments about the quality of what such people do end the regress. This is seen most clearly when phenomena are controversial – then everything that went into producing them can come under scrutiny and be challenged.

Empirical studies of experimentation show that experiments may settle disagreements but that such closure is neither permanent nor is it due solely to experiment. This result coincides importantly with Nickles' argument for a generative view of justification. We can put this succinctly in terms of the familiar distinction between 'knowing how' and 'knowing that'. Traditional philosophical methodologies assume that knowing *how* to produce a phenomenon or datum is irrelevant to showing *that* it is the case. All that matters is the theoretical interpretation of phenomena and data. However, if 'closed' debates are always liable to be reopened on all fronts (for example, by new empirical and theoretical results and instrumental possibilities), then the 'how' never ceases to be important. The competences which enable generative forms of justification are never as far away as consequentialist philosophies assume.

Realising nature

Experimental strategies are epistemologically compelling when they rely on methods and assumptions which are not themselves being challenged by the audience to which they are addressed. It would be wrong to suppose that being embedded in a successful experimental tradition (Faradayan lines, or Wilsonian fogs, for example) is tantamount to being entrenched in reality. Bachelard's phenomenotechnics, for instance, despite its claim that natural events must be realised through practices in order to be treated scientifically, nevertheless preserves a distinction between these two positions.[7] But doesn't Nature have a say? Compare what Gooding and Pickering say about the place of Nature. For Gooding, Nature has a cognitive role in shaping

7. Bachelard (1938).

representations in the early development of magnetic field theory. The cognitive role of experiment, in which it was used to learn how to manipulate and represent new aspects of Nature, gradually changed to the familiar epistemologically significant role in which experiment was used to defend theoretical claims about Nature. But if phenomena are always elicited and rendered, then the distinction between observation and experiment begins to blur.[8] With it go the reassuring notions that scientists produce natural phenomena and laws *and* that they can claim that their representations of those facts correspond to the way things really are. The notion that representations correspond to things, processes or relationships is easy to state but notoriously difficult to justify. That is not our concern here. Pickering offers an engaging compromise between realism (as correspondence) and pragmatism (as a form of instrumentalism). He denies that scientists produce correspondences but he is reluctant to deny that they do engage a real world. What happens is that scientists aim for and sometimes achieve stable and coherent accounts of their activities, rather than, say, the certainty that they have discerned the truth about field lines or fractional charges. Stability depends on skills, instruments, theories and the mutually reinforcing judgments of other scientists about all three. But according to Pickering stability also depends on how the natural world is. That is never shown directly, but rather by how nature is implicated in the material culture of instruments and skills with which scientists engage nature. Gooding and Pickering agree that experiment contributes to coherent, intelligible renderings of natural phenomena. Both stop short of a correspondence – realist explanation of the success of experiment.

Skills and transparency

This introduces a paradox that occurs in several chapters, though for some of our readers it may prove to be apparent rather than real. It is well known that scientists take some of the outcomes of their activities (but not others) to be real (natural) phenomena. Many chapters show how important observational activity is to the production of significant phenomena and data. Then why do scientists' own accounts treat phenomena as natural, leaving out the human contribution? The problem is implied by Schaffer and discussed in Gooding's chapter. Nickles identifies this as the paradox of transparency. If it is assumed that Nature can have some decisive role in the outcome of experiments, then there is another problem as well, namely, explaining how scientists discern

8. Gooding (1989).

natural fact from artifacts of human procedures, instruments or theoretical predilections. Once the body of skills is brought into focus, however, the ability to discern natural phenomena from human artifact becomes less mysterious and, if anything, more impressive. The institutionalisation of skills, practices and assumptions through scientists' education may dissolve the paradox that phenomena come to seem independent of the methods and instruments by which, according to their published accounts, they are produced. After all, scientists are trained to present their findings as phenomena or data rather than as the production of artifacts and instruments. As Gooding puts it, some of the experiments that begin as exercises in cognition, as ways of eliciting phenomena, eventually become epistemologically important, as exemplars for showing the way things are. However, as a matter of individual psychology, a problem remains: we cannot dismiss the possibility that for some scientists – Faraday amongst them – there will be a tension between the personal experience of the constructedness of natural phenomena and the overtly natural, objective appearance accorded to them in reconstructed public accounts. Newton, by contrast, does not seem to have paused in his confidence that his trials could be made to lose all their artifactual accompaniment. We have indicated in Part II that these styles of presentation have their own history. Part IV demonstrates that they are also consequent upon the variations in the constituencies who witness, share and appeal to experimenter's work.

PART IV: THE CONSTITUENCY OF EXPERIMENT

Prima facie the experimental laboratories described by John Krige and Jim Secord in their two chapters could not be more dissimilar. The high energy accelerators located at CERN in Geneva have become one of the landmarks of today's scientific terrain. The science carried out at CERN – particle physics – is very much 'big science', involving large financial contributions from a number of European countries, large teams of scientists, and an infrastructure of support facilities the size of a small city. CERN even has its own official team of historians, of which John Krige is a part. Andrew Crosse's science, on the other hand, has long passed into relative obscurity. His experimental work was carried out in the music hall of his own home and required little funding, often relying on materials to be found at home, such as the melted-down family plate and chunks of brick and old flower pots. Crosse worked alone in poetic communication with Nature and had little interest in the public reception of his claims. He had no in-house historians.

Experiment as a resource

Both studies show that the genesis of experiments no matter how important or how marginal, and the use to which they are subsequently put, are dependent upon the wider social context. In the case of CERN, a number of key groups were mobilised to provide funds and support for the new experimental facility. In the case of Crosse different groups played an important role in the interpretation of his results and in the long-term fate of his experimental claims. Both the genesis and interpretation of experiments can be contested as different constituencies seek to shape experiments to their own particular interests. Experiments can be seen as the resources which give different constituencies within and outside science the means to advance distinct interests or particular research programmes.

John Krige's paper on why Britain joined CERN warns against the danger of viewing the genesis of a big experiment as being the result solely of a series of rational decisions. The constituency for experiment in this case involved a number of actors with different perceptions of the long-term needs of British experimental physics and the final outcome was reached only after a protracted series of negotiations. As Krige notes, it is also important to pay close attention to when these actors took the stage, because at different points in the negotiations their strategic interests changed. For instance, Chadwick and Thomson were initially opposed to the European venture in 1949, as Britain was at that time pre-eminent in the technology of accelerators and there seemed little to gain from a new, costly and (from their viewpoint) scientifically dubious, project. However, by 1952, the development of the strong-focusing principle at the Brookhaven Cosmotron in the USA made large accelerators technically more feasible and less costly, and, at a stroke, Britain lost its lead in accelerator technology. Also important in the decision to join CERN was the establishment of a new group of younger experimental physicists and engineers around the facilities at Harwell. This group was very vocal in pushing for British physicists to have access to larger machines. The decision to join CERN can only be understood against the background of the interests of these different constituencies.

Krige demonstrates that an important part of the conflict between the younger group of particle physicists, led by Donald Fry, and the older generation of physicists, such as Chadwick and Thomson, was their different conception of the appropriate environment in which experimental physics could flourish. The style of science entailed by a large facility shared amongst a number of countries was alien to Chadwick and Thomson's own experiences. They had seen Britain's pre-eminence arise largely from relatively small-scale research, carried out in a university context with a few technicians (C.T.R. Wilson's development of the cloud chamber in pursuit of his special interest in 'condensation physics', discussed by Galison and Assmus, typifies this individual style). The new generation of physicists, on the other hand, were familiar with team work and saw the scientific potential which a larger facility would offer.

It would be all too easy to see the eventual decision to join CERN as a triumph for the rationality of the new generation of physicists. However, as Krige shows, the situation was much more complex. Chadwick and Thomson were first persuaded to pursue a European option. This would have involved the British aligning themselves with Bohr and Kramers, who had their own reasons to reject Auger's plans for a new European laboratory. The British offered the use of the accelerator at Liverpool in return for pushing for a new European theoretical centre to be based at Bohr's Institute for Theoretical Physics in Copenhagen. As Krige follows the twists and turns of the negotiations between the different groups within the European physics elite and the negotiation in Whitehall as the physicists pressed their case with the British government, it becomes clear that these experiments did not come into being solely on the grounds of technical merit or for the purposes of testing theories.

Of course, Krige's study is of 'big science'. It may be tempting to say that in view of the cost and size of high energy physics experiments, it is inevitable that they will depend upon the capriciousness of the political process, rather than being decided by the dictates of good scientific method. However, Secord's study of Crosse shows that for small experiments, too, we can understand developments only by drawing upon the wider constituencies within which Crosse worked and which ultimately dictated the fate of his experimental claims. Crosse's experimental programme stemmed from his adaptation of a new nineteenth-century technology, the voltaic cell. Crosse as an experimenter also has some resonances with C.T.R. Wilson. Wilson was influenced by the Victorian tradition of mimetic experimentation and Crosse, too, was motivated by a desire to reproduce natural phenomena in the laboratory. Crosse hoped that by passing current from a voltaic cell through chemical solutions he could reproduce the crystalline patterns he had observed when out walking in Holwell cavern, a famous romantic beauty spot. It was in the course of such electrical experiments that, quite unexpectedly, mites (acari) appeared.

One of the important features to which the new sociology of science has drawn attention is the credibility of experimenters. Credibility is scarcely in doubt in the case of work produced by CERN – one does not invest millions of pounds and employ the cream of the physics community in order to doubt the results. However, credibility is clearly an issue when the experimenter is working alone in Somerset; when he has deliberately cut himself off from his colleagues and isolated himself from the latest scientific developments; when he has very crude equipment, works largely for his own pleasure, and publishes no results; and then to cap it all, has found a completely unexpected result which challenges one of the sacred boundaries of Victorian science – that between the study of natural philosophy and the study of life. Why should anyone take Crosse and his acari seriously? The fact that Crosse was taken seriously can only be understood against the wider audience for his experiments. Secord shows that it was this constituency which first thrust Crosse to prominence against his own wishes and which then subsequently took up the acari in a variety of different ways even though Crosse himself seems to have found the appearance of acari only to be of passing interest in relation to his main goal of understanding the electrical basis of natural phenomena.

The public and the popular press

One important constituency in the acari affair was that of the popular press. The development of the steam press at the start of the nineteenth century meant that suddenly a whole range of newspapers and periodicals could have large print runs. The development of 'popular science', or 'steam intellect', as it was known, meant that there was a wide audience for scientific discoveries. This popular audience represented in some ways a threat to the Victorian gentleman scientists. Organisations such as the British Association for the Advancement of Science provided an invaluable platform from which these scientists tried to reassert control and shape popular science. It was at a British Association meeting that Crosse first received attention. Because of his well-known detachment, he was seen as being above the petty squabbles of the time. He was thus held up to be a romantic scientific hero, a man whose work was to be revered. His experiments on the application of electricity to geology also appealed greatly to the constituency of geologists who were hopeful that geology could, like other physical sciences, be set on a firm experimental basis. Crosse was heralded as the instigator of a scientific revolution in geology. And this was all *before* acari had even appeared in the laboratory. Thus, despite his lack of connections with mainstream science, Crosse was made into an influential figure by his fellow geologists and then in the world of popular science. Indeed, the news of acari first appeared as a popular science item reported in the *Somerset County Gazette*. The story was soon taken up by *The Times* and numerous other newspapers and periodicals.

By contrast, the constituency served by the media is notably absent from the CERN case. The crucial negotiations over Britain's future stake in CERN were nearly all carried out within and amongst the European and British science establishment and the British political establishment. Krige explicitly notes that arguments to do with the need for European co-operation which might be seen as having some political cachet among the wider public (and have proved to be important in the development of projects such as *Concorde* and the Channel tunnel) played no part at all in Britain joining CERN. The lack of an influential popular constituency for science can perhaps be seen in the current debate over whether Britain should withdraw from CERN. The government can contemplate withdrawal with little risk of public outcry. Things in nineteenth-century Britain were very different. The popular constituency could take over scientific ideas, giving them a life of their own, almost independent of leading scientific opinion. When no less an authority than Faraday attempted to intercede in the acari debate he was misquoted as having actually endorsed Crosse's claims. This was ironic, since Faraday's stature was due partly to his having done so much to enlist public interest in science through demonstrations at the Royal Institution.

The end of debate

The debate over acari exhibits many of the characteristics of other episodes of 'extraordinary science'. Replications of Crosse's claims were particularly problematic and Secord's study shows once again that replication is not the simple procedure for warranting scientific belief which many philosophers have claimed. Here, Secord's chapter parallels Schaffer's on Newton's prism experiment. Initially, a ritual refutation of Crosse's work was sought – one that could be quoted at the next British Association meeting, in order to discredit Crosse's acari in public eyes. However, acari would not vanish quite as easily as many hoped. The controversy gained new momentum when a positive replication was reported. Secord shows how different communities judged closure differently: acari survived for some electrical researchers and the popular press, but were rejected (and discredited) almost at once for some natural philosophers.

Crosse's claims touched on some of the fundamentals of Victorian attitudes towards science and life, and it is no surprise to find that they were bitterly contested. Crosse found himself the subject of much vitriol. As Secord notes, to do work on such a controversial phenomenon required a major investment of time, reputation, money, philosophy and faith. Although this was not 'big science', in terms of the personal investment, the stakes could scarcely have been higher.

However, before acari faded, they had become enmeshed (albeit temporarily) with the interests and practices of a number of different scientific groupings. The constituencies for acari were many and varied. For some they provided a resource in founding a new specialist society for the study of electricity, for others they gave reason to found a new discipline of electrobiology, for yet others acari were of most interest for the taxonomic puzzle they presented. In view of the sacred divide between the study of life and study of the physical world, it was no surprise to find that acari became part of a wider platform upon which an argument for atheism was launched. Acari became an actor in the mid-Victorian campaign for natural law, development and the 'science of progress'. All these different constituencies provided the wider social context within which acari lived.

Although Krige does not examine the reception of results from CERN in his chapter, other studies of modern science show that experimental claims, even within high energy physics, can provoke controversy. Although in modern science the constituency of interests has changed (for instance, religion and the press no longer play such an important role) different interest groups within the physics community associated with different competence and instrumental traditions can contest experimental claims.

The battle to define the appropriate constituency for experimental knowledge is every bit as important today as it was in the nineteenth century. Ultimately, as Gooding and Nickles argue in chapters 7 and 10, experiments only acquire their meaning in a particular context, for a particular constituency. Studies in this Part show that negotiations over the appropriate constituency are inseparable from the establishment of experimental fact itself. Once established, the facts acquire an existence independent of the constituency from which they emerged.

This can make them less susceptible to challenges to the methods, beliefs and competence of the practitioners from whose work they first developed.

Reliability

What makes an experiment, or a set of experiments, believable? Several chapters take this as their concern, including those by Gooding and by Galison and Assmus. This is also the question which Allan Franklin and Donald MacKenzie attempt to answer. Franklin draws on his experience as an experimental physicist and his work in the history and philosophy of science to delineate, for a number of cases in physics, the factors which convinced experimenters that particular experiments were reliable. Franklin is concerned with philosophical issues of experiment but shares with many of the authors in this volume a commitment to the study of how science is actually carried out. MacKenzie's paper also concentrates on why a particular set of experiments is held to produce reliable information for a particular community of specialists. His case comes from technology rather than science. The experiments he examines test the reliability and accuracy of nuclear weapons. MacKenzie takes up the argument developed in recent work on technology that ideas drawn from the sociology of science can be extended to the domain of technology.[9] Indeed there is a growing awareness amongst historians of technology that science and technology are much less disparate activities than was once assumed. MacKenzie is concerned to show how the types of argument encountered in the testing of missiles parallel arguments to be found over the testing of experiment in science.

Franklin outlines a number of 'epistemological strategies' or arguments designed to establish the validity of an experimental result or observation. These include: looking at the same phenomena with different pieces of apparatus; prediction of what will be observed under specified circumstances; regularities and properties of the phenomena themselves which suggest they are not artifacts; explanation of observations with an existing accepted theory of the phenomena; the elimination of all sources of error and alternative explanations; calibration and experimental checks; predictions of a lack of phenomenon; and

9. Bijker, Hughes & Pinch (1987); Latour (1987), p.174.

statistical validation. These strategies are shown to have been important in a variety of cases drawn from both contemporary physics and from the history of physics.

Franklin's chapter is littered with examples. Millikan's oil drop experiment, for instance, is used to show the strategy of 'properties of the phenomena as validations'. In this case, it was the consistency of Millikan's results (whereby he found 'one definite invariable quantity of electricity, or a very small multiple of this quantity') which argued for their validity and against their interpretation as experimental artifacts. Another example, drawn from infrared spectroscopy of organic molecules, is used to illustrate the strategy of 'calibration'. In infrared spectroscopy it is not always possible to prepare a pure sample and the substance to be analysed is often put in an oil paste. In such cases, one finds the spectrum of the oil superimposed on the spectrum of the substance, and this gives confidence that the observations are genuine.

Rational belief or cultural practice?

Franklin does not claim that his list of strategies is exclusive or exhaustive, nor that they or any subset of them form necessary or sufficient conditions for what he calls 'rational belief' in an experiment. His examples demonstrate that one or more of these strategies tend to convince practising scientists that particular experiments are reliable. Franklin contrasts his *epistemological* strategies with the explanations of belief in experiments offered by sociologists of science such as Collins and Pickering, who tend to treat such strategies as culturally accepted practices.[10] The warrant for such practices, according to Franklin, is to be found in rationality, which he does not regard as a cultural category.

In so far as it excludes any social basis for the validation of judgment about rationality and believability, Franklin's juxtaposition of 'rationality' with the beliefs of practising scientists is a powerful reassertion of one of the main planks of traditional philosophy of science. According to Franklin, scientists act rationally all the time, it is just that philosophy of science has lost touch with what scientists actually do. A close look at what convinces working scientists is found to be the saviour of rationality. In many ways, Franklin adopts the most traditional epistemology to be found in this book. For him, social context plays no part in experiment because there are epistemological rules which can be applied straightforwardly in the field to separate the wheat of a

10. Pickering (1984); Collins (1985); Franklin (1986), pp. 104–5, 244.

genuine result from the chaff of error. His conclusions should be compared with those of Nickles, who argues just the opposite.

Rationality

The two chapters in Part V raise in different ways the issue of why experiments gain credibility. For Franklin, this process has the hallmark of rationality, while for MacKenzie, the process is irretrievably social. We might be wary of Franklin's usurpation of the term 'rationality'. Few historians or sociologists of science would want to *oppose* social processes to rational ones in the way he suggests. As Nickles argues in chapter 10, the social negotiation of knowledge can be seen as a rational activity in that it is rule-governed and the rules include criteria of consistency, accuracy and reliability. Indeed, one of the merits of MacKenzie's chapter is that it draws out similarities in the way that experimental knowledge is shaped in a variety of contexts in both science and technology. The finding of such widespread regularities scarcely suggests that science and technology are merely 'irrational' or that 'anything goes'. No one seriously disputes that the criteria Franklin points to are the strategies which matter. The issue for MacKenzie, and for other social constructivists, is rather: how do these particular strategies do their work, and from where do they gain their epistemological warrant? MacKenzie shows how reliability is the outcome of a social process; Franklin shows that it is reliability which counts. Franklin takes as *explanans* what MacKenzie takes as *explanandum*.

Experiment and persuasion

Of all our contributors, Franklin is the most optimistic about the self-sufficiency of epistemological strategies as forms of persuasion. This is because he considers them in contexts where, apparently, criticism was not faced explicitly. In Part III, Gooding, Galison and Assmus, and Pickering are also concerned with experiment in the construction of reliable empirical claims, rather than the argumentative role exposed during controversy. However, their studies suggest that the construction of phenomena *always* takes place against a back-drop of potential or actual rival interpretations which are incompatible in some respect with the one they examine. Gooding points out that there were fundamental theoretical differences between Faraday and Ampère and that the existence of a rival explanation of Faradayan facts affects our explanation of the influence of his ideas. Different attitudes to the epis-

temological warrant carried by small-scale reproduction of natural processes are central to Galison's and Assmus's explanation of the eventual
dissolution of Wilson's 'condensation physics' into particle physics and
meteorology. The assimilation of Wilson's cloud chamber and its rein-
vention as the bubble chamber show that new epistemological strategies
developed around these instruments. It should also be remembered
that Wilson did not accept the epistemological legitimacy of Thomson's
experimental strategies. Was he irrational to resist the assimilation of
condensation physics? Consider the wider constituency of experiment,
discussed in Part IV. The electric mites that emerged from Crosse's trials
of the effect of electricity on crystallisation were no more extraordinary
for the relevant audience than the odd properties of electromagnetic
action discovered towards the end of 1820 were for their appropriate
community. Secord shows that the insects were unacceptable on moral
and theological grounds to many Victorians and to many gentlemen
of science in particular. Gooding shows that the Oersted effect was
welcomed, despite its anomalous aspects. The epistemological strategies
adopted by supporters of acari carried little weight against the opposition of the majority of scientists. It is not possible to explain these
different receptions of equally 'extraordinary' phenomena without
recourse to wider assumptions and the values of Victorian scientists
and their publics. Was this majority irrational to reject these legitimate
experimental strategies?

An equally instructive comparison is with Pickering's discussion of
Morpurgo's quark hunt. Where Franklin takes the sufficiency of tradition for granted, Pickering shows the potential vulnerability of accepted
assumptions. His example underlines a main point of Cantor's chapter
and a moral we can draw from Secord's: all experimentation addresses
a human audience as well as nature, and the presentation of nature is
affected by whether that audience is sympathetic or sceptical. Pickering
shows that the dynamics of experimentation are shaped by social as
well as natural parameters, pointing out that the classical laws of electrostatics are amongst the many things taken for granted by Morpurgo's
team. These laws were also taken for granted by everyone Morpurgo
addressed. This need not have been the case, but the fact that it was
so is far from trivial. Had Morpurgo situated his results with respect to
some other theory of the way charges behave in electrostatic fields,
this would have required a different set of epistemological strategies,
because it would have changed his quark detector from an *acceptably
black* box into a controversial Pandora's Box of challengeable assumptions.

Accuracy

Like Pickering, MacKenzie looks at one example in detail. His chapter considers the testing of the US intercontinental ballistic missiles launched from Vandenberg Air Force Base in California towards Kwajalein atoll in the Marshall Islands. He also discusses what is known about the Soviet testing of ballistic missiles and how this information is assessed by the American missile designers and intelligence analysts. From his in-depth interviews with a variety of US engineers, scientists, and nuclear strategists, MacKenzie documents how the test results have over the years been subject to a number of challenges. For instance, one early challenge concerned whether missiles flown with dummy warheads would perform differently from those flown with the real thing. MacKenzie shows that his argument can be understood in terms of different actors making different judgments about similarity and difference. Such judgments are crucial in determining whether two experiments are to be counted as similar and, as such, they are often challenged in controversies over the replication of experiments.[11] Parallel cases are considered in the chapters by Schaffer and Secord.

Much of the challenge to the test results has centred on the facticity of missile accuracies; are the accuracies deduced from performance over test ranges in the Pacific reliable guides to what could be achieved in the operational use of missiles, such as in a nuclear first strike? This debate encompasses many different aspects of missile testing, and includes such factors as how accuracy is defined (including complex procedures for the assessment of the relative contributions of random and systematic error in the measure of accuracy), whether specific features of the Pacific test range mean that results cannot be extrapolated to missiles fired at the Soviet Union, and whether engineering modifications, made in the course of preparing missiles for test firing, are significant.

What Gooding, Galison and Assmus show for results – that they depend on the context of practices – MacKenzie here shows for judgments about accuracy, precision and similarity. Throughout his analysis, MacKenzie shows that arguments drawn from the sociology of scientific experimentation are equally applicable to the case of testing a technology. As well as the key role played by similarity and difference judgments, he shows how modalities are added to and dropped from claims, in order to establish their facticity and how the testing of missiles rests

11. Barnes (1982); Collins (1985), chapter 2; Shapin & Schaffer (1985), chapter 6.

upon the craft skills and tacit knowledge of the missile engineers.[12] Some of the challenges levelled at the results of missile tests can be seen to stem from the strategic interests of particular groups. Thus, the argument put forward that tests of missiles without real warheads were inadequate, was, until 1964, mounted by the manned bomber lobby. MacKenzie outlines a variety of different interest groups who have mounted subsequent challenges. Again, as in Part IV, constituencies for experiment are seen to be important in understanding the dynamics of the debate.

MacKenzie concludes his chapter by drawing out some of the wider political implications of missile testing. He points out that as the testing of missiles is not a neutral activity but has an inherent political dimension, a comprehensive test ban treaty would not end pressure for, and conflict over, nuclear weapon development. Such a treaty would simply inject a new powerful element into the social processes whereby our knowledge of nuclear weaponry emerges. Perhaps a sociology of technological testing can make important contributions towards the wider political debate. There can scarcely be a more poignant case where the detailed negotiation of pieces of experimental knowledge can be tied up directly with the politics of mass destruction.

References

Bachelard, G. (1938). *La Formation de l'Esprit Scientifique*. Paris: Vrin.
Barnes, B. (1982). *T.S. Kuhn and Social Science*. London: Macmillan.
Bijker, W., Hughes, T. & Pinch, T. (ed.) (1987). *The Social Construction of Technological Systems*. Cambridge, Mass.: MIT Press.
Collins, H.M. (1985). *Changing Order*. Beverly Hills and London: Sage.
Fleck, L. (1979). *Genesis and Development of a Scientific Fact*. Chicago and London: University of Chicago Press.
Franklin, A. (1986). *The Neglect of Experiment*. Cambridge: Cambridge University Press.
Gooding, D.C. (1989). *The Making of Meaning*. Dordrecht: Martinus Nijhoff. (in press).
Hacking, I. (1983). *Representing and Intervening*. Cambridge: Cambridge University Press.
Heilbron, J.L. (1979). *Electricity in the Seventeenth and Eighteenth Centuries*. Berkeley: University of California Press.
Latour, B. (1987). *Science in Action*. Milton Keynes: Open University Press; Cambridge, Mass.: Harvard University Press.
Latour, B. & Woolgar, S. (1979). *Laboratory Life*. Beverly Hills and London: Sage.

12. Latour & Woolgar (1979), pp. 81–6.

Pickering, A. (1984). *Constructing Quarks*. Edinburgh: Edinburgh University Press; Chicago: University of Chicago Press.

Pinch, T. (1986). *Confronting Nature*. Dordrecht: Reidel.

Price, D.J. de S. (1984). Of sealing wax and string. *Natural History, 93,* 48–56.

Shapin, S. & Schaffer, S. (1985). *Leviathan and the Air Pump: Hobbes, Boyle and the Experimental Life*. Princeton: Princeton University Press.

PART I

INSTRUMENTS IN EXPERIMENT

1

SCIENTIFIC INSTRUMENTS: MODELS OF BRASS AND AIDS TO DISCOVERY

W.D. HACKMANN

INTRODUCTION

Scientific instruments became indispensable in collecting and dissecting natural phenomena in the seventeenth century, leading to the development of techniques that are at the roots of modern science. Yet the impact of these devices on the emerging natural, or experimental philosophy has until recently been little studied. They made visible what could not be seen by the unaided senses. But what was *actually* seen, and how the instrument-induced phenomena were interpreted, was often determined by contemporary intellectual frameworks.[1] Instruments also led to the questioning of the validity of observations: were these phenomena real or fictitious? Some instruments were tools of measurement, others, such as the armillary sphere and orrery, were models of how the natural world was perceived.[2]

Central to the study of instruments is how they were used in making experiments. These were often laboratory replications of 'real world' natural processes; visual demonstrations that had a powerful didactic role and propagandised the particular view being advanced. The intellectual framework was the mechanical philosophy, based on 'common sense' everyday experience, making use of causal structural explanations. This allowed for deductive testing and for formulating new predictions.[3] The mechanical philosophy was particularly suited to investigating those phenomena that could be isolated, reproduced, or modelled in the laboratory.

---

1. See Hackmann (1979), taken up by D'Agostino & Ianniello (1980).
2. For a development of this distinction, see Price (1980).
3. Gabbey (1985), pp. 9–84, especially pp. 11–12.

There has been little analysis of the various and complex uses made of scientific instruments.[4] They have mostly been studied as antiquarian objects or as cultural artifacts and even as heroic devices responsible for particular scientific breakthroughs. Examples of heroic instruments that readily come to mind are Boyle's air pumps or Herschel's giant telescopes. How they interacted with experiments, and their impact on scientific method, are key problems which have received little attention. Scientific instruments were (and still are) used as arbiters between contending theories, and sheer size could give a psychological advantage. A celebrated example from the eighteenth century is the giant electrostatic generator made for the Dutch natural philosopher Martinus van Marum by the English instrument maker John Cuthbertson.[5] It fulfilled all the requirements of a heroic device (Fig.1.1): it was large and constructed to the limits of what was technically possible. This awesome machine with its two glass discs five feet in diameter, produced discharges twenty-four inches long; in modern terms, between 300 000 and 500 000 volts. It is not difficult to understand why Van Marum's fellow scientists could be persuaded by his experimental results. These were one of the main factors for the early acceptance of Lavoisier's combustion (oxidation) theory in the Netherlands.[6]

No apparatus is, however, self-evidently superior. Its value, like that of the experiment, lies in its power of persuasion. Bachelard has likened an instrument to *'un théorème réifié.*[7] Sociologists of science, interested in experimentation, have started to examine the part played by instruments in the scientific and psychological strategies developed to reach a consensus of opinion about a particular theory. Although these analyses are, generally, rather ahistorical, they have at least focused

4. According to Price, the role of scientific instruments and techniques in the development of science has been grossly underestimated, see his 'Sealing wax and string: a philosophy of the experimenter's craft and its role in the genesis of high technology', George B. Sarton Memorial Lecture at Annual Meeting AAAS, May 27, 1983, and Price (1984). This is a viewpoint with which I agree entirely. Other neglected areas are the relationship between precise instrumental measurement and quantification of theory, the reliability accorded to measurements, the estimates of experimental error, and the uses of numbers in tables and graphs (see Heilbron, 1980). There is little treatment of the development of electrostatic apparatus in Heilbron (1979), which in other respects is the most comprehensive work on the early history of electricity.

5. See my Oxford prize essay (Hackmann, 1973) on the Cuthbertsons; also discussed in Hackmann (1978a), pp. 154–170.

6. For Van Marum's chemical and electrochemical experiments, see Levere (1969), and also Hackmann (1971).

7. See Bachelard (1933, 1938, 1951). For a review of the viewpoints on natural philosophy, but stressing the cultural aspects, see Schaffer (1983a), and on Bachelard, see Gaukroger (1976) and Schaffer (1980).

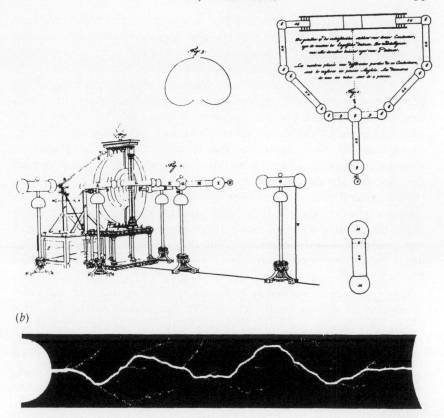

(b)

Figure 1.1. (a) Martinus van Marum's large electrostatic generator at Teyler's Tweede Genootschap (now Teyler's Museum), and (b) its 24 inch discharge taken by Van Marum as a visual proof of Franklin's single electric fluid theory, 1784. (Van Marum, *Verhandelingen Uitgegeven door Teyler's Tweede Genootschap*, 1785, vol. 3, Plates II and III)

on a key issue largely ignored by historians of science: What is the role of instruments in bringing about changes in scientific beliefs? This question is not at all easy to answer.

The interaction between instruments, experiments, and the development of scientific concepts is extremely complex. To study this we require a variety of approaches, from single detailed historical case studies to sociological analyses of experimentation as a group activity. Instruments (as embodiments of experiments) have recently begun to be taken more seriously. Let us hope that historians interested in this aspect will take the trouble to learn how these instruments functioned by examining the actual objects and not just the flat representations in textbooks and laboratory reports, and also what could be expected from the experiments. Thus, a great deal can be learned from repeating Newton's prism or Galileo's 'inclined plane' experiments.

Van Helden has pointed out that scientific instruments bridged the gap between the mathematical and experimental sciences which developed together from the seventeenth century.[8] The origins of the experimental, or mechanical philosophy are as complex as the historiography that has grown up around this subject.[9] Crombie has traced the experimental method back to the Middle Ages; observing Nature with the unaided senses and performing simple experiments that did not violate natural processes have been carried out since antiquity.[10] In the sixteenth century, trade and exploration encouraged navigation, resulting in the improvements of such instruments as the mariner's astrolabe, the quadrant, and the compass.[11] These devices were based on earlier angle-measuring instruments developed for astronomy. Probably the oldest scientific instrument is the sundial.

Angle-measuring surveying and topographical instruments were also developed in the sixteenth century. The increased market in these devices required by the military engineer and land surveyor led to the establishment of instrument-making centres.[12] Here we see the coming together of the skills of the mathematician and the engraver in brass and copper. Mathematical practitioners, skilled in devising and using new instruments, became involved in those subjects vital to European colonial and economic expansion, such as navigation, surveying, horology, cartography, astronomy, and fortification. In England, such a group centred on Gresham College, founded in London in 1597. This is how the techniques and empirical approach of the craftsmen and artisans found their way into the emerging experimental sciences – stressed in the late sixteenth century by Bernard Palissy and Pierre Ramus, and in the following century by Francis Bacon, Robert Boyle, and Robert

8. An interesting resume of the early period is Van Helden (1983), pp. 49–83, especially p. 60; a subject that still requires much more analysis
9. For a recent selection of opinions, see Kuhn (1976, reprinted in 1977); Harré (1980) pp. 11–54; Schaffer (1980), pp. 55–81; (1983a), pp. 285–314; Bennett (1986).
10. See Crombie (1953), especially his commentary on Lynn White's 'What accelerated technological progress in the western Middle Ages?', (pp. 272–91), also Crombie (1963), pp. 316–23, and his paper in Crombie (1961), pp. 13–20, on the meaning of measurement in the natural and social sciences.
11. Maddison (1963); Waters (1966); McConnell (1980); Turner (1983a), pp. 243–58, especially p. 245.
12. Taylor (1954, 1971); Waters (1958); Stimson (1976); Turner (1983b), pp. 93–106.

Hooke.[13] This tremendous shift in both social and political attitudes should not be underestimated. It is symbolised in the painting by Johannes Stradanus, or Jan van der Straet (1570), depicting the Grand Duke of Tuscany, Francesco I, not with a book, but in his private laboratory, surrounded by assistants, wearing an artisan's leather apron, making chemicals. This is the earliest painting of a chemical laboratory with no alchemical or allegorical overtones.[14]

INSTRUMENTATION AND THE EXPERIMENTAL METHOD

Key features of the pragmatic research routines developed in the seventeenth century were the experimental method, the experiments themselves, the instruments, and the experimenter's attitudes and expectations. The experimental method delineated what was thought to be possible in the laboratory, and what validity to place on the results. Instruments became an essential aid to discovery. Central to the development of this experimental philosophy were debates about the limits of knowledge, and the validity of sense experience or perception. Francis Bacon argued in *Novum organum* (1620) that nature should be interpreted through the senses, aided by experiments 'fit and apposite'.[15] Properties of phenomena examined in this way should be listed in 'natural histories'. His inductive method, and the empirical techniques utilised by Robert Norman in his *The Newe Attractive* (1581) and in William Gilbert's *De magnete* (1600), were translated into more complicated laboratory strategies by Boyle and Hooke, culminating in Newton's *Opticks* (1704). This book was cast in the rational mathematical structure of his *Principia* (1687), but configurations of apparatus replaced the synthesis, and the proof was the *experimentum crucis*.[16] These early instrumental techniques focused on the importance of the senses in observations – both in terms of philosophical debates about the nature of perception and the examination of the practical limitations

---

13. Boyle admitted freely his indebtedness to the practices of 'tradesmen', see Birch (1772), pp. 396–9, 415, 442; both he and Diderot stress the importance of science in improving the crafts, which was one of the purposes of the *Encyclopédie*, see Harré (1980), pp. 39–40.
14. Hackmann (1985), p. 110. This painting is in the Studiolo of Francesco I in the Palazzo Vecchio, Florence. Probably the same laboratory is depicted by Stradanus in Plate 7 entitled 'Distillatio' in the series *Nova Reperta*. This plate was engraved at about the same time.
15. Bacon's *Works*, (1857–74), pp. 61,83.
16. For a discussion of Newton's problems with the *experimentum crucis*, see Lohne (1968) and Schaffer (Chapter 2 this volume).

of the senses. It led Hooke, for instance, to study the physiological mechanism of vision.[17]

If sensations could not be trusted, then neither could the phenomena induced by instruments. On the whole, these deeper philosophical questions were ignored or laid aside by the pragmatic experimenter during the laboratory phase of his work. Yet their implications could not be so easily disregarded when attempting to explain the underlying *mechanisms*, or causes, of these observations. As Ian Hacking has pointed out, the experimental philosopher could never get outside his representations to be sure that they corresponded to the 'real' world as observed. Every test of a representation is just another representation – as Bishop Berkeley had it: 'Nothing is so much like an idea as an idea.' Bacon, with his lawyer's mind, made the distinction between what was directly perceptible and those invisible events which could only be 'evoked', but the subtlety of this was not appreciated until much later.[18]

The 'Newtonian style'

Newton distinguished between a mathematical formulation (such as the inverse-square law of gravitation which exists in 'real world' observations) and the invention of a physical explanation, or model, for such a force (for instance, in terms of an unobservable aether). I.B. Cohen has identified this as the 'Newtonian style' which found its clearest expression in the *Principia*.[19] The implications of Newton's methodology were not well understood by his contempories. They became further blurred in the eighteenth century because of the strict experimental approach advocated by such popular teachers and textbook writers as Desaguliers and Benjamin Martin. Such experimenters shared less with Newton than with his disciple, the Scottish common-sense philosopher Thomas Reid.[20]

In this context it is instructive to compare the attitudes of two great Dutch eighteenth-century experimentalists, Willem Jacob's Gravesande and Peter van Musschenbroek, who together spanned seventy years of natural philosophy. Their textbooks and teaching methods shaped

---

17. Waller (1705), p. 98. For a modern discussion on the concept of observation, see Shapere (1982). He points out that the information has to be transformed by instruments into human-accessible information (p. 509).
18. Hacking (1983), pp. 169, 273.
19. Cohen (1980), pp. xii–iv, 16. The loss of this distinction in the late eighteenth century is described by Heilbron (1980), pp. 370–5, in the case of electricity.
20. Laudan (1970), pp. 103–31; (1981), pp. 86–109; Schofield (1970).

the experimental sciences more than any others at that time – an influence that lasted well into the nineteenth century. A mute testimony to their impact is the antique physics teaching apparatus that has survived in many of the old university collections. Both emphasised different aspects of Newton's methodology.

's Gravesande, who taught in the early years of the century, was the more cautious, mathematical, and strictly phenomenalistic. He tried to adhere strictly to Newton's 'hypotheses non fingo'. He did not use his experimental results to hypothesise about underlying mechanical causes, such as the *nature* of light or electricity. His successor, Van Musschenbroek, on the other hand, used his experimental data to speculate about such *untestable* entities as atoms or sharp acid particles, although he did warn against the introduction of unverifiable hypotheses. He was willing to tackle such typically eighteenth-century subjects as magnetism, heat, and other phenomena which were difficult to quantify or cast into general laws. For this reason these phenomena had been scarcely investigated by his predecessor.[21] Both 's Gravesande and Van Musschenbroek stressed the didactic use of experiments, for which they developed many pieces of demonstration apparatus. Newton's eighteenth-century followers were particularly influenced by the *Opticks*, with its impressive sequence of experimental 'proofs', especially by the way this technique was successfully transferred to other branches of physics by 's Gravesande and Van Musschenbroek. It was founded on the belief that instruments reproduced 'real' natural processes: the rainbow spectrum produced by Newton's prism was not an artifact of this device, nor was the large spark produced by Van Marum's generator, taken by him as 'proving' the existence of a single electric fluid.[22]

Natural symmetries, natural processes, and common sense experience

Underlying the experimentalist's faith in his laboratory procedures was the concept of the 'principle of the uniformity of nature', formulated, for instance, by Newton in his 'third rule of reasoning':

The qualitites of bodies, which admit neither classification nor remission of degrees; and which are found to belong to all bodies within the reach of our experiments, are to be esteemed the universal qualities of all bodies whatsoever.[23]

21. De Pater (1979), pp. 57–121; Hackmann (1985), pp. 110–12. For a further discussion on the observational *(macro-)* level and the theoretical *(micro-)* level, see note 53.
22. Hackmann (1971), p.336
23. Newton (ed. A. Motte & F. Cajori, 1966), vol. 2, pp. 398–400, for quotation and details.

The same concept had already been formulated by Aristotle, and in the eighteenth century was generally known as the 'principle of the simplicity of nature', according to which similar effects whether produced in the laboratory or by Nature, were due to the same causes. Perhaps at the root of this idea was the ancient concept of symmetry, based on every-day observations of the natural world. The Greek word 'symmetry' has evolved with time, taking on its more precise modern meaning in the eighteenth century. The Greeks did not distinguish between symmetry, proportion, harmony, 'commensurable' (in geometry), or 'indifference' (in cosmology). According to Aristotle, Anaximander argued that the disc-shaped Earth was in equilibrium in the universe because of its perfect symmetry or 'indifference' with respect to every direction in space. This concept may also be the reason why the later Greek astronomers were committed to a spherical Earth and cosmos, and to the idea of perfect motion being circular.

The Pythagoreans studied symmetries and harmonies in mathematics, music and cosmology. They are reputed to have discovered the five regular solids: cube, pyramid, octahedron, icosahedron, and dodecahedron, which much later were to figure largely in the cosmology of Kepler. 'Polar symmetries' (antitheses and contrarities) were also important in their description of the world. These included hot and cold, right and left, male and female, odd and even, the opposing forces of Love and Hate, and the four primary elements (fire, air, water and earth). To Archimedes, symmetry kept a balance in equilibrium. Similarly, Leibniz formulated his 'principle of sufficient reason': such balance would remain in equilibrium unless there was a sufficient reason for it to tilt. Galileo and Descartes both made use of geometrical symmetry. Thus, symmetry in all its forms has been implicit in describing natural processes since antiquity, although we have to wait until the late eighteenth century for its precise mathematical description in geometry and algebra.[24] This concept became a vital element not only in the construction of physical theories and in the processes of analysis, prediction, and problem-solving, but also in the design of laboratory experiments and instruments capable of replication.

These intuitive *a priori* ideas were also expressed in the emerging seventeenth-century natural theology. The hallmarks of God's universe were harmony and economy, so that natural phenomena created in the laboratory had the same causes as those in the world outside. Research in the laboratory was an act of worship; it brought the experimenter closer to his Creator. Uniformity of Nature, economy, symmetry,

24. Roche (1988).

and harmony, were at the core of the conservation laws. These were implicitly accepted by eighteenth-century experimentalists, such as Franklin with his single-fluid theory of electrostatic behaviour, and later stated explicitly by Faraday and others. It has been argued by some historians that Faraday's theology was crucial for his phenomenalism, his postulation of lines of force, and his *a priori* acceptance of the conservation principle.[25] These laws could not be 'proven' by instrumental evidence. The ideal could only be approached by increasingly more sensitive and accurate instruments and experiments.[26]

INSTRUMENTS: 'PASSIVE' AND 'ACTIVE' EXPLORERS OF NATURE

Instruments can either be passive or active when exploring Nature. Among the former are tools of measurement such as clocks, chemical balances, electrometers, galvanometers, and graduated astronomical angle-measuring instruments. Increasing their precision made 'breakthroughs' in theory, or new discoveries, possible but not inevitable. For instance, Tycho Brahe's new angular measurements were necessary for Kepler's work, and Flamsteed's for Newton's.[27] To give another example, by the 1720s the angular resolving power of these observational instruments had improved so much that it was possible to discover the aberration of light, and nutation. Similarly, the accuracy of the chemical balance was improved steadily during the eighteenth century when experimental chemists such as Lavoisier became interested in the quantitative use of this instrument. This increase in sensitivity must have given them confidence in their *a priori* concept of the conservation of (chemical) mass, and certainly in the quantitative analysis of the chemical reactions they were investigating.

A second group of significant passive observational instruments

25. Gooding (1980; 1982) and Cantor (1985), pp. 69–81. Heilbron does not accept that Faraday's concepts came from philosophical or metaphysical considerations, but argues that they have their roots in contemporary physics, in Heilbron (1981), pp. 187–213, especially pp. 187–9, and note 113.

26. For an early example, see Naylor (1980). He and Tom Settle showed by means of careful reconstruction of the 'inclined plane' experiment that Galileo could not have discovered the law of fall solely by experimental evidence, *contra* to the claim made by Stillmann Drake (1974/5, 1975); see also Van Helden (1983), p. 79, note 66, who shares Drake's opinion.

27. For a fuller discussion of this aspect of instrumentation, see Hackmann (1985), pp. 104–5.

included the telescope and the microscope. These revealed hitherto unsuspected phenomena and structures not observable with the naked eye. Discoveries with these devices were usually not expected, because they could not be related to past experiences. Galileo demonstrated that the images revealed by his telescope existed in the real world and were not artifacts of the instruments or figments of his imagination. His newly-discovered satellites of Jupiter, or the mountains and valleys on the moon really existed. It was, of course, not at all easy for the early telescopists and microscopists to interpret these unfamiliar images. Galileo's trained artist's eyes helped him to identify the shifting patterns of the moon's surface with the features of the Earth's crust. It took until the middle of the seventeenth century before they became routine research instruments and the observations made with them no longer regarded with mistrust.

Hooke fully appreciated the importance of these optical instruments in enhancing the imperfect senses. He described this process in his *Micrographia* (1665):

The next care to be taken, in respect to the Senses, is a supplying of their infirmities with Instruments, as it were, the adding of artificial Organs to the natural; this in one of them has been of late accomplished with prodigious benefit to all sorts of useful knowledge, by the invention of Optical glasses. By the means of the Telescope, there is nothing so far distant but may be represented to our view; and by the help of Microscopes, there is nothing so small, as to escape our inquiry; hence there is a new visible World discovered to the understanding. By this means the Heavens are open'd, and a vast number of new Stars, and new Motions, and new Productions appear in them, to which all the ancient Astronomers were utterly Strangers. By this the Earth it self, which lyes so near us, under our feet, shews quite a new thing to us, and in every little particle of its matter we now behold almost as great a variety of Creatures as we were able to reckon up in the whole Universe it self.[28]

However, these instruments were still *passive* in that they did not actively interact with Nature in the same way as the new 'philosophical' instruments, such as the air pump or the electrical machine. These made it possible to isolate phenomena in a controlled laboratory environment, such as in the bell-jar of the air pump. It became generally accepted that these devices reproduced 'real' natural processes. The

28. Hooke (1665), preface. For a brief analysis of his use of instruments, see Bennett (1980), and for his scientific method, see 'Espinasse (1956), pp. 16–41; Centore (1970), pp. 16–40. These views are also discussed in Hackmann (1988), concentrating on attitudes to instrumentation at the time of Halley, Hooke, Boyle and Newton. The contemporary attitude at the Royal Society is described by Heilbron (1983).

most common debates were about the validity of the observations. Others were concerned with the underlying theoretical framework to which these experiments gave credence. A typical eighteenth-century example is the fluid theories of electrical action. At the deepest level – notably the controversy between Hobbes and Boyle – they were about the experimental method, but by the eighteenth century these conflicts had been almost entirely resolved by the experimental community.[29]

Instrumental replicability

At the instrumental level, the debate centred on replicability. Boyle and Hooke, for instance, were well aware of the importance of this as a standard or test of laboratory-induced knowledge. When in 1821, Faraday discovered the rotation of a current-carrying wire around a permanent fixed magnet, he overcame the problem of replicability by making a pocket version of his experimental apparatus which he sent to his scientific colleagues. Experimental philosophers stressed the importance of controlling the factors that could affect the results, otherwise these could not be replicated – the crucial test of a successful experiment. This became increasingly difficult to achieve as laboratory apparatus and experiments grew in complexity from the eighteenth century onwards. Van Musschenbroek, for instance, advised that great care should be taken in the design of apparatus and experiments. Yet like all other experimentalists, he was beset by two fundamental problems which influenced replicability: (a) mechanical design factors, and (b) unknown external influences. Thus, he observed that the weighings made with his balance were affected by changes in the moisture contents of the strings by which the pans were suspended. This could be the reason, he suggested, why Florentine academicians decided that cold bodies were heavier than warm ones. Detailed laboratory records should be kept of all the factors that might affect the experiment, including where it was made, in which season, the hour, the direction and strength of the wind, barometric pressure, humidity and temperature. The experiment should be repeated in different ways and under varying conditions. It could take a long time to appreciate which par-

---

29. The electrical case has been expanded in Hackmann (1979). On the controversy between Hobbes and Boyle, see Shapin & Schaffer (1985), pp. xi, 110–54, and on the closure of debate necessary for the development from prototype to standard laboratory instruments (pp. 274–6). Gooding (Gooding & James, 1985, pp. 120–3) describes how Faraday sent a pocket rotation apparatus to fellow scientists in order to overcome the problem of replicability.

ticular factors were important. To cite just one example, electrometers could not become accurate or reproduce standard results until at least the 1830s, when Faraday realised that they had to be screened from the minute extraneous charges of electricity in the air. The fact that many of the laboratory devices were 'theory-laden' could also influence what was perceived. Gilbert, for instance, in 1600, recognised electrical attraction, but not repulsion in the fairly ambiguous movements of his versorium.

This distinction between passive (observational) and active (phenomena-interactive) instruments may help us to understand the basic features of experimental philosophy as implied in research strategies. As the trade in scientific instruments developed in the eighteenth century, the makers distinguished between mathematical, optical, and philosophical instruments. This distinction broadly accords with the categories in this chapter. Experimental apparatus began to be likened to the telescope and microscope, in that the more powerful they became, the more they would 'magnify' natural phenomena and lead to new discoveries. This was the driving force behind the development of ever larger electrostatic generators in the eighteenth century, of which Van Marum's machine was the ultimate example. Priestley well understood the active role of philosophical instruments in the exploration of Nature:

The Instructions we were able to get from books is, comparably, soon exhausted; but philosophical instruments are an endless fund of knowledge. By philosophical instruments, however, I do not mean the globes, the orrery, and others, which are only the means by which ingenious men have hit upon, to explain their own conceptions of things to others and, which, therefore, like books, have no uses more extensive than the view of human ingenuity; but such as the air pump, condensing engine, pyrometer, etc. (with which electrical machines are to be ranked) and which exhibit the operations of nature, that is of the God of nature himself, which are infinitely various. By the help of these machines, we are able to put an endless variety of things into an endless variety of situations, while nature herself is the agent that shows the results. Hereby the laws of her action are observed, and the most important discoveries may be made; such as those who first contrived the instruments could have no idea of.[30]

Priestley's main concern was in discovering descriptive laws from a mass of laboratory observations. His distinction between models of brass (globes, orreries, etc.) and philosophical (experimental) apparatus with which new discoveries were possible, is very pertinent. Some of

30. Priestley (1775), vol. 1, xxi–iii.

these models began life as observational instruments but then 'degenerated' into demonstration devices. A case in point is the astrolabe, which evolved from an instrument for plotting star positions into a teaching device and a means of casting horoscopes. Priestley did not, however, distinguish further between these passive brass models and model experiments in the laboratory, or lecture–demonstration (didactic) devices. These are all crucial distinctions if we want to understand the development of natural philosophy. Experimental apparatus, such as von Guericke's air pump or Hauksbee's 'electrostatic' generator, started their evolution towards standard laboratory devices as experimental prototypes. They only became non-controversial laboratory instruments after the debates about what was observed with them had been resolved, and the replications of these observations had been accepted.

Lecture–demonstrations and the illustrative experiment

Experimental apparatus was also adapted specifically for lecture–demonstrations. Hauksbee and Whiston were early pioneers of the illustrative experiment for teaching and for communicating novel results. 's Gravesande and Van Musschenbroek were the first to write comprehensive physics textbooks based on this system. To ensure that their didactic apparatus could be copied they took immense care over the engravings and descriptions. A typical late eighteenth-century pedagogic device was the fall machine of George Atwood used to demonstrate the laws of uniformly accelerated motion (Fig.1.2). Previously, Galileo's 'inclined plane' experiment had been standardised for pedagogic use, but Atwood's intention was to devise a more flexible arrangement 'to subject to experimental examination' Newton's laws of motion. The instrument was developed with the able assistance of the instrument makers Benjamin Martin and George Adams the Younger, and it remained a popular class-room device until the early twentieth century.[31] Thus, this type of instrument, although developed from previous experimental work, was not designed for further research but for demonstrating established knowledge more effectively. The electromagnetic and electrodynamic discoveries made by those such as Oersted, Faraday, Ampère, Barlow, and de la Rive were immediately translated into didactic devices so that the phenomena could be easily

---

31. Atwood (1784); Greenslade (1985). For the didactic apparatus of 's Gravesande and Van Musschenbroek, see Crommelin (1951); and Rowbottom (1968) and Schaffer (1983b) on the popularity of lecture demonstrations.

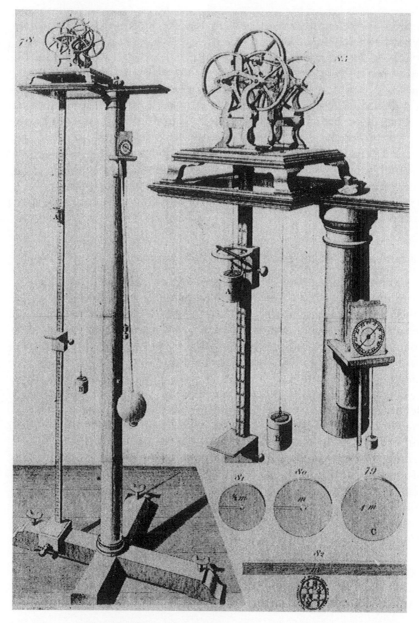

Figure 1.2. Atwood's fall machine for demonstrating the laws of uniformly accelerated motion, 1784. (Atwood, *Treatise on the Rectilinear Motion and Rotation of Bodies,* 1784)

visualised. The Italian scientist Leopoldo Nobili, for instance, devised a portable teaching laboratory for demonstrating these novel discoveries (Fig.1.3).[32]

ANALOGIES AND MODEL EXPERIMENTS

The intuitive feeling about the underlying regularity of natural processes became the basis for two powerful techniques in scientific research: (a) the use of the analogous argument in the framing of physical theories, and (b) the 'model experiment' in which natural processes were 'recreated' in the laboratory by means of scale models, when the strict inductive procedures could not be applied, or when no direct experimental intervention was possible. The use of the metaphor became a crucial feature in eighteenth-century experimental philosophy. John North has analysed the development of the concept of scientific analogy from the seventeenth until the mid-nineteenth century.[33] Newton used this technique when exploring the properties of light and the colour spectrum produced by a prism; the former he based on the analogy with sound, the latter on the mixing of pigments.[34] William Harvey likened the action of the heart to a force-pump circulating the blood, and the dominant eighteenth-century view of electricity was in terms of an hydrostatic model in which all could be explained by one or two imponderable fluids. This technique had a great bearing on the kind of experiments that would be undertaken, and on the design of laboratory apparatus.

The analogous argument became a key element in the research strategy. To give one example, in a series of experiments much lauded by late eighteenth-century plan physiologists, Van Marum argued that the structure of plant vessels and animal arteries must be similar as both were affected in the same way by electrical discharges. He had to invoke the argument by analogy as the structure of plant vessels, unlike that of the arteries, could not be determined even by the best contemporary microscopes.[35]

32. Brenni and Hackmann (1984), pp. 29–100, and my Introduction in this volume (pp. 7–10). For Faraday's use of illustrative experiments, and his capacity to visualise natural phenomena in the laboratory, see D.C. Gooding in ed. D.C. Gooding & F. James (1985), pp. 105–35, especially p. 106.

33. North (1981), pp. 115–40; also Hesse (1966) and Leatherdale (1974).

34. Cantor & Hodge (1981), p. 20; also North (1981), pp. 118–21; Biernson (1972) and Cohen (1980), p. 205.

35. Hackmann (1972), pp. 24–5. For Faraday's use of analogy, see, for instance, Gooding (1981, 1985), and for Rowland's magnetic analogy to Ohm's law, see Miller (1975). Rowland was very much guided in this use of analogy and models by Faraday's *Experimental Researches in Electricity*, published in 1839–55.

Figure 1.3. Nobili's kit demonstrating the electromagnetic properties of a current, circa 1830 (In the Isitituto e Museo di Storia della Scienza in Florence, inventory number 1553. The photograph was taken by Franca Principe)

The model experiment, often with scaled-down apparatus, was evolved from abstract conceptual models formulated to illustrate a particular property or the supposed mechanism of natural phenomena. As Crombie has pointed out, Descartes' innovation was to conceive nature operating according to principles of the engineering-model.[36] Just as Harvey developed a force-pump analogy for the heart, so in his *Les Météores* (1637) Descartes used a model raindrop, in the form of a spherical glass flask filled with water, to investigate the properties of the rainbow. Here he followed, without acknowledgement, the work of Theodoric of Freiburg three centuries earlier. In *La Dioptrique* (1637), Descartes compared the eye with a model made from a camera obscura, and then showed (by means of an experiment already described by Christopher Scheiner) how the eye could be examined like an optical instrument. He removed the back of the eye (the retina) of a freshly dead ox or man and cast the image instead on a piece of thin white paper or eggshell.

Descartes admitted that a particular natural phenomenon could usually be explained by several conceptual models or hypotheses. The procedure which he developed was a kind of *experimentum crucis* in which he devised experiments which allowed him to choose between these models. He contrasted his models, or assumptions, with those used in astronomy, which were not true physically but were framed to yield predictive conclusions and to 'save the appearances'. He was a capable observer and an ingenious experimenter, as demonstrated by his analysis of the properties of lenses, which allowed him to give a scientific account both of the eye and the telescope.

Cohen has contrasted Descartes' use of conceptual models with Newton's.[37] Unlike Newton, Descartes did not separate the analysis of natural phenomena into their mathematical and physical aspects. His use of models, too, was quite different from Newton's approach. For instance, in his *Dioptrique* he illustrates the transmission of light by means of three models which not only conflict with each other, but

36. Crombie (1985), pp. 192–201. Hooke followed this example. See, for instance, the telling quotation in Bennett (1980), p. 44 taken from Waller (1705), p. 61:
A most generous Help of Discovery in all kinds of Philosophical Inquiry is, to attempt to compare the workings of Nature in that particular that is under Examination, to as many various, mechanical and intelligible ways of Operations as the Mind is furnisht with.
Kelvin was one of the prominent late nineteenth-century physicist who used this aid in teaching physics.

37. Cohen (1980), pp. 103–5.

also with what he considers to be the fundamental properties of light. Each model highlights a *specific* property which, by means of the argument from analogy, may suggest further experiments or provide useful information. There is no single model for all the properties. Newton, on the other hand, uses two kinds of models: his observations lead him to conceive mathematical analogues of nature simplified and idealised – such as the law of gravitation – and then he provides conceptual models with which to *explain* gravity or the 'spring' of air.

Experiments to isolate and to imitate nature

Direct observations with instruments, and the analogous argument illustrated by conceptual models, yielded two basically different types of experiments. The first consisted of laboratory procedures isolating certain phenomena and determining their properties; the second of laboratory models with which to imitate the phenomena as perceived – what Van Marum in 1781 called 'imitative experiments'.[38] This mode of conducting laboratory research became common in the eighteenth century, when the arsenal of laboratory apparatus also improved drastically. The growth of electrical theory in this century necessitated the development of instrumentation, unlike the preceding century, when there was little electrical theory and correspondingly little electrical instrumentation, as Priestley wrote so eloquently:

While nothing more than electrical attraction and repulsion were known, nothing that we should now call an *electrical apparatus* was necessary. Every thing that was known might be exhibited by means of a piece of amber, sealing-wax, or glass; which the philosopher rubbed against his coat, and presented to bits of paper, feathers, and other light bodies that came his way, and cost him nothing.[39]

Typical examples of eighteenth-century experiments of the first kind are the elucidation of the properties of electrical attraction or the electric spark, in order to answer such questions as: does electrical attraction follow the inverse-square law of gravitation, or can the electric spark ignite inflammable material? This last question was also considered to be relevant to chemistry. For if the electric spark was found to be hot, then it might be identified with the phlogiston or material fire of chemical reactions. Thus, newly discovered properties by improved apparatus could lead to unforseen directions of study, in this case of electrochemical phenomena.

38. Hackmann (1970), pp. 225–6, and *idem* (1979), p. 222; Van Marum (1781).
39. Priestley (1775), 2, pp. 86–7.

Hauksbee's 'chafing machine'

Francis Hauksbee, a strict adherent to Newton's dictum not to frame hypotheses, followed this experimental procedure in the early 1700s. He tried to discover the reason for the 'mercurial phosphor', earlier observed by Jean Picard in his barometer. From a sequence of carefully devised experiments, he drew the conclusion that this strange phenomenon was caused by the friction of mercury on the glass, and that the same effect could be produced by rubbing glass with wool. As he put it more carefully, 'the Light produc'd proceeds from some Quality in the Glass (upon such a Friction or Motion given to it) and not from the Mercury upon any other Account, than only as it is a proper Body, which, by beating or rubbing on the Glass produces Light', but added, 'nothwithstanding all this, the Matter is doubtful; and there may (for all that we know) be a Luminous Quality in the Mercury, as well as in the Glass or other Bodies'.[40] This conclusion demonstrates that he could not divorce himself entirely from Descartes' theory of light according to which the luminosity of a body is due to an internal motion of its particles. This theory was widely held by Hauksbee's contemporaries Boyle, Hooke and Newton.

In the course of his laboratory investigations he devised two pieces of apparatus: the first was his 'mercurial phosphor' chafing apparatus in which what we now call electroluminescence was produced by rubbing amber or glass in the partial vacuum of the bell-jar of an air pump. The second was his rubbed glass globe machine for investigating the action of electricity on threads suspended inside and outside the globe (Fig.1.4). Hauksbee's experiments, significantly, did not result in a general recognition that the luminescence produced by the mercury barometer was electrical. In his *Physico-Mechanical Experiments* (1709), for which the *Opticks* was obviously the model, the mercury experiments and those dealing with electricity are kept in separate sections. The possible electrical nature of both sets is only made by inference. Those repeating these experiments in the early part of the century, such as the German rationalist philosopher Christian Wolff, Du Fay and Desaguliers, also kept separate those experiments dealing with barometric luminescence from the electric ones with rubbed glass. They are dealt with in the same way in the first major textbook on physics, 's Gravesande's *Mathematical Elements of Natural Philosophy, Confirm'd by Experiments; or; An Introduction to Sir Isaac Newton's Philosophy* (1720–21), in which Hauksbee's two sets of experiments are summarised in a

40. Hauksbee (1709), p. 49; Freudenthal (1981), p. 209.

(a)

(b)

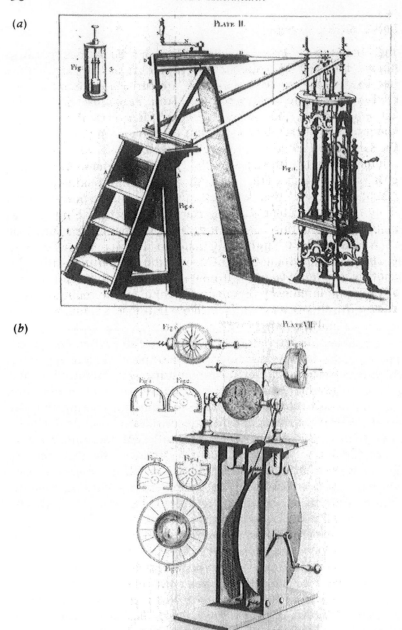

Figure 1.4. *(a)* Hauksbee's 'mercurial phosphor' chafing device, and *(b)* globe
'electrical' machine, 1705. (Hauksbee, *Physico-Mechanical Experiments*, 1709,
Plates II and VII)

chapter under the heading of 'The Fire adheres to Bodies, and is contain'd in them; where we shall also speak of Electricity'.[41] The first convincing demonstration by means of laboratory apparatus that (electrical) attraction occurred on the outside of the barometer tube at the same time that the mercury glowed within was not made until 1741, and the best experimental proof was published four years later by the Berlin Royal Academy.[42] This proof was accepted because it fitted in with the theoretical framework that had grown up around the subject at that time.

A careful analysis of Hauksbee's research programme shows how his concepts were modified to keep step with his experimental results, and how this led to the construction of the first true frictional electrical machine (a designation never used by him). This prototype electrostatic generator had much in common with the sulphur ball apparatus of Otto von Guericke, developed in the 1660s, but superficial similarities can be very misleading in the history of instrumentation. These two instruments, and Hauksbee's 'mercurial phosphor' chafing device, are with hindsight all electrostatic generators, but at the time they were developed in accordance with different theories. All three were 'theory-laden' or reified theorems; the metamorphoses of theories into tangible scientific instruments.

Von Guericke's apparatus was a model experiment to demonstrate that the nature of gravitation was not magnetic but electrical, caused by the friction of the air on the rotating Earth. The sulphur ball represented the Earth, which was made up mainly of 'sulphurous particles', and the hand with which it was rubbed while being rotated represented the atmosphere. According to an anonymous reviewer (probably Boyle or Hooke), with this globe the 'Impulsive, Attractive, Expulsive and other vertues of the Earth . . . . may be ocularly exhibited'.[43] Thus, in this experiment von Guericke had replaced the magnetic terrella model of Gilbert by an electrical one.

Hauksbee's 'mercurial phosphorus' chafing machine was derived from the physical theory that this glow required a vacuum and motion, while his prototype electrostatic generator has its origins in contemporary chemical theory linking phosphorescence with electricity. Amber was considered to be a natural phosphor composed largely of sulphur, and from this there was an easy transition to electrical experiments with rubbed glass. This transition from phosphor to the study of elec-

41. 's Gravesande (1720/1), 2, pp. 2-13.
42. Hackmann (1979), p. 209.
43. Hackmann (1978a), p. 25; anonymous review in the *Philosophical Transactions of the Royal Society*, 7 (1672), pp. 5103–5.

tricity also occurred in the simultaneous research itineraries of Hauksbee's contemporaries Samuel Wall, in London, and Pierre Polinière, in Paris, which reinforces the view that all three experiments were guided by the same physical and chemical theories.[44] Wall and Polinière both connected their small-scale laboratory researches on phosphors with the large-scale natural phenomenon of lightning, by suggesting that they were manifestations of the same sulphureous principle.

CONCEPTUAL MODELS OF BRASS AND WOOD

In the eighteenth century many attempts were made with laboratory scale-models to reproduce naturally occurring phenomena, such as earthquakes, the formation of waterspouts and whirlwinds, Northern lights, and the behaviour of lightning (Fig.1.5). Most of these were theoretically very trivial, but made dramatic subjects for the ever popular lecture–demonstrations. In one model, devised by Priestley in the late 1760s, a powerful discharge from a Leyden battery passed through a wet plank floating in a basin caused wooden blocks placed on top to tumble. This demonstrated the electrical nature of earthquakes: the wet plank represented the Earth and the blocks, buildings. In 1785 Priestley wrote to Van Marum encouraging him to improve on this model experiment with his huge generator.[45]

Many eighteenth-century model experiments dealt with atmospheric electricity, in which most theories could only be 'tested' by means of laboratory models. It was generally assumed that cloud formation was an electrical phenomenon, although there was much debate about the details of the mechanism. Most atmospheric phenomena were thought to be caused by Nature attempting to restore imbalances in the amount of electricity in different regions of the atmosphere. One of Van Marum's most striking experiments with 'artificial clouds' was performed before the members of his scientific society in Haarlem in 1787. It was typical of the mixture of theory and showmanship found in many of these experiments. He made two 'artificial clouds' from large placentas filled with hydrogen gas. Both were charged by his electrical machine; one positively and the other negatively. This caused them to rise and also to approach each other slowly. When a short distance apart, a spark jumped across the gap and both clouds immediately began their descent.

44. Freudenthal (1981), pp. 225–6. On his experimental philosophy in the context of the French attitudes, see Home (1985).
45. Hackmann (1970), pp. 239–40 for the primary references.

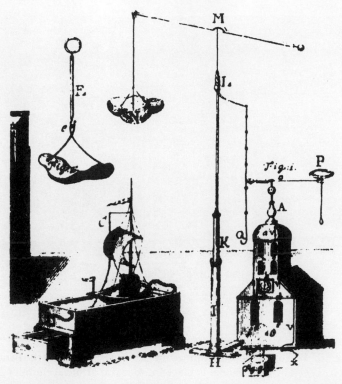

Figure 1.5. Model lightning experiments, 1780. (J. Langenbucher, *Beschreibung einer Beträchtlich Verbesserten Elektrisiermachine*, Augsburg, 1780)

Van Marum argues that this experiment demonstrated the observed behaviour of clouds before and during a thunderstorm. To make it more spectacular for the unscientific members of his audience, he arranged for a third balloon filled with a mixture of hydrogen and air to be placed between the two clouds so that it exploded with a resounding bang when the spark occurred.[46]

Although, as we have seen, others had noted the similarity between an electric spark and lightning, it was Franklin who, by means of small-scale model experiments suggested the essential similarity between artificially-produced electricity in the laboratory and the lightning flash, and proposed the famous kite experiment. An attempt to replicate the latter gave Professor Richmann the dubious honour of being the

46. Ibid., pp. 223–4. For other examples, see Heilbron (1980), pp. 373–4. The development of the air condenser aided simulations of the behaviour of electrified clouds in the laboratory.

Figure 1.6. Wilson's model lightning conductor experiments at the Pantheon, London, 1777. The scale model is placed under the artificial thundercloud of pasteboard and tinfoil. (Wilson, *An Account of Experiments made at the Pantheon on the Nature and Use of Conductors*, 1778)

first martyr to the new science of electricity, when he lost his life in St Petersburg in 1753, because of insufficient knowledge about earthing his conductor.

The most impressive of the model lightning experiments was performed in 1777 by Benjamin Wilson at the Pantheon in London, in the presence of King George III (Fig.1.6). His electrical machine was of conventional size, but its conductor was extremely large, made of pasteboard tubes coated with tinfoil, 155 feet long and about 16 inches in diameter, and suspended by silk cords from the theatre's ceiling. The discharge could be augmented further by connecting to it a second conductor consisting of 3,900 yards of copper wire. This arrangement, although less efficient than a good-sized Leyden battery, represented an electrified cloud, under which he placed a model house, representing the Purfleet arsenal, complete with lightning conductor. Wilson's aim was to demonstrate that the lightning conductor terminating in a ball was less dangerous than Franklin's design ending in a sharp point, as his model ball-shaped conductor did not attract the lightning discharge as readily from the artificial cloud as the pointed version. In reality the phenomenon is much more complex than was then appreciated, and it makes little difference in the real world whether a lightning conductor terminates in a sharp point or in a small ball.[47] George III's acceptance of Wilson's demonstration had more to do with political considerations. The American War of Independence had made even Franklin's scientific opinions unpopular. Wilson's public experiment had the hallmarks to make it acceptable. Its impresive size made it more persuasive; it was clearly an exercise in propaganda. The results were reasonably visible to the audience; after all, 'seeing is believing'. Furthermore, it fitted squarely into the contemporary theoretical framework of electrical action and the behaviour of the lightning conductor, according to which electricity behaved like a fluid, and the lightning conductor like a pipe – what Sir Oliver Lodge in 1892 was to call the 'rainpipe analogy'.[48] Franklin's experiments followed the same conceptual model.

Cavendish's model electric fish

Model experiments ranged from the flamboyant, with perhaps little scientific content, performed before fashionable audiences, to those which made significant contributions. A particularly elegant and fruitful model experiment was performed by Cavendish in 1776 in order to solve the vexing problem of how the torpedo fish could give shocks without producing sparks.[49] He demonstrated with conventional laboratory apparatus (Leyden battery, wires, and a Henley (repulsion) and Lane (spark) electrometer) that the fish's ability rested on two conditions; (a) on the nature of 'divided circuits', in which the path taken by the discharge depended on the different resistances of these circuits, and (b) on the relationship between the charge or quantity of electricity and its intensity. He confirmed his findings by replicating the results by means of an artificial torpedo (Fig.1.7). This he constructed of wood and leather, and placed in a trough filled with salt water of the same salinity as sea water. Its electric organ was represented by two pewter discs which could be connected to a large Leyden battery of 49 jars. This battery still produced shocks as powerful as the fish's

47. Wilson (1778); Hackmann (1979), p. 221.
48. Lodge (1892, chapter 28, especially p. 372). In the early 1800s Sir William Snow Harris could state that a lightning conductor was perfectly safe in a barrel of gunpowder!
49. Cavendish (1776), and Maxwell (1921), 1, pp. 9, 194–210, 284, 419–28.

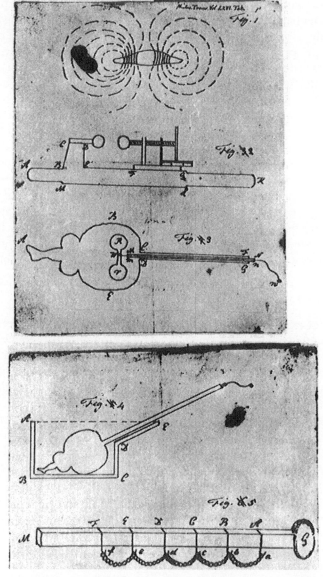

Figure 1.7. Cavendish's drawing of his artificial torpedo fish, 1776. These were copied in his *Philosophical Transactions* paper. The sheepskin body is given the same shape as the real fish, and the electrical organ is represented by the pewter discs (Rt) in his Fig. 3. Fig. 1 depicts the conjectured path of electricity in the water; Fig. 2 his spark-electrometer; and Fig. 5 the five links of a brass chain through which the shock of the model fish would pass. (Reproduced by permission of The Council of the Royal Society)

even though it was so weakly electrified that its spark could only be seen with a microscope. Priestley and Lane, who had doubted the torpedo's electrical nature, received shocks to their complete satisfaction. This experiment gained additional significance when Volta constructed his pile in 1800. He used the same arguments to show the electricity of this device, and called it an 'artificial electric organ', which was fundamentally the same as the torpedo. Neither the pile nor the torpedo give electrostatic signs because they operate at too low a tension.

Although experiments based on analogy and on models could be immensely fruitful, they could also be misleading. A purely practical problem of model experiments was that the laboratory results might well be different from real-world phenomena because of differences in scale. A more basic difficulty was that logically there was no reason why the phenomena recreated in the laboratory with models should be the same as the natural ones. This was the fundamental weakness of the analogous argument. Observations of natural symmetries led Gilbert to postulate that his spherical loadstone was surrounded by a uniform field of magnetic power or action (his *'orbis virtutis'*); a concept that was to have important consequences in the history of magnetism. It also induced him to construct his magnetic terrella with which he claimed that he could demonstrate the variation of magnetic dip with latitude. He argued that the magnetic compass was not attracted to the Pole Star but that this was a property of the Earth. But this concept of the Earth as a gigantic magnet also caused him to suggest that the Earth's magnetic axis was the same as the axis of rotation. We saw that the same type of analogous argument almost two hundred years later convinced Van Marum erroneously that plant vessels and arteries had similar structures. Gilbert's diagram of the observed behaviour of the magnetic needles in the *'orbis virtutis'* is far too regular, and one wonders to what extent this was based on preconceived ideas of symmetry. The same is true of his diagram depicting magnetic dip, which would require a much bigger magnetic terrella than the one illustrated for this effect to show up sensibly.

The main problem with the approach of eighteenth-century natural philosophy was the stress on data-gathering and the lack of mathematical idealisation or synthesis. Experiments were interpreted in terms of macroscopic observable forces and not synthesised to idealised forces between macroscopic particles. The shift in emphasis began in the late eighteenth century with the new generation of mathematical physicists, such as Cavendish, in England, and Coulomb, in France. Their attitudes (very much based on the *Principia*), and their disregard for serious macroscopic explanatory models (such as the behaviour of electrostatics in terms of one or two fluids), were not fully appreciated by the older experimental philosophers like Volta, who remained too close to their experiments to be able to formulate general idealised laws. Instead, this older generation of experimentalists expressed specific rules – such as those of Van Musschenbroek on magnetism. Faraday, too, was in this sense a natural philosopher and not an experimental physicist. His wonderful capacity for seeing the pattern of the phenomenon in his mind's eye was one he shared with some of his eighteenth-century forebears.[50] Their experimental data were reworked and synthesised in the nineteenth century by Maxwell and others.

CONCLUSIONS

Physical phenomena were studied and described at two levels; instrumental and philosophical. At the shallower instrumental *macro-level*, laboratory-induced properties of the natural phenomena were investigated by means of experimental apparatus. At this level properties and processes were more or less directly observed and measured. Predictions were made concerning their behaviour under different conditions and these were tested with further apparatus. The development of new and improved experimental apparatus and laboratory procedures was essential in this process which could also lead to totally unexpected discoveries, such as the Leyden jar in 1745, which set off an entirely new train of ideas and experiments. This device had three important consequences. First, it made possible the study of much more powerful electrical phenomena, which led to further discoveries. Second, it helped to formulate new concepts about electrostatic action. These included the electric circuit, intensity or level of charge, capacity or area of coated surface electrified, quantity of charge, and indirectly, electrical resistance. Third, it led to the design of new electrical devices such as the electrophorus (1775) and the parallel-plate condenser (1778), both in essence Leyden jars with movable coatings. These were consequences of Franklin's one-fluid theory of the action of the Leyden jar, and they in turn led to the discovery of 'adhesive' or 'contact electricity' in the 1790s, resulting in the invention of the Voltaic pile in 1800. This extraordinary device started off another train of discoveries in both electricity and chemistry. Thus, the interaction between

50. De Pater (1979), pp. 122–7; Hackmann (1985), p. 112. On Faraday's capacity to visualise phenomena and translate this into illustrative experiments, see Gooding (1981); (1985), chapter 6.

instrumentation and scientific exploration and discovery flowed in both directions: advances in instruments led to new discoveries (or highlighted specific properties), and this in turn influenced instrument design.

At this level, anomalies in the expected behaviour of the apparatus not only caused philosophical debates about the validity of the experimental method, but also led to new theories. When startled natural philosophers were first confronted by the Voltaic pile (the first source of electric current) it appeared to them like a perpetual Leyden jar. Attempts to explain it by traditional principles failed. It was soon discovered that other factors such as current intensity and electrical resistance were required for an adequate theory. Price has focused on this aspect of instrumental science. He suggests that Kuhn's revolutionary changes, as embodied in his 'paradigms', might be due more to unexpected discoveries made in advances in contemporary technology than to leaps of theoretician's genius.[51] The link between successful instruments and the formulation of theories is indeed a complex one. We have seen that instrumental replicability was related to the development of an effective laboratory strategy. The factors that affected the operation of instruments often had to be isolated empirically. Such understanding took time to evolve.[52]

At the deeper philosophical *micro-level*, attempts were made to use instrumental results in theorising about the underlying mechanisms of phenomena, such as the corpuscular (or fluid) nature of light or electricity. Here no direct experimental intervention was possible, and the description was based on analogies and models. No clear-cut distinction was made between these conceptual models in the laboratory and the 'real world' phenomena experienced by the senses. It was at this level that most 'circular arguments' occurred. An *ad hoc* premise was formulated to explain a specific phenomenon and this was then used as a 'proof' for the original premise. According to George Adams, in 1787, Franklin's concept of a single electric fluid, of which an accumulation gave a positive charge and a deficiency a negative charge, was such a circular argument. The shape of the Leyden jar gave visual evidence for the hydrostatic model of a single fluid coursing through the circuit, but Franklin could only prove the existence of both electrical conditions (plus and minus) by demonstrating experimentally that they attracted

51. Price (1980, 1984). For an interesting paper on Kuhn's predecessor, Ludwick Fleck, see Buchdahl (1983).
52. On the problems of replicability in the case of electrometry, see Hackmann (1978b), pp. 10–14. New materials are also important in the evolution of instruments, see for instance, Burnett (1986), pp. 217–38. Much more work needs to be done on this subject.

each other. He then used the same phenomenon of attraction as a proof of the existence of positive and negative electricity. As Adams pointed out: 'Of this however there is not one proof, and all the attempts that have hitherto been made to prove it, are only arguing in a circle, or proving the thing by itself'.[53] A fundamental anomaly in Franklin's theory was the repulsion of negative electricity; to answer this was the impetus behind much of the progress made in this subject in the latter half of the eighteenth century. Priestley accepted Franklin's single-fluid theory rather than the two-fluid theory because it explained electrical phenomena more *simply*, but he made the important admission that neither could be proven.[54] Mathematical physicists like Coulomb could see this type of intellectual invention for what it was worth. The real value lay in inspiring new avenues of exploration.

In natural philosophy a complex link was established between instrumentation, theory and the observer. Instruments are tangible signposts to the technical barriers reached at a particular time. Examining their capabilities allows us to determine what was technically possible, and gives us insights into the problems faced by the early experimenters. Hence these artifacts should be regarded as valuable source material in the history of science.

53. Adams (1787), pp. 45, 79–81, 117–18, 142–6, 150–2. In the 5th edition (1799), enlarged by W. Jones, all these arguments are brought together (see pp. 107–31). Laudan (1981), pp. 21–5 distinguishes between the *macro-sciences* which deal with processes and properties which can be more or less *directly* observed and measured, and the *micro-entities* which are the unobservable entities (corpuscles, etc.) postulated by the theories formulated to explain the observable phenomena. Physical theories were not derived *from* experience but were tested in the laboratory *against* experience, and then entered the knowledge system. His analysis has highlighted the problems faced by the mechanical philosophers when defining the relationship between theoretical knowledge and sensory experience. In this chapter I have concentrated on the relationship between instrumentation and experiment (sensory experience).
54. Priestley (1775), 2, pp. 41–52.

References

Adams, G. (1787). *An Essay on Electricity*, 3rd edn. London: Hindmarsh.
d'Agostino, S. & Ianniello, M.G. (1980). Elettrometri e galvanometri dell'ottocento evoluzione strumentale e contesti teorici. *Annali dell'Istituto e Museo di Storia della Scienza di Firenze*, **5**(2), 69–82
Atwood, G. (1784). *A Treatise on the Rectilinear Motion and Rotation of Bodies, with a Description of Original Experiments Relative to the Subject*. Cambridge.
Bachelard, G. (1933). *Les Intuitions Atomistiques*. Paris: Presses Universitaires de France.

Bachelard, G. (1938). *La Formation de l'Esprit Scientifique.* Paris: Presses Universitaires de France.

Bachelard, G. (1951). *L'Activité Rationaliste de Physique Contemporaine.* Paris: Presses Universitaires de France.

Bacon, Francis, (1857–74). *Works,* 14 vols., ed. R.E. Ellis, J. Spedding & D. Heath. London: Longman.

Bennett, J.A. (1980). Robert Hooke as Mechanic and Natural Philosopher. *Notes and Records of the Royal Society,* **35,** 33–48.

Bennett, J.A. (1986). The mechanics' philosophy and the mechanical philosophy. *History of Science,* **24,** 1–28.

Biernson, G. (1972). Why did Newton see indigo in the spectrum? *American Journal of Physics,* **40,** 526–33.

Birch, T. (1772). *The Works of the Honourable Robert Boyle,* 6 vols., new edn. London.

Brenni, P. & Hackmann, W.D. (1984). Gli strumenti scientifici. In *L'Eredità Scientifica di Leopoldo Nobili,* pp. 29–45. Reggio Emilia: n.p.

Buchdahl, G. (1983). Styles of scientific thinking. In *Using History of Physics in Innovatory Physics Education,* ed. F. Bevilacqua & P.J. Kennedy. Pavia: Centro Studi per la Didattica della Facoltà di Scienze Matematiche, Fisichi e Naturali of the Università di Pavia and the International Commission on Physics Education.

Burnett, J. (1986). The use of new materials in the manufacture of scientific instruments, c. 1880–c. 1920. In *The History and Preservation of Chemical Instrumentation,* ed. J.T. Stock & M.V. Orna, pp. 217–38. Dordrecht and Boston: Reidel.

Cantor, G.N. (1985). Reading the book of nature: the relation between Faraday's religion and science. In *Faraday Rediscovered,* ed. D. Gooding & F.A.J.L. James, pp. 69–82. London: Macmillan; New York: Stockton Press.

Cantor, G.N. & M.J.S. Hodge. (1981). Introduction: major themes in the development of ether theories from the ancients to 1900. In *Conceptions of Ether, Studies in the History of Ether Theories 1740–1900,* ed. G.N. Cantor & M.J.S. Hodge, pp. 1–60. Cambridge: Cambridge University Press.

Cavendish, H. (1776). An account of some attempts to imitate the effects of the torpedo by electricity. *Philosophical Transactions of the Royal Society,* **66,** 196–225.

Centore, F.F. (1970). *Robert Hooke's Contributions to Mechanics.* The Hague: Nijhoff.

Cohen, I.B. (1980). *The Newtonian Revolution.* Cambridge: Cambridge University Press.

Crombie, A.C. (1953). *Robert Grosseteste and Origins of Experimental Science 1100–1700.* Oxford: Clarendon Press.

Crombie, A.C. (1958). Some aspects of Descartes' attitudes to hypothesis and experiment. *Actes du 2ème Symposium International d'Histoire des Sciences,* Florence.

Crombie, A.C. (1961). Quantification in medieval physics. In *Quantification: a History of the Meaning of Measurement in the Natural and Social Sciences,* ed. H. Woolf, pp. 13–30. Indianapolis: Bobbs-Merrill.

Crombie, A.C. (1963). Commentary on Lynn White's 'What accelerated technological progress in the Western Middle Ages?' In *Scientific Change*, ed. A.C. Crombie, pp. 316–23. London: Heinemann.

Crommelin, C.A. (1951). *Descriptive Catalogue of the Physical Instruments of the Eighteenth Century in the National Museum of the History of Science at Leyden.* Leyden: National Museum of the History of Science.

Drake, S. (1974, publ. 1975). Galileo's work on free fall in 1604. *Physis*, 16, 309-22.

Drake, S. (1975). The role of music in Galileo's experiments. *Scientific American*, 232, 98-194.

Espinasse, M. (1956). *Robert Hooke*. London: W. Heineman.

Freudenthal, G. (1981). Early electricity between chemistry and physics: the simultaneous itineraries of Francis Hauksbee, Samuel Wall, and Pierre Polinière. *Historical Studies of the Physical Sciences*, 11, 203–29.

Gabbey, A. (1985). The mechanical philosophy and its problems: mechanical explanations, impenetrability, and perpetual motion. In *Change and Progress in Modern Science*, ed. J.C. Pitt, pp. 9–84. Dordrecht: Boston; London: Reidel.

Gaukroger, S. (1976). Bachelard and the problem of epistomological analysis. *Studies in the History and Philosophy of Science*, 7, 189–244.

Gooding, D. (1980). Metaphysics versus measurement: the conversion and conservation of force in Faraday's physics. *Annals of Science*, 37, 1–29.

Gooding, D. (1981). Final steps to the field theory: Faraday's study of magnetic phenomena, 1845–1850. *Historical Studies in the Physical Sciences*, 11, 231–75.

Gooding, D. (1982). Empiricism in practice: teleology, economy, and observation in Faraday's physics. *Isis*, 73, 46–67.

Gooding, D. (1985). Experiment and concept formation in electromagnetic science and technology in England in the 1820s. *History and Technology*, 2, 151–76.

Gooding, D. & James, F.A.J.L. (ed.) (1985). *Faraday Rediscovered*. London: Macmillan; New York: Stockton Press

's Gravesande, W.J. (1720/1). *Mathematical Elements of Natural Philosophy*, 2 vols. London: Senex.

Greenslade, T.S. (1985). Atwood's machine. *The Physics Teacher*, 23, 24–8.

Hacking, I. (1983). *Representing and Intervening*. Cambridge University Press.

Hackmann, W.D. (1970). The electrical researches of Martinus Van Marum (1750–1837). Unpublished M.A. thesis, Queen's University, Belfast.

Hackmann, W.D. (1971). Electrical Researches. In *Martinus Van Marum: Life and Work*, vol.3, ed. R.J. Forbes, *et al.*, pp. 329–78. Leiden: Noordhoff for the Hollandsche Maatschappij der Wetenschappen.

Hackmann, W.D. (1972). The researches of Dr. Martinus Van Marum (1750–1837) on the influence of electricity on animals and plants. *Medical History*, 16, 11–26.

Hackmann, W.D. (1973). *John and Jonathan Cuthbertson: The Invention and Development of the Eighteenth-Century Plate Electrical Machine*. Leiden: Museum Boerhaave.

Hackmann, W.D. (1978a). *Electricity from Glass: The History of the Frictional Electrical Machine 1600-1850*. Alphen aan den Rijn: Sijthoff and Noordhoff.

Hackmann, W.D. (1978b). Eighteenth century electrostatic measuring devices. *Annali*, **3**, 3–58.

Hackmann, W.D. (1979). The relationship between concept and instrument design in eighteenth-century experimental science. *Annals of Science*, **26**, 205–24.

Hackmann, W.D. (1985). Instrumentation in the theory and practice of science: scientific instruments as evidence and as an aid to discovery. *Annali*, **10 (2)**, 87–115.

Harré, R. (1980). Knowledge. In *The Ferment of Knowledge. Studies in the Historiography of Eighteenth-Century Science*, ed. G.S. Rousseau & R. Porter, pp. 11–54. Cambridge: Cambridge University Press.

Hauksbee, F. (1709). *Physico-Mechanical Experiments on Various Subjects, Containing an Account of Several Surprising Phenomenon Touching Light and Electricity*. London.

Heilbron, J.L. (1979). *Electricity in the Seventeenth and Eighteenth Centuries*. Berkeley: University of California Press.

Heilbron, J.L. (1980). Experimental natural philosophy. In *The Ferment of Knowledge. Studies in the Historiography of Eighteenth-Century Science*, ed. G.S. Rousseau & R. Porter, pp. 357-87. Cambridge: Cambridge University Press.

Heilbron, J.L. (1981). The electrical field before Faraday. In *Conceptions of Ether, Studies 1740–1900*, ed. G.N. Cantor & M.J.S. Hodge, pp. 187–214. Cambridge: Cambridge University Press.

Heilbron, J.L. (1983). *Physics at the Royal Society during Newton's Presidency*. Los Angeles: William Andrews Clark Memorial Library, UCLA.

Hesse, M. (1966). *Models and Analogies in Science*. Notre Dame, Ind.: University of Notre Dame.

Home, R.W. (1985). The notion of experimental physics in early eighteenth-century France. In *Change and Progress in Modern Science*, ed. J.C. Pitt, pp. 107–32. Dordrecht and Boston: Reidel.

Hooke, R. (1665). *Micrographia: or Some Physiological Descriptions of Minute Bodies Made by Magnifying Glasses*. London.

Kuhn, T.S. (1976). Mathematical versus experimental traditions in the development of physical science. *Journal of Interdisciplinary History*, **7**, 1–31, reprinted in his (1977) *The Essential Tension*. Chicago and London: University of Chicago Press.

Laudan, L.L. (1970). Thomas Reid and the Newtonian turn of British methodological thought. In *The Methodological Heritage of Newton*, ed. R.E. Butts & J.W. Davis, pp. 103–31. Oxford: Blackwell.

Laudan, L.L. (1981). *Science and Hypothesis*, vol. 19, University of Western Ontario Series in Philosophy of Science. Dordrecht, Boston and London: Reidel.

Leatherdale, W.H. (1974). *The Role of Analogy, Models and Metaphor in Science*. Amsterdam and Oxford: North Holland; New York: American Elsevier.

Levere, T.H. (1969) Martinus Van Marum and the introduction of Lavoisier's

64   W.D. Hackmann

Lohne, J.A. (1968). Experimentum crucis. Notes and Records of the Royal Society, 23, 169–99.
McConnell, A. (1980). Geomagnetic Instruments Before 1900. London: Harriet Wynter.
Maddison, F.R. (1963). Early astronomical and mathematical instruments. Brief survey of sources and modern studies. History of Science, 2, 17–50.
Maddison, F.R. (1969). Medieval Scientific Instruments and the Development of Navigational Instruments in the XV and XVI Centuries. Coimbra: Agrupamento de Estutos de Cartografia Antiga, series sep. XXX.
Marum, M. Van (1781). Antwoord op de vraag: door proeven aan te toonen, welke luchverhevelingen van de werking der natuurlijke electriciteit afhangen; dezelve 'er door worden voortgebracht; en welke de bekwaamste middelen zijn om onze huizen, schepen en personen tegen de schadelijken invloed derzelven te beveiligen. Verhandelingen Betaafsch Genootschap der Proefondervindelijke Wijsbegeerte, 6, 1–18, 23–64.
Maxwell, J.C. (ed.)(1921) The Electrical Researches of the Honourable H. Cavendish, 2 vol., rev. J. Larmor. Cambridge: Cambridge University Press.
Miller, J.D. (1975). Rowland's magnetic analogy to Ohm's law. Isis, 66, 230–41.
Naylor, R.H. (1980). The role of experiment in Galileo's early work on the law of fall. Annals of Science, 37, 363–78.
Newton, I. (1966). Mathematical Principles of Natural Philosophy, 2 vols. trans. A. Motte, rev. F. Cajori. Berkeley and Los Angeles: University of California Press.
North, J.D. (1981). Science and analogy. In On Scientific Discoveries, ed. M.D. Grmek, R.S. Cohen & G. Cimino, pp. 115–40. Dordrecht and Boston: Reidel.
Pater, C. de. (1979). Petrus van Musschenbroek (1692–1761): een Newtoniaans Natuuronderzoeker. Utrecht: Printed Ph.D. thesis.
Price, D. de Solla. (1980). Philosophical mechanism and mechanical philosophy. Annali, 5(1), 75–85.
Price, D. de Solla. (1984). Of sealing wax and string. Natural History, 93, 48–56.
Priestley, J. (1775). The History and Present State of Electricity, with Original Experiments, 3rd edn, 2 vols. London.
Roche, J. (1988). A critical study of symmetry in physics from Galileo to Newton. Barcelona Conference on the History of Ideas, 1983. Barcelona.
Rowbottom, M. (1968). The teaching of experimental philosophy in England, 1700–1730. Actes du XIe Congrès internationale d'Histoire des sciences, vol. 4. Warsaw.
Schaffer, S. (1980). Natural Philosophy. In The Ferment of Knowledge. Studies in the Historiography of Eighteenth-Century Science, ed. G.S. Rousseau & R. Porter, pp. 77– 91. Cambridge: Cambridge University Press.

Schaffer, S. (1983a). History of physical sciences. In *Information Sources in the History of Science and Medicine,* ed. P. Corsi & P. Weindling, pp. 285–316. London: Butterworth.

Schaffer, S. (1983b). Natural philosophy and public spectacle in the eighteenth century. *History of Science,* 21, 1–43.

Schofield, R.E. (1970). *Mechanism and Materialism.* Princeton: Princeton University Press.

Shapere, D. (1982). The concept of observation in science and philosophy. *Philosophy of Science,* **49,** 485–525

Shapin, S. & Schaffer, S. (1985). *Leviathan and the Air-Pump.* Princeton: Princeton University Press.

Stimson, A. (1976). The influence of the Royal Observatory at Greenwich upon the design of 17th and 18th Century angle-measuring instruments at sea. *Vistas in Astronomy,* **20,** 123–30.

Taylor, E.G.R. (1954). *The Mathematical Practitioners of Tudor and Stuart England.* Cambridge: Cambridge University Press for the Institute of Navigation.

Taylor, E.G.R. (1971). *The Haven Finding Art: a History of Navigation from Odysseus to Captain Cook,* rev. edn. London: Hollis and Carter for the Institute of Navigation.

Turner, G. L'E. (1983a). Scientific instruments. In *Information Sources in the History of Science and Medicine,* ed. P. Corsi & P. Weindling, pp. 243–58. London: Butterworths.

Turner, G.L.E. (1983b) Mathematical instrument-making in London in the sixteenth century. In *English Map-making 1500–1650,* ed. S. Tyacke, pp. 93–106. London: The British Library.

Van Helden, A. (1983). The birth of the modern scientific instrument 1550–1700. In *The Uses of Science in the Age of Newton,* ed. J.G. Burke, pp. 49–84. Berkeley, Los Angeles and London: University of California Press

Waller, R. (1705). *The Posthumous Works of Robert Hooke.* London.

Waters, D.W. (1958). *The Art of Navigation in England in Elizabethan and Early Stuart Times.* London: Hollis and Carter

Waters, D.W. (1966). *The Sea- or Mariner's Astrolabe.* Coimbra: Agrupamento de Estudos de Cartografia Antiga, Secqão, XV.

Wilson, B. (1778). *An Account of Experiments Made at the Pantheon on the Nature and Use of Conductors.* London.

2

GLASS WORKS:
NEWTON'S PRISMS AND
THE USES OF EXPERIMENT

SIMON SCHAFFER

'Instruments are in truth reified theorems.'
(Gaston Bachelard, 1933)

'Perhaps, said the Marchioness, Nature has reserved the
Merit of demonstrating Truth to the English prisms; that is,
to those by whose means she at first discovered herself.'
(Francesco Algarotti, 1737)

Experimental controversy involves contest about authority. The acceptance of a matter of fact on the basis of an experimental report involves
conceding authority to the reporter and to the instruments used in the
experiment. In the seventeenth century, experimental philosophers
used a wide range of means to make authority for their work. Conviction was thought to result from a long series of trials or from a single
decisive experiment. It might result from being present as a witness at
such a trial, achieving a replication of such a trial, or by reading a
report given in so much circumstantial details that such direct witnessing was obviated. Authority might be held to lie in the credit of a
single experimenter or in the communal assent of the experimental
community. In their controversies, experimental philosophers often
strenuously debated these differing ways of making conviction. Such
fights show some of the uses of experiment in reaching agreement
among disputants. Furthermore, they show how experimental instruments play a central role in these usages, and are resources which
experimenters deploy in their struggles to achieve authority.[1]

1. For examples of experimenters' ambiguities about the significance of many trials or
   of uniquely decisive ones, and about the relative importance of direct or 'virtual'
   witnessing, see Boyle (1664), sigs. A2-A4; Shapin (1984).

The controversy discussed in this paper centred on Newton's work on light and colours between the 1660s and the 1720s. His trials with prisms, notably the celebrated *experimentum crucis*, were emblematic of experimental philosophy. They became so in at least two ways. After 1704, Newton claimed his trials had been replicated by competent experimenters, making facts which natural philosophers must acknowledge and use in their own work. Second, Newton also claimed that amongst these trials it was possible to pick out those which were 'crucial', and which would decisively settle dispute. The authority of these emblems was retrospectively located in the events surrounding the first public announcements Newton made in the early 1670s. But these emblematic uses were developed over several decades – they were not swiftly achieved, nor were they ever uncontested. Newton's programme was a site at which natural philosophers debated the boundary between experiments, 'concluding directly and without any suspicion of doubt', and hypotheses, 'conjectured by barely inferring 'tis thus because not otherwise or because it satisfies all phaenomena'. To some of his critics, Newton seemed to violate the rules of the experimental life. He was attacked as dogmatical, overestimating the authority due to his reports, providing too few trials to license his conclusions, and reporting experiments which could not be replicated. Thus, in contrast to Robert Boyle's celebrated emphases on 'histories' of many trials, Newton interrupted his first published account of optical trials to state that 'the historicall narration of these experiments would make a discourse too tedious & confused, & therefore I shall rather lay down the *Doctrine* first and then, for its examination, give you an instance or two of the *Experiments*, as a specimen of the rest'. Comments such as these highlighted differences over proper conduct in experimental work and reportage.[2]

Both the problem of the 'cruciality' of experiments and that of their 'replicability' are typical of the experimental sciences. It is misleading to treat the authority of such experiments as self-evident, for this obscures the detailed character of experimental controversy. The ground of such authority was often the matter in dispute. The resolution of such disputes masks the process by which agreement is accomplished. Agreement includes consensus about the conduct and meaning of a particular trial. Where experiments are interpreted as conveying unarguable lessons about the contents of Nature, this indicates that a controversy has already reached a stage of provisional closure. Only then will experiments be defined through an exemplary method, standardised tools and an agreed matter of fact. This paper examines the

2. Newton (1959–77), 1, pp. 96–7, 209. For comments see Bechler (1974); Dear (1985).

career of one of Newton's experiments, a trial with two prisms which he first recorded in a notebook in 1666 as the 'forty-fourth' of a long list of experiments on light and colour. In significantly changed format, this trial was made into an *'experimentum crucis'*. Newton's *experimentum crucis* was the object of considerable debate during the 1670s and has remained a central topic of philosophical and historical attention. The term was not used in his notebooks, drafts or lectures before 1672, nor did it appear in the *Opticks* in 1704. Nevertheless, the label remained current among Newton's readers and disciples. However, the *reference* of the term changed markedly. There was no consensus among the experimenters on the lesson which its author intended should be taught by this trial, nor on the proper method for conducting it. Some tried their version of this experiment, obtained results different from those which they held Newton had reported and rejected Newton's account of light and colour. These experimenters treated the trial as crucial but used that cruciality to undermine Newton's theory. Others, notably Robert Hooke, replicated Newton's trial, but then argued that the trial was not crucial, and denied that this replication licensed Newton's account. 'Cruciality' was an accomplishment which varied with outcomes of attempted replication.

The character of this accomplishment was intimately connected with issues of instrumentation, specifically, with the evaluations experimenters gave of the quality and arrangement of their *prisms*. There was no uncontroversial way of making these evaluations authoritative. For one community of experimenters during one period of time, Newton's *experimentum crucis* could be associated with an obvious procedure, involving complex arrangements of specially crafted prisms and lenses and a self-evident matter of fact, involving the chromatic homogeneity and the fixed refrangibility of primitive colour-making light rays. In the crucial experiment, a prism was used to make 'primitive' rays, and then one of these rays subjected to a second refraction in a second prism. Newton sometimes claimed that if white light were transmitted through a prism it could be separated into a set of 'primitive' colour-making rays. A properly separated 'primitive' ray could not then be further divided by transmission through another prism. It was a standing difficulty that many of Newton's critics reported that they could split putatively 'primitive' rays into further colours. But for Newton and his allies, a 'primitive' ray could be simply defined as a ray which could not be split by a second refraction. Then experimenters who managed to split such a ray could be criticised by Newton for their failure to produce 'primitive' rays. This argument established a troubling circle, akin to what H.M. Collins calls the 'experimenter's regress'. The criterion of a good experiment was that it produced the matter of fact which Newton sought to establish. Experimenters had to be convinced of this matter of fact before they could share this criterion. Once conviction had been achieved, then this criterion seemed unchallengeable. After closure, the procedures for making 'primitive' rays became self-evident. This paper documents the process by which this self-evidence was accomplished.[3]

The unarguable meaning which individuates an experiment is not achieved without struggle. How, then, do experiments acquire their identity? Scientific instruments play a decisive role in this process. Newton's arguments suggested that good prisms were those which made 'primitive' rays. His critics were told they were using bad prisms. Instruments help make experiments compelling, because the self-evidence which is attached to instrumental procedures after closure links complex experiments to agreed matters of fact. This closure makes instruments into what are seen as uncontestable transmitters of messages from nature, that is, it makes them 'transparent'. This process is comparable to what Trevor Pinch calls the 'black boxing' of instruments: he argues that after closure 'the social struggle over a piece of knowledge has become embedded in a piece of apparatus'. Such pieces can then be treated as if they 'regularly produce reliable and uncontentious information about the natural world'. Prisms have become so uncontentious that it is now hard to recapture the sense of their contingent and controversial use. Yet it is that contingent and controversial use which must be recovered in order to understand how 'transparency' is accomplished.[4]

It must also be stressed that 'transparency' is not necessarily achieved permanently. The 'transparency' of instruments may vary during controversy. Protagonists in disputes may engage in the 'deconstruction' of provisionally achieved 'transparency'. On occasion it may be useful to emphasise the specific complexities of an instrument in order to defend observation reports against criticism. Newton sometimes argued that the failings of prisms explained troubled experimental results. At other times, Newton and his critics minimised the role of their instruments in order to highlight what they claimed were basic conceptual disagreements. Furthermore, the accomplishment of 'transparency' sus-

---

3. For 'experimenter's regress' and replication, see Collins (1985), pp. 79–100, 129–30. For discussion of the process by which 'open' settings reach closure in experimental dispute, see Latour (1987).
4. Pinch (1986), pp. 212–14.

tains a realist history of experimental argument. Outcomes of earlier debates are then attributed to the 'obvious virtues' of instruments, rather than to the complex of practices and presuppositions which govern organised experimentation. When Newton's supporters claimed that his critics had chosen the wrong instruments, this allowed the further claim that Nature clearly spoke of the truth of Newton's theory of light and colour. But a more considered history of the prisms used in these arguments shows how this claim was accomplished as part of the provisional, local closure of the optical controversy.[5]

THE PRISM BECOMES 'THE USEFULLEST INSTRUMENT' (1660–66)

Newton's reports of trials with lenses, mirrors and prisms, first delivered to the public after 1670, were connected with the active interest of his contemporaries in telescope and microscope design. This instrumental context will be considered first, before analysing the means by which the 'cruciality' of Newton's trials was connected with the troubles of such optical devices. This connection with glass working gave Newton's reports much of their immediate impact in London. When he started lecturing on light and colour at Cambridge in January 1670, Newton began by suggesting that attempts to grind conical glasses to avoid spherical aberration were as futile as efforts 'to plough the seashore'. He then indicated that even if such conics could be produced, there was a 'property inherent in the nature of light' which prevented the 'perfection of dioptrics'. This property was the specific refrangibility of primitive colour-making rays.[6] In 1672, Newton repeated these views in his letter to the Secretary of the Royal Society, Henry Oldenburg. In a carefully crafted reminiscence, Newton claimed that in early 1666 he had been working on the grinding of non-spherical glasses, when the understanding that 'Light it self is a *Heterogeneous mixture of differently refrangible rays*' prompted his abandonment of 'my aforesaid Glassworks'.[7] The tangible products of Newton's glass-works were versions of a reflecting telescope which would avoid some of the difficulties of aberration. The letter of February 1672 followed hard upon the demonstration of one version of this telescope at the Royal Society in January and Newton's election to the Society's fellowship.[8]

5. For the 'deconstruction' of experimental set-ups in controversy see Pickering (1981); Pinch (1981).
6. Newton (1967–81), 3, pp. 438–9; Newton (1984–), 1, p. 49.
7. Newton (1959–77), 1, p. 95. For Wren's contemporary work on nonspherical lenses see Bennett (1982), pp. 34–8.
8. Birch (1756–57), 3, p.4; Newton (1959–77), 1, pp. 3–4, 73–76; Mills and Turvey (1979); Newton (1984–), 1, pp. 427–428.

The range of optical instruments which Newton discussed posed different problems for his colleagues. The status of prisms in experimental optics was rather different from that of telescopes and microscopes. There was little technical work on the design and improvement of prisms, but optical instruments using lenses and mirrors were recognised as troublesome and in need of improvement. Newton could rely on his colleagues' interests in the latter, but he had to assume they had little interest in the former. Following the argument of Descartes in his *Dioptrique*, natural philosophers were aware of the problem of spherical aberration and they debated methods by which non-spherical lenses could be formed.[9] Yet it is unlikely that any parabolic lens was produced in the seventeenth century.[10] There was also much debate between Robert Hooke and others such as the French astronomer Adrien Auzout on the relative quality of various kinds of glass, since optical blanks were supposed to be as free as possible of veins and bubbles.[11] In the 1660s and 1670s, Venetian glass remained the standard against which other glass was to be compared. The role of glass-workers in Italy and elsewhere was also important. Newton's relations with the London glass-maker Christopher Cock were a significant part of his efforts to improve telescopes.[12] In general, however, the secretive practices of glass-makers were not easily subjected to enquiry. During the 1670s, glass workers were beginning to produce 'flint' glass by the addition of lead oxide, which had a high refractive index and was suitable for optical display, such as chandeliers, but which also tended to crack too easily. Most lenses of the period were full of air bubbles and flaws.[13]

The immediate response to Newton's views was partly governed by this work on the glass and metal technology of optical instruments. He debated technical issues with such as Hooke, Auzout and the Scottish mathematician James Gregory.[14] The controversy with Hooke explicitly raised the problem of whether further improvements could be expected

9. For Hooke and Newton on the making of corrected lenses, see Hooke (1665), sig. e2^r; Hooke (1666); Newton (1959–77), 1, pp. 4, 53–4, 95; Bechler (1975), pp. 104–6.
10. Van Helden (1974).
11. Hooke (1665), sig. e1^v; Oldenburg (1965–86), 2, pp. 383–9, 420, 468–9; for early eighteenth-century work on variations in glass quality see van der Star (1983), pp. 145–59.
12. For improvements of Venetian glass see Charleston (1957), pp. 218–21; Pedersen (1968), pp. 148–9; for Newton and instrument makers see Birch (1756–57), 3, pp. 4, 8, 19, 43; Newton (1959–77), 1, pp. 82–3, 85–7, 123–5, 185; Newton (1962), pp. 402–4.
13. Peddle (1921–22).
14. Newton (1959–77), 1, pp. 126–9. In March 1672, answering Auzout's worries about the performance of his mirrors, Newton proposed the inclusion of a crystal prism along the axis of the reflector.

from glass works. Hooke claimed priority for his own techniques for the improvement of telescopes by compound lenses. He told the Royal Society that 'I am a little troubled that this supposition should make Mr. Newton wholy lay aside the thoughts of improving microscopes and telescopes by Refractions'.[15] Newton composed a series of drafts in answer to Hooke which denied that he had given up the hope of improving optical instruments by lens designs.[16] He claimed he had a method for correcting aberration without conic sections. He now told Oldenburg that 'I examined what may be done not onely by *Glasses alone, but more especially by a complication of divers successive Mediums,* as by two or more *Glasses* or *Chrystalls* with *water* or some other fluid between them, all wch together may performe the office of one *Glasse'.* Such instruments were obvious and visible matters of dispute between Newton and his fellow experimenters and instrument makers.[17]

Unlike lenses and mirrors, prisms did not figure significantly in the construction of the instruments of astronomy and microscopy. Indeed, it is not clear why prisms were commercially available to Restoration natural philosophers. It has been suggested that they were used in chandeliers, or as toys. There is some evidence that well into the seventeenth-century prisms were seen as playfully deceitful and that the production of prismatic colours was indeed a common entertainment. The prismatic phenomena appeared in texts of natural magic, some of which Newton read in the 1650s. There were several anecdotes of 'a most pleasant and delightfull experiment' using 'a three square cristall prisme' to cast coloured images. For the Catholic natural philosopher Thomas White, writing in exile in Paris in the 1650s, it was still a commonplace that in 'Prismaticall glasses . . . we are pleas'd to know our selves delightfully cosen'd'. His colleague Kenelm Digby reported that triangular prisms were commonly known as 'Fools Paradises'. The transformation of prisms into instruments of experimental philosophy would have been a marked displacement of their use and significance.[18] On the other hand, there were several references to the use of crystalline prisms in medieval and Renaissance texts, particularly in association with the production of the rainbow. The 1660s saw a widespread deploy-

15. Birch (1756-57), 3, pp.4, 8, 10–15; Newton (1959–77), 1, p. 111.
16. Hall (1955); Newton (1967–81), 3, pp. 442–3, 512–13 n. 61; Bechler (1975), pp. 109–13; Shapiro (1979); Newton (1984–), 1, p. 429.
17. Newton (1959–77), 1, pp. 172, 191-2.
18. Dollond (1758); King (1955), pp. 144–50; Bechler (1975), pp. 125–6; Cantor (1983), pp. 64–9 (for optical instruments and lenses). Birch (1756–57), 3, p.41; Lohne (1961), pp. 393–4; Mills (1981), p. 14, for Newton's work on prisms. Peacham (1634), p. 140; Bate (1654), pp. 150–1; della Porta (1658), pp. 355–70; White (1654), p. 181; Digby (1669), p. 323; Huxley (1959), for comments on prism trials in other texts.

ment of prisms in the detailed investigation of the production of colours. As Alan Shapiro has suggested, they did so partly because of important challenges to scholastic theories of light and colour mounted by Cartesians. The schoolmen often distinguished two kinds of colour. *Emphatic* or *apparent* colours were those displayed through prisms or in the rainbow, where light was changed into colours through adjacent darkness, and *real* colours, disclosed in bodies by light but not produced by that light. In his *Dioptrique* and *Météores*, collected in an Amsterdam edition of 1656 and carefully studied by Newton during 1664, Descartes effaced this distinction. All colours were apparent, and thus all colours were produced the way prismatic colours were.[19]

This gave prisms a key new role in the analysis of colour. Prismatic colours could be seen as representations of the production of many other kinds of colour. In his analysis of the rainbow, Descartes reported a set of experiments using a prism to show how colours were produced in one refraction at the boundary between light and darkness. Such Cartesian texts provided a warrant for analysing colours with prisms, and also provided a new and influential target for criticism. It was reported that Newton bought his first prisms in the mid-1660s in order to attack Descartes's theories. Experimenters such as Robert Boyle and Robert Hooke began to write about prisms. But they did not yet make prisms into privileged instruments. In his *Micrographia*, Hooke coined the term *experimentum crucis* to describe an experiment he claimed decisively refuted Cartesian doctrine, but this was not a prism trial. Newton read *Micrographia* very carefully during 1665. Boyle's *Experiments and considerations touching colours* mainly reported trials with lenses, mirrors, and chemical tinctures and dyes. He used prisms sparingly. Newton's notes of the autumn of 1664 contain extensive comments drawn from Boyle, principally upon the varying appearance of colours in different situations and lights. Boyle reported just four experiments using prisms. Two of these involved the production of as many as four sets of emphatical colours, or 'irises', as Boyle called them, from rays of sunlight falling upon an equilateral crystal prism in a darkened room. In another pair of trials, Boyle sought to sustain the view that there was no difference between real and emphatic colours by casting the prismatic iris upon a 'really' coloured object. He sought to show that emphatic colours combined with real ones just as real colours did with each other. Prismatic blue shone on red cloth made it seem purple. Attempts to use prisms tinted with real colours were very troubled,

19. For the Cartesian and scholastic accounts of colour, see Westfall (1962), p. 343, 347; Lohne (1967); Lohne (1968), pp. 174–9; Sabra (1981), pp. 60–8; Newton (1983), pp. 246–9, 432, 434; Nakajima (1984); Newton (1984–), 1, p.4.

C A P V T VIII. 215

aquam egrediendo iterum fracti tendebant ad E. Nam fimul ac *oculum per-*
corpus aliquod opacum & obfcurum alicui linearum A B, B C, *veniunt*
C D, vel D E opponebam, rubicundus calor evanefcebat; & *poſt duas refractio-*
licet totam pilam, exceptis duobus punctis B & D obnuberem, *nes & u-*
& corpora obfcura ubivis circumponerem, dummodo nihil actio- *nam refle-*
nem radiorum A B C D impediret, lucide tamen ille refulge- *xionem: ex-*
bat. Poftea eodem modo inveftigatâ causâ rubri illius coloris, *autem ſeve*
qui apparebat in K, inveni, illum effe à radiis Solis, qui venien- *ſecundariâ ex radiis*
tes ab F ad G, ibi refrangebantur verfus H, & in H reflexi ad *poſt duas*
I, rurfufque ab I reflexi ad K, tandemque iterum fracti in pun- *refractiones*
cto K, tendebant ad E. Atque ita primaria Iris fit à radiis poft *& duas re-*
duas refractiones & unam reflexionem ad oculum venientibus: fe- *ad oculum*
cundaria verò à radiis, qui non nifi poft duas refractiones & duas *pervenien-*
reflexiones eodem pertingunt. Ideoque hæc femper alterâ minus *tibus; quo fiat ut illa*
eft confpicua. *ſit debilior.*

Sed fupererat adhuc præcipua difficultas, in eo quod, etiamfi po- IV.
fito alio ejus pilæ fitu, radii etiam poft duas refractiones & unam *Quomodo etiam ope*
aut duas reflexiones, ad oculum poffint pervenire, nulli tamen nifi *vitrei pri-*
in eo fitu, de quo jam locuti fumus, ejufmodi colores exhibeant. *ſinatis co-*
Atque ut hanc amolirer, inquifivi an non aliqua alia res inveniri poffet, *lores Iridis videantur.*
cujus ope colores eodem
moco apparerent, ut factâ ejus
comparatione cum aquæ guttis,
tanto facilius de eorum causâ ju-
dicarem. Et commodum recor-
datus, per prifma vel triangulum
ex Cryftallo fimiles videri, u-
num confideravi, quale eft
M N P, cujus duæ fupeficies
M N, & N P funt omnino
planæ, & una in alteram ita in-
clinata, ut angulum 30 vel 40
circiter graduum contineant;atque
ideo fi radii Solis A B C,
penetrent M N ad angulos rectos, aut fere rectos, ita ut nul-
lam notabilem refractionem vi-
trum ingrediendo patiantur, fatis magnam exeundo per N de-
beant

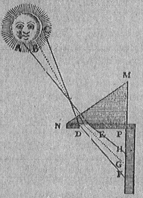

Figure 2.1. Descartes's prism trial: 'a prism or triangle of crystal' inflects rays
such that they 'paint all the colours of the rainbow' on a sheet of cloth or
white paper. Descartes argues that there must be at least one refraction and a
shadow to produce these colours. (Descartes, 'Meteors' in *Opera philosophica*,
1656, 3rd edn, p. 215. Amsterdam. This is the version Newton annotated in
1664–65. By permission of the Syndics of Cambridge University Library)

because the tints rendered the glass rather opaque. However, despite these difficulties, and the limited use Boyle made of his instrument, he emphasised the prism's status: it was 'the usefullest Instrument Men have yet imploy'd about the Contemplation of Colours' and 'the Instrument upon whose Effects we may the most Commodiously speculate the Nature of Emphatical Colours, (and perhaps that of Others too)'. Newton soon took up this suggestion, so changing the place of the prism in experimental optics.[20]

NEWTON TRANSFORMS THE USES OF THE 'GLASS-PRISM' (1666–72)

In his 1664 notes, Newton sought to emulate some of Boyle's colour-mixing trials with prisms but his experiments did not involve any prismatic projection, that is, they did not yet involve the casting of an iris upon a screen or wall. Instead, he examined coloured bands and threads by looking at these objects through a prism. These experiments, particularly that with a bicoloured thread examined through a prism, prompted the thought that blue-making rays were refracted more than red-making ones. Such an examination made the blue and red parts of the thread seem to separate from each other. Newton attributed this phenomenon to differing refrangibility. He did not record a projection until his manuscript 'Of colours', written in 1666 after his reading of Hooke's *Micrographia* and the beginning of his 'glass works'.[21]

In contrast with the strategies of Boyle and Hooke, Newton's manuscript of 1666 marked an important change in prism techniques. He used at least three different prisms separately and in combination. He also noted the use of a prism made of a 'four square vessell' of polished glass filled with water and a device constructed of two prisms tied together 'basis to basis'. Newton began changing the commercial 'triangular glass prism' into a complex experimental instrument. He recorded a long series of 'Experiments wth ye Prisme', two of which are particularly important. The seventh experiment involved a prismatic projection of an image across a space of at least 21 feet in a darkened room. This was designed to show that even when the prism was set so that light passed through it symmetrically, the prismatic image was oblong rather than circular, for from this Newton would argue that the shape of the image was due to the different refrangibility of different

20. For Newton's reading in Descartes, Boyle and Hooke, see Descartes (1656), p. 215; Boyle (1664), pp. 191–3, 224–9; Hooke (1665), p. 54; Lohne (1968), p. 179. Newton's notes are discussed in Hall (1955), pp. 27–8, 36–7; Westfall (1962), pp. 345–7; Mamiani (1976), pp. 81–94; Newton (1983), pp. 440–2, 452–62, 481.
21. Newton (1983), pp. 430, 440, 434, 467–8.

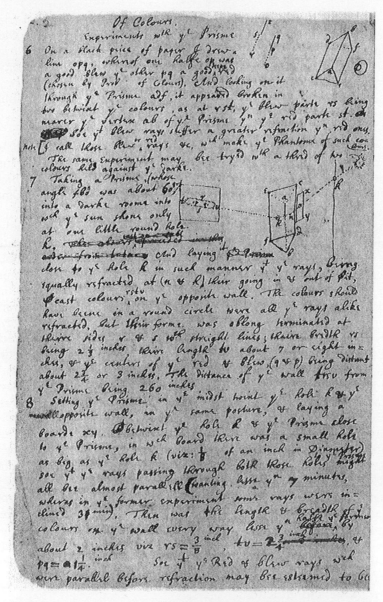

Figure 2.2. Newton's notes on the prismatic projection of light and the formation of a broadened spectrum, recorded in his notebook 'Of Colours' in 1666. 'The colours should have beene in a round circle . . . . but their forme was oblong'. (Cambridge University Library MSS Add 3975 f. 2. By permission of the Syndics of Cambridge University Library)

rays. This experiment appeared, carefully rewritten, as the first trial presented both in 1670 at Cambridge and in 1672 in his letter to Oldenburg, the 'celebrated phaenomena of colours'. The forty-fourth trial involved the use of a second prism to refract light rays again after their emergence from the first one. This experiment was designed to show that each ray had a specific refrangibility and made a specific colour. It was later to be substantially reworked as the celebrated *experimentum crucis*.

Both experiments fulfilled their role because of a set of claims about the way in which prisms worked. In order to understand these transformations in presentation and meaning and the persuasive role which Newton designed them to serve, it is necessary to consider the role and use of the prisms deployed in these trials.[22]

The claims about prisms which Newton made in these early notes remained tacit in the initial publication of his experiments. The dramatic innovation in the tactics of prism trials and the challenges to the utility of common prisms were not made visible. But controversy prompts protagonists to expose such tacit knowledge. During his trials of the 1660s and the controversies of the 1670s, Newton specified more details of how prisms should properly be prepared and used. He gave experimenters instructions about the *differences* between prisms which were commonly available and those which could best display the phenomena he reported. This implies that the provenance of Newton's own instruments is an important factor. However, none of the extant prisms associated with Newton seems to correspond to any of those whose use he describes. It is reported that Newton bought a prism at Stourbridge Fair in Cambridge in August 1665. Since 'he could not demonstrate' his hypothesis of colours against Descartes without a second one, he bought another there in 1666. But Newton left Cambridge before the 1665 Stourbridge fair and none was held in 1666. It has been suggested that Newton was recalling the fairs held at midsummer in those two years. A notebook also records the purchase of three prisms and collections of 'glasses' in London and Cambridge during 1668. As we have seen, the essay 'Of colours' suggests that Newton already had three prisms available to him during 1666. Further manuscripts of the period from 1668 record the purchase of optical machinery and work on lens grinding.[23]

22. Newton (1959–77), 1, pp. 92, 94; Newton (1983), pp. 478, 470, 472, 468.
23. For the provenance of Newton's prisms see Westfall (1980), pp. 156–8; Mills (1981), pp. 14–16, 27–32.

Two implications of these stories are of interest. First, prisms were evidently the sort of objects purchased at commercial fairs and in the City. They were correspondingly priced; prisms were relatively cheap tools for Newton's expanding programme of practical natural philosophy.[24] Second, while it would rapidly become clear to Newton that it was necessary to prepare prisms carefully for his optical trials, to concentrate attention on glass quality and prism design, nevertheless he gave no details of these protocols in his first communications with the London experimenters. As he put it bluntly in February 1672, 'I shall without further ceremony acquaint you that in the beginning of the year 1666 . . . I procured me a Triangular glass-Prisme'. Later in the letter he gave the dimensions of the prism and the refractive power of its glass (a value which indicates that this prism was unusual in containing some lead). Rather few indications were added in the body of the paper, which itself contained but four trials, save the instruction that the prism should be 'clear and colourless'. A second prism was invoked without any specification of its quality or geometry.

Newton's instructions proved insufficiently detailed for his audience. In experiments designed to show the important and controversial fact that 'uncompounded' rays could not be changed, Newton did not provide a recipe for making these 'primitive' or 'uncompounded' rays. Instead, he said that 'there should be perfecter separation of the Colours, than, after the manner above described, can be made by the Refraction of one single Prisme'. Evidently the perfect separation of an 'uncompounded' ray relied on special techniques in handling prisms. The separation of such rays was a novel feature of experimental optics. The existence of such rays was a novel feature of optical theory. Yet Newton relied on the familiarity of the common prism, and merely added that 'how to make such further separations will scarce be difficult to them, that consider the discovered laws of Refractions'. But these laws were precisely the matter of dispute. The subsequent career of Newton's *experimentum crucis* and the detailed interpretations of its author and critics show how vulnerable was the 'obviousness' of Newton's account and how important were the 'difficulties' of his instruments.[25]

24. For the prices of prisms see Newton (1936), pp. 52–3; Newton (1967–81), 1, pp. xii–xiii. Newton recorded that his three prisms of 1668 cost three shillings for the lot. This compares with purchases of glasses costing 14 shillings in Cambridge and another 16 shillings in London, eight shillings for a chemical furnace and £2 for chemicals during the same period.
25. Newton (1959–77), 1, pp. 92, 93, 100, 102.

NEWTON'S PRISMS AND HIS AUDIENCES (1670–72)

When he chose to give his first published account of his new doctrine of light and colours in early 1672, Newton helped himself to the rich resource of prism experiments which he had described in his Lucasian lectures at Cambridge. The contrast between these lectures and the version Newton released to his audience helps reveal how he sought to persuade that audience. In the lectures Newton described experiments using several prisms to show that light rays were differently refrangible and that differently refrangible rays displayed different colours. Each refracted colour-making ray was sent successively through a second prism onto a screen. In his sixth lecture, he drew the same conclusions from set-ups where the refracted rays were made to undergo total internal reflection in the second prism. Very few of these trials were then summarised for the Royal Society during the 1670s. In the lectures, no one trial appeared to be especially significant and most experiments seemed to need special equipment and technique. But in his communications with his fellow experimenters Newton made one trial 'crucial' and also suppressed most of the details of the procedures he had used.[26]

Nowhere in the earlier version of his lectures did Newton provide a clear demonstration of his doctrine of the immutability of the colour displayed by 'primitive' rays. There was no 'crucial' experiment. Instead, he rehearsed variations in the placing of screens, the illumination of the chamber in which the experiments were to be performed and the movement, position and quality of the prisms themselves. He proposed moving the first prism from its original place behind the first screen to a position between the sun and the screen. This was designed to remove the suspicion that the different angles of refraction of different rays might be due to different angles of incidence of sunlight at the first prism. He tried covering the leading side of the second prism with black paper pierced with a single hole, in order to admit only a few rays to the second refraction. In his sixth lecture he also changed the orientation of the two prisms so that sometimes they were crossed, and at others parallel.[27]

While enriching the possible tactics of prismatic trials, Newton also addressed the problem of prism quality and design. He was making prisms into experimental instruments. These instruments were supposed to demonstrate a novel and complex doctrine of the origin of

26. Newton (1984–), 1, pp. 95–9, 133–9.
27. Newton (1984–), 1, pp. 96–7, 134–5.

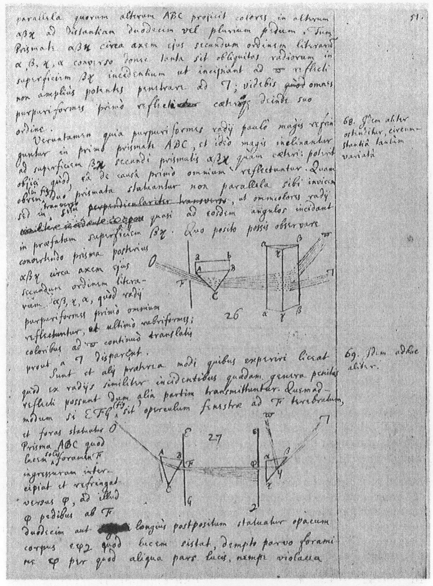

Figure 2.3. Newton's notes in the Lucasian lectures on variants of trials with two prisms, including experiments using total internal reflection (Fig. 27) to demonstrate the specific refrangibility of different 'primitive' rays. (Sixth optical lecture, Cambridge University Library MSS Add 4002 p. 51. By permission of the Syndics of Cambridge University Library)

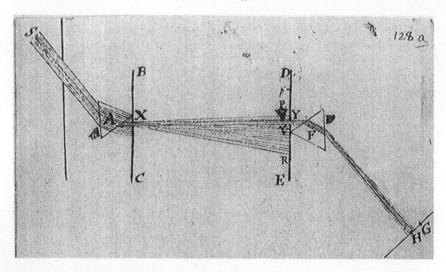

Figure 2.4. A loose sheet recording a preliminary version of the two prism experiment from the manuscript of the Lucasian optics lectures. (Cambridge University Library MSS Add 4002 f. 128a. By permission of the Syndics of Cambridge University Library)

colour. Colours were not generated by modifications of light inside prisms. Refractions analysed light into its constituent rays. This doctrine would not stand if it could be shown that irregularities in Newton's prisms were the cause of dispersion.[28] Furthermore, his prisms had to be capable of separating 'primitive' colour-making rays. He identified three troubles with common prisms: their small angles, their refractive powers and, most importantly, the fact that common prisms were often tinged with colour and vitiated with bubbles and veins. He lectured that

instead of the glass prisms commonly sold (which are too slender) you must use broader ones, such as those you can make from glass plates highly polished on both sides and joined together in the form of a small prism-shaped vessel; the vessel should be filled with very clear water and sealed all around with cement . . . Those prisms, moreover, that are made wholly of glass are often tinged with some colour, such as green or yellow.

He repeated this advice in the sixth lecture when discussing total internal reflection and later recommended that the best glass was that 'used

28. From 1672, Newton reported attention to 'unevenness in the glass, or other contingent irregularity' in his prisms. He summed up these irregularities, including 'veins, an uneven polish, or fortuitous position of the pores of glass'. Newton (1958), p. 224; Newton (1959–77), 1, p. 93.

to make mirrors'.[29] Therefore, much of Newton's work centred on ways of telling whether prisms were working 'properly'. Only properly working prisms could show that his doctrine was right. To persuade his audience of this doctrine, he would have to persuade them to change the way they used prisms and to change the prisms they used.

This task proved troublesome during the 1670s. There was an ambiguity about the lessons Newton claimed his experiments taught. In the forty-fourth experiment of his notebook of 1666, Newton had derived two important consequences: first, blue-making rays were refracted more than red; second, these rays could not be split into further colours. But during the 1670s these two consequences were often separated from each other. The second lesson raised more trouble, because it was not easy to make 'uncompounded colours' with common prisms. Special instruments and protocols were needed. All colours 'proper to bodys' were mixed, as Newton noted in 1666. Furthermore, even in his lectures Newton had trouble demonstrating immutability. In his sixth lecture he briefly commented that he would 'show afterward' that 'no light of any simple colour can be changed in its colour . . . in refractions'.[30] But he did not honour this promise in the original lectures. He did so only in the revised version he completed in October 1674. Writing in the midst of his disputes with critics, he conceded that the two prism trial 'is not yet perfect in all respects'. In this trial, a red-making ray from the first prism did not display further colours when refracted through a second prism placed transverse to the first. However, if the second prism was placed parallel to the first, then the red-making ray also displayed yellow colours after the second refraction. This seemed to challenge the fundamental doctrine of immutability. To ward off this possible challenge, made all too obvious in the responses he was receiving from his correspondents in London and elsewhere, Newton changed the experimental protocol. He now stipulated that the holes through which the light was transmitted should be made as small as possible, while he also claimed that the best way to perform the experiment was to subject each ray to *many successive refractions*, not just two. He held that immutability would be proven if 'the apparent changes [in colour] would become smaller by repeated refractions, because simpler colours would arise at every step'. As Alan Shapiro has pointed out, this was just the strategy Newton used much later in his published *Opticks*. But he did not make these details clear to any of his colleagues during the seventeenth century.[31]

29. Newton (1984–), 1, pp. 105, 131, 153.
30. Newton (1984–), 1, p. 143.
31. Shapiro (1980), pp. 215–16; Newton (1984–), 1, pp. 453–5, 145 n. 33.

The differences in presentation fomented dispute between Newton and his audiences. In the letter to Oldenburg of February 1672, Newton selected some of his earlier trials, rewrote his autobiography, omitted many important details, notably those on prism quality and design, and revised some of the lessons these experiments were supposed to teach. The *experimentum crucis* was a simplified and revised form of large numbers of experiments given in the third and sixth Cambridge lectures. The first prism was placed between the Sun and the first screen and then turned slowly by hand. The lesson Newton derived here was the existence of differing refrangibility, not any consequence about the specificity or immutability of colour. But elsewhere in the letter Newton did discuss his commitment to immutability of colour. He claimed boldly that 'when any one sort of Rays hath been well parted from those of other kinds, it hath afterwards obstinately retained its colour, notwithstanding my utmost endeavours to change it'. He mentioned efforts involving prisms, coloured reflectors and thin films, but gave no 'history' of these attempts.[32]

The technique for making a 'well parted ray' was not spelt out. The criterion for a 'well parted ray' seemed tautological to some of Newton's audience. 'Well parted rays' were only recognisable as just those which did not display further colours after a refraction, yet the doctrine in question was whether the colours displayed by such rays could be changed. In May 1673, for example, Newton told the Dutch natural philosopher Christiaan Huygens that he had given sufficient details for 'them who know how to examin whether a colour be simple or compounded', while in June he wrote again that the proposition of immutability 'might be further proved apart by experiments, too long to be here described'. This reticence was important, for several reasons. Ronald Laymon has suggested that Newton's *experimentum crucis* only works if 'idealized descriptions of the experiments are used'.[33] Newton's 'idealization' of his group of trials demanded that his prisms be seen as commonplace and the lay-out of his trials be treated in a highly abstracted form. This was part of his effort to win assent to his new doctrine. Furthermore, many of his readers assumed that the experiments reported by Newton to Oldenburg from 1672, particularly the *experimentum crucis*, were designed to demonstrate immutability *of colour*. Thus if the experimenters could show immutability was false, they held that Newton's doctrine would fall. Since Newton held that the

32. Newton (1959–77), 1, p. 97.
33. Newton (1959–77), 1, pp. 265, 294. For an analysis which notes the role of idealisation in Newton's reports see Laymon (1978), pp. 51–3.

demonstration of immutability, or of specific refrangibility, demanded prisms handled in special ways, the details of experimental tactics and of instruments were fundamental items in this dispute.

The response to Newton's first paper showed that the techniques for handling prisms were an important part of the dispute about his claims. This response varied a great deal. Many found Newton's work dramatic and compelling. For them, Newton became 'our happy wonder of ingenuity and best broacher of *new light'*. But not all were persuaded. These differences evince contrasting experimental technologies and philosophies.[34] In London, Hooke replicated trials with two prisms but denied their decisive role. Newton was also countered by a group of English Jesuits at Liège, including the mathematics professor Francis Line, his student John Gascoines and the theology professor Anthony Lucas. The Jesuits initially proffered challenges to the presuppositions of Newton's trials and series of experiments of their own. When told by Newton that the *experimentum crucis* was the only trial to be examined, they reported a failure to replicate his alleged result. There was here no agreement on proper use and design of prisms. Nor was there agreement on the meaning and authority of the *experimentum crucis*.

Critics often contested Newton's implied claim that the *experimentum crucis* rendered the performance of long series of trials unnecessary. For example, Newton criticised Lucas for his effort to perform large numbers of optical trials. 'Instead of a multitude of things', Lucas should 'try only the *Experimentum Crucis*. For it is not number of Experiments, but weight to be regarded; and where one will do, what need many?' Yet this was not always the view which Newton and his interlocutors expressed. Newton himself denied in 1677 that 'I brought ye *Experimentum Crucis* to prove all'.[35] Hooke denied that this so-called 'crucial experiment' proved anything decisive: 'it is not that, which he soe calls, will doe the turne, for the same phaenomenon will be salved by my hypothesis as well as by his'. Indeed, Hooke implied that one of Newton's principal failings was the small number of trials the Cambridge professor reported, in contrast with the 'many hundreds of tryalls' which Hooke himself had performed and described. Hooke said he had

34. Fairfax (1674), p. 51.
35. Newton (1959–77), 2, pp. 79-80, 258.

not erected an hypothesis 'without first trying some hundreds of expts'.[36]

Hooke was writing a week after reading Newton's first paper on colours. Hooke, Boyle and the mathematician and divine Seth Ward were appointed by the Society to report on Newton's letter when it was read in London on February 8 1672. Hooke also recalled his 'crucial experiment' of 1665, one which did not use a prism. While he attributed great weight to the simple 'Experiment or Observation of Crystal', which, he held, decisively proved his own hypothesis of colour, he did not privilege the prism in the way Newton sought to achieve. His favoured experiments, used tellingly against Descartes in 1665, involved mica, thin plates and glasses filled with variously coloured liquids. He did not pay attention to the detailed flaws and corrections of commercial prisms which Newton charted in his Cambridge lectures. Newton sought to minimise what he saw as the defects of these objects: but bubbles, veins and tints provided Hooke with opportunities for further mechanical ingenuity. Hooke and Newton had different experimental technologies in their treatment of these devices and they drew very different conclusions from their trials.[37]

In April and May 1672, Hooke showed the Society a series of trials with two prisms, including one which demonstrated that 'rays of light being separated by one prism into distinct colours, the reflection [sic] made by another prism does not alter those colours'. By June, he had also replicated the *experimentum crucis*: yet he still insisted that 'I think it not an *Experimentum crucis,* as I may possibly shew hereafter'. He told the Royal Society's President that this trial might prove that 'coloured Radiations' maintain fixed refrangibilities: it did *not* prove what Hooke claimed Newton wanted to prove, that there was a 'coloured ray in the light before refraction'. Indeed, Newton did not seem consistent in his account of what this trial showed. In February, he said that it demonstrated that there were differently refrangible rays in light without reference to colour; in June, when publicly answering Hooke, he said that it demonstrated that 'rays of divers colours considered apart do at equall incidences suffer unequall refractions', so raising the issue of specific colour.[38]

36. Newton (1959-77), 1, pp. 110–11. Contrast Leibniz's views on Newton's experiments in Leibniz (1965), pp. 488–9.
37. For the exchanges between Hooke and Newton see Birch (1756–57), 3, pp. 10–15; Newton (1959–77), 1, pp. 110–14. For Hooke's views see Hooke (1665), pp. 47–54; Hooke (1705), p. 54.
38. Birch (1756–57), 3, pp. 47, 50; Newton, (1959–77), 1, pp. 195, 202–3.

Throughout their subsequent exchanges on prismatic colours, which continued to 1678, Hooke accepted what he took to be matters of fact in Newton's trials, and freely acknowledged that his replications had worked. But Hooke read Newton as arguing for specific refrangibility and for immutability in the 'crucial' experiment, and he denied that this trial was persuasive. Thus when Newton was at a meeting of the Society in March 1675, he apparently heard Hooke confirm that the trials reported three years earlier had been replicated. However, at the same meetings, Hooke developed his own vibration theory of light and colour, citing new experiments on diffraction to show that 'colours may be made without refraction' and that his own doctrine could successfully save all the phenomena of colour. By the end of 1675, Newton was prepared to claim that Hooke had 'accommodated his Hypothesis to this my suggestion' of the origin of colours. Hooke, in his turn, was reported as believing Newton had plagiarised his 'suggestion' from the *Micrographia*. Despite attempts at reconciliation between the protagonists, it appeared that Newton's 'crucial' experiment had not acquired authority, nor a fixed meaning. As late as 1690, Hooke told the Royal Society that he was aware of no 'Better' theory of colour than his own, thus writing Newton out of the history of optics.[39]

Newton and his Jesuit critics also discussed rival prism techniques and the meaning of the *experimentum crucis*. In spring 1672, Newton already found himself compelled to give fuller details of his trials than those presented in his initial paper. Newton sent Oldenburg a diagram of the experiment and conceded that 'I am apt to beleive [sic] that some of the experiments may seem obscure by reason of the brevity wherewith I writ them wch should have been described more largely & explained with schemes if they had been then intended for the publick'.[40] The reaction of Line and the colleagues who continued his work from autumn 1674 showed how hard it was for Newton to achieve authority over his 'publick'. It demonstrated the problems of achieving agreed replication.[41] During 1675, Newton appealed to the Royal Society in order to authorise his claim that the 'crucial' experiment had been replicated in London and was, allegedly, easy to perform anywhere: 'it

39. For Newton's relation with Hooke's criticisms, see Hall (1951), p. 221; Westfall (1963); Hall and Westfall (1967). The exchange and Hooke's historiography of optics are documented in Hooke (1705), pp. 186–90; Birch (1756–57), 3, pp. 194–5; Hooke (1935), pp. 148–9, 153; Newton (1959–77), 1, pp. 357, 360, 362–3, 408, 412.
40. Newton (1959–77), 1, pp. 166–7, 205–6, 212; Oldenburg (1965–86), 9, pp. 132–3.
41. For initial exchanges with Line, see Newton (1959–77), 1, pp. 318, 336, 329; Westfall (1966), pp. 303–4. Line claimed that he had read the optics of Newton's predecessor, Isaac Barrow and had performed optical trials in the 1640s with Kenelm Digby. For these earlier prism experiments see Digby (1669), p. 341.

may be tryed (though not so perfectly) even wthout darkning a room, or ye expence of any more time then a qter of an hower'.[42]

As the argument with the Liège experimenters became angrier, during 1676, Newton was told that they had often done their own optical experiments before witnesses at Line's house: 'we think it probable he hath tried his experiment thrice for Mr. Newton's once'. Gascoines suggested that only some unreported difference in the arrangement or type of prism could explain this conflicting result. In January and February 1676, Newton responded to this challenge with many more details of his own trials. He gave the dimension of the holes used to admit light. He re-emphasised the placing of the prism at minimum deviation and said the trial worked best when the sky was clear. He advised on the best way of darkening the room. Importantly, he said Gascoines should 'get a prism with an angle about 60 or 65 degrees . . . If his prism be pretty nearly equilateral, such as I suppose are usually sold in other places as well as in *England*, he may make use of the biggest angle'.[43]

These specifications were designed to elicit a replication of the experiment in Liège. 'Ye business being about matter of fact was not proper to be decided by writing but by trying it before competent witnesses'. Newton implied that if the Jesuits could not make these experiments work, then it must be due to their wilful incompetence rather than to subtle differences in technique. Newton wanted pictures, because 'a scheme or two . . . will make the business plainer'. The trial was demonstrated at the Society in April 1676 and Oldenburg told Gascoines and his colleagues of this allegedly decisive success.[44] However, the immediate response of the Liège natural philosophers showed that this result was not compelling. On the contrary: Lucas immediately answered Oldenburg's letter with the comment that 'I was much rejoyced to see the tryalls of that Illustrious Company, agree soe exactly with ours here, tho in somewhat ours disagree from Mr. Newton'. Lucas and his colleagues interpreted the Royal Society's experiment as reconcilable with their own. They continued to produce evidence which they held refuted Newton's doctrine. Newton was astonished: the Royal Society 'found them succeed as I affirmed'. Exchanges with Lucas continued until 1678 when Newton violently withdrew from all such dispute.[45]

42. Newton (1959–77), 1, pp. 335, 357–8, 410–11.
43. Newton (1959–77), 1, pp. 394, 409–10.
44. Birch (1756–57), 3, pp. 313–14; Newton (1959–77), 1, pp. 423–4 and 2, p. 6.
45. Newton (1959–77), 2, pp. 12, 76–81. For the dispute with Lucas see Westfall (1966); Gruner (1973); Guerlac (1981), pp. 89–98.

In this correspondence the status of the *experimentum crucis* was challenged at once. Lucas reported a set of new experiments designed to show that different colour-making rays did not differ in refrangibility. Newton told Lucas to try only the *experimentum crucis*. Lucas did so in October 1676, but without conceding its privileged role or the lesson which Newton claimed it taught. Lucas appealed to the precedent of Boyle's pneumatics, which had been supported by 'a vast number of new experiments'. He asked why he should accept Newton's stricture that the controversy must hinge on the outcome of one trial. Lucas also argued that the experiment, if successful, would not show intrinsically different refrangibilities in different colour-making rays. Lucas read Newton as seeking to prove a doctrine about colour with his 'crucial' trial, not merely a simpler result about unequal refraction; and he read Newton as illegitimately basing his authority on a single trial, rather than a mass of evidence. Newton, once again, was furious with what he saw as a failure to grasp the sense, or the authority, of his key experiment. He contemplated a publication of a major treatise on his optical work. The alternative was silence.[46]

Newton was drawn into a final exchange with Lucas. For when Lucas did try the *experimentum crucis* in autumn 1676, he reported a result different from that attributed to Newton. He reported that even though he had worked 'exactly according to Mr. Newtons directions' he found as 'a result of many trialls' that violet rays displayed a 'considerable quantity of red ones' after the second refraction. Newton's delayed reply re-emphasised the meaning he wished to give this trial: 'you think I brought it to prove that rays of different colours are differently refrangible'. Newton held that this thought was mistaken. Yet there were grounds for Lucas's reading. Recall Newton's public statement of June 1672 that the 'crucial' experiment proved that 'rays of divers colours considered apart do at equall incidences suffer unequall refractions'. But Newton told Lucas that 'I bring it to prove (without respect to colours) yt light consists in rays differently refrangible'.[47] Here Newton insisted that the crucial experiment taught nothing about colours. He did so because he had to discredit Lucas's version of this experiment. Lucas's report suggested that Newton was wrong about the constant colour displayed by truly uncompounded rays. So Newton replied that the experiment was not designed to prove the homogeneity of 'uncom-

46. Newton (1959–77), 2, pp. 8–11, 79, 182–3, 183–5; Oldenburg (1965–86), 13, pp. 99–100; Westfall (1966), p. 311; Gruner (1973): pp. 318–21.
47. Newton (1959–77), 2, p. 257; Oldenburg (1965–86), 13, p. 101; Lohne (1968), pp. 185–6. Newton's original formulation is in Newton (1959–77), 1, p. 187.

pounded' rays. This answer was directly linked with the problem of
the quality and design of prisms. Newton's other tactic in his attack on
Lucas was to challenge the Jesuit's instruments. Newton alleged that
they were incapable of producing 'uncompounded' colours.

The controversy demonstrates a central trouble of replication and
instrumentation. Further work was necessary to establish whether the
two men were discussing the 'same' experiment. Newton said that
since Lucas's experiment was different from the 'crucial' experiment,
its different result did not discredit that experiment. Lucas insisted that
he had 'but follow'd the way which [Newton] himselfe had track'd out
for me'. At several points in this exchange, Newton drew attention to
the need for proper prisms. In August 1676, he reported trials which
compared changing prism angles with changes in the length of the
spectrum produced. He then advised that the *experimentum crucis* must
be made with 'Prisms which refract so much as to make the length of
the Image five times its breadth, and rather more than less; for, otherwise Experiments will not succeed so plainly with others as they have
done with me'. He also pointed out the need for prisms with plane or
convex sides when making spectra: Newton suspected that Lucas was
using a concave instrument. Lucas confirmed that this was so in
October; later, he also considered 'the difference of glasse the prismes
are made of'.[48]

These details affected the debate on changes of colour displayed by
refracted rays, since Newton claimed that the ability to make a wide
spectrum affected the ability to separate genuinely 'uncompounded'
rays. He instructed Lucas on the character of 'compound' rays. Lucas
was wrong 'to take your ordinary colours of ye Prism to be my [un]com-
pounded ones'. So Newton emphasised that Lucas had not obtained
properly 'uncompounded' rays because he had the wrong prisms and
used them badly. Lucas's believed that none of these differences in
prisms could explain why he had managed to change the colour of light
rays by refraction. He told Newton that 'if all rayes differently coloured
had an unequall refrangibility', as Newton apparently believed, then
'the variety of prismes could no more refract different colours *equally*,
that it can change the nature of rayes'. To reach closure here, Newton
would have had to persuade Lucas to change his prisms and then, as
Newton himself did, to interpret these changes as the correction of

48. Newton (1959–77), 2, pp. 80–1, 189–91. As R.S. Westfall has shown, there is no
    evidence that the Jesuits were using prisms made of different glass and considerable
    evidence that they had trouble measuring their angles and detecting the concavity
    of the prisms' sides. See Westfall (1966), p. 309.

important defects. In March 1678 Lucas did report difficulty in getting prisms with good glass. But closure was not accomplished: instead, as we have seen, from summer 1678 Newton broke off any further debate on the issue.[49]

Newton's arguments with Hooke and Lucas show that the *status of the experimentum crucis* was hard to fix. There was no agreed criterion for a competent prism experiment or for a good prism. Only when the status of the experiment was fixed did this criterion become available. Replicability and meaning both hinged on the establishment of this emblem. Some of Newton's audience read the experiment as a claim about colour immutability. Newton sometimes provided them with grounds for this reading. By the 1720s, in fact, he seemed to have come to agree with this reading. That is, the experiment which Newton now counted as his decisive one had a new and fixed meaning: 'refracted light does not change its colour'. This slogan appeared on a technically defective but important emblem of Newton's programme, the vignette of the *experimentum crucis* which Newton designed for the 1722 French edition of his magisterial *Opticks*. 'Cherubs and spectators' were excised from this design so as to give pride of place to the prisms, which testified to the truth of this incontrovertible fact of immutability of colour.[50] Once this fact was established and firmly wedded to the instruments, a means then existed for discriminating good prisms and competent experimental arrangements. The process by which this criterion acquired self-evidence will be examined in the final sections of this chapter.

'AN UNHAPPY CHOICE OF PRISMS'; THE ACHIEVEMENT OF 'TRANSPARENCY' (1704–22)

Those who eventually accepted the emblematic status of the *experimentum crucis* and Newton's prisms produced a story which explained why the experiment and the instruments had not swayed critics in the 1670s. In popular texts such as Voltaire's *Elements of Sir Isaac Newton's Philosophy* (1738) and Algarotti's *Newtonianism for Ladies* (1737) it was claimed that those who had not succeeded in replicating Newton's trials 'had not been happy enough in the Choice

49. Newton (1959–77), 2, pp. 252, 254–5, 269; Oldenburg (1965–1986), 13, p. 101; Gruner (1973), p. 327.
50. For the vignette to the French *Opticks* see Newton (1952), pp. 73, 122; Newton (1959–77), 7, pp. 155, 179, 201, 213; Lohne (1968), pp., 193–6; Guerlac (1981), pp. 156–63.

of . . . Prisms'. They were recording a rather common view.[51]
Experimenters who had reported trials which differed from those of
Newton were now dismissed from consideration because their instruments must have been defective. This claim depended on a prior consensus on the status of Newton's trials and his instruments. After assuming the Presidency of the Royal Society in 1703, producing the *Opticks*
and working closely with the experimental philosopher J.T. Desaguliers,
Newton was in a position to claim that any optical experiment, if
performed with the right prisms, would guarantee the truth of his
doctrine. In London, the prism had become a 'transparent' instrument.
This was an accomplishment of Newton and his allies. It demanded a
reconstruction of the record of the optical controversies. This reconstruction involved both the public exposition of new prism techniques and
a reinterpretation of previous failures to replicate Newton's claims. As
Newton took power over the key resources of experimental philosophy,
Newtonian optics acquired a disciplinary history and a standardised
technology.

The appearance of Newton's *Opticks* was a key event in this process.
Consideration of its opening sections shows how Newton reconstructed
his trials to make his authority inside the experimental community.[52]
Initial passages described ways of separating 'uncompounded' rays.
This had been a key trouble of the 1670s. Then, Newton had faced a
dilemma: he could, apparently, only sway his colleagues with prisms
they were used to employing. Yet he reckoned that with these, an
'uncompounded' ray could not easily be made. So it was hard to spell
out a decisive experiment to prove that such rays could be made and
did not display further new colours when passed through a second
prism. The work of Lucas or of the French experimenter Edmé Mariotte
made this problem only too clear. Like Lucas, Mariotte had performed
trials in the 1670s which purported to challenge the *experimentum crucis*
and demonstrate that the doctrine of colour immutability was false.
He had considerable expertise in experimental optics. Within a year of
Newton's first paper on light and colour, he had conducted trials in
Paris on the mutability of colours which challenged Newton's claims
about the hues of 'uncompounded' rays.[53].

51. Voltaire (1738), p. 101; Algarotti (1742), 2, p. 60.
52. For the impact of the book see Guerlac (1981), pp. 106–11; Hall (1975).
53. Leibniz mentioned these trials to Oldenburg in February 1673. They were repeated
    at the Paris Academy of Sciences in 1679 and published in Mariotte's *Traité des couleures*
    in 1681. Newton owned a 1717 edition of this text and marked the passage which
    challenged his doctrine. For Mariotte's work see Oldenburg (1965–86), 9, p. 485;
    Shapiro (1980a), pp. 283–4; Guerlac (1981), pp. 98–9. Newton's notes are in Harrison
    (1978), pp. 21, 25.

In his book, Mariotte developed an anti-Cartesian version of the modification hypothesis, supposing that the colour of a light ray could be changed in refraction. In his version of Newton's experiment with two prisms, he did not place the first prism at minimum deviation, nor did he place a screen immediately after this prism to collimate the rays produced. He used a white card to separate out a single ray after the first refraction and then examined what happened to this ray when it was refracted a second time. Mariotte was confident that this arrangement allowed him to make well separated rays. His card was at least 30 feet from the first prism, displaying a spectrum of a similar width to that Newton reported. He also ensured that 'the room is very dark and no sensible light passes through the slit in the card apart from that which is coloured'. Yet he reported that a purely violet ray displayed red and yellow tinges after the second refraction. Assuming that Newton's whole theory was supposed to stand or fall by this experiment, Mariotte concluded that 'the ingenious hypothesis of Mr. Newton must not at all be accepted'.[54] This single report of a single refutation of a view Newton had not quite expressed in print was an important resource in European responses to Newton's optical doctrine. Mariotte was often cited, notably by Leibniz, as providing an important challenge to Newton's theory. Leibniz, Mariotte's 'old friend', repeatedly reminded correspondents of the challenge to the matter of fact of colour immutability which the trials of the 1670s suggested.[55]

Newton's *Opticks*, painstakingly assembled during the 1690s, provided new resources with which to respond to these worries. Neither Lucas nor Mariotte was mentioned in the new book. Hooke, recently deceased, received only a cursory reference. Nor did Newton use the name *experimentum crucis* to sanctify the sixth experiment of the book, which used two prisms to prove constant refrangibility. Constancy of colour became at least as important a feature of his scheme. Thus Newton shifted the weight of his argument. The whole of the fourth proposition was devoted to a description of the way to make 'uncompounded' rays. At last, Newton gave a relatively full public account of the instruments 'sufficient for trying all the experiments in this book about simple light'.[56] Notable techniques included the positioning of a lens before the first prism to diminish the incident image. He also detailed the kinds of glass to be used: recalling remarks made in the Cambridge lectures, he specified 'Glass free from Bubbles and Veins'.

54. Mariotte (1740), pp. 226–31.
55. Leibniz (1965), pp. 488–9; Guerlac (1981), p. 116.
56. Newton (1704), pp. 30–2, 49.

The prisms must have 'truly plane' sides, not convex as he had suggested to Lucas. The polish should be 'elaborate', not 'wrought with Putty' which produced 'little convex polite Risings like waves'. The edges of prisms and lenses should be covered in black paper. Yet he conceded that 'it's difficult to get Glass prisms fit for this Purpose', referring to his own practice of using vessels made of 'broken Looking-Glasses' filled with rain water and a lead salt to increase the refraction. Elsewhere, he discussed ways these water-filled vessels should be used and reported the failings of a prism 'made of a dark coloured Glass inclining to green.' When he replaced this with a prism of 'clear white Glass' he still found 'two or three little Bubbles' and covered the defective parts of the Prism with black paper. After 1704, Newton annotated his own copy of the book with further changes in specified prism angles and in details of liquids with which to fill prismatic vessels. Once again, these remarks changed the way prisms were to be handled and then assessed. The stage was set for the claim that unsuccessful replicators were using bad prisms. The book provided a range of such resources and Newton set out to get the Royal Society's experimenters to use them.[57]

Newton used his power over the Society against his critics, notably what he perceived as a conspiracy headed by Leibniz and the writers of the Leipzig journal *Acta eruditorum*. This strategy exploited the resources of the Presidency and the expertise of the experimenters, Hauksbee and Desaguliers, in a campaign in which the allegedly superior quality of English prisms soon became important. As Steven Shapin has argued, this was a campaign with very important political resonances. One of its most dramatic aspects was the assertion of the authority of the experimenters of Augustan London. In late 1707, Leibniz heard about French interest in the *Opticks* and urged experimenters there to replicate the troubled Mariotte report. At the same moment, Newton ordered Hauksbee to begin trying the *Opticks* experiments before the Royal Society. These trials were then reported in the *Acta eruditorum*. As the war with Newton began to consume his attention, Leibniz decided to publish his views on Mariotte in the Leipzig journal. So from the summer of 1714, Newton directed Desaguliers, Hauksbee's successor, to show that uncompounded colours could be made, that Mariotte's instruments were defective and that Leibniz was wrong.[58]

---

57. Newton (1704), pp. 49–51, 55, 63–4. Newton argued that rays which displayed violet and indigo were hard to make 'uncompounded', because of scattered light from 'the Inequalities of the Prism'. This was an obvious resource to use against Mariotte, whose trials had produced colour changes in the violet.

58. Hall (1980); Guerlac (1981), pp. 110–11, 116–17; Shapin (1981); Heilbron (1983), pp. 90–1.

Desaguliers had to address the problem of *replicability*. To destroy Mariotte's credit, he had to show experiments which resembled those of the 1670s. If he used too many of the new protocols outlined in the *Opticks* it would appear that Mariotte had been reading Newton's initial reports correctly. If he did not use these new protocols, then he would fail to produce 'uncompounded' colours. Desaguliers turned to Newton for aid: the President helped draft the paper which appeared under Desaguliers's name in the *Philosophical Transactions*. The two men emphasised that previously printed reports were sufficient to allow replication. 'Some Gentlemen abroad' had 'complained that they had not found the Experiments answer, for want of sufficient Directions in *Sir Isaac Newton's Opticks*; tho' I had no other Directions than what I found there'. But Desaguliers also accepted that much new information was necessary to allow these trials to be successfully repeated. He allowed that Newton's original papers of the 1670s were inadequately detailed. A technique for separating monochromatic rays 'was not published before Sir Is. Newton's *Opticks* came abroad' in 1704. This explained why Lucas and Mariotte had 'reported that the [crucial] Experiment did not succeed'. Furthermore, Desaguliers added many significant details even to the fuller descriptions of the *Opticks*, including the lenses and prisms to be used in the optical trials. Prisms should be made of the green glass used for the object glasses of telescopes. 'The best white prisms', it emerged, were inadequate for the purpose, being 'commonly full of veins'. Desaguliers' tactic, following Newton's advice, was to marry the techniques for making uncompounded colours described in the fourth proposition of the *Opticks*, including the careful treatment of clear glass prisms and the use of a collimating lens, with the lay-out of experiment six, that which most closely resembled the original *experimentum crucis* of 1672. Desaguliers now did what Newton had not done. He revived the name *experimentum crucis* for this completely reconstructed trial. Part of this reconstruction involved the claim that the 'crucial' experiment demonstrated colour homogeneity rather than specific refrangibility. Desaguliers tailored his experiments for effective witnessing. Spectators were each given a hand-held prism through which to view the spectrum cast on the final screen. Desaguliers made all these important changes and conceded that 'several have confessed to me that they at first used to fail in this experiment'. But it was essential that he stipulate that he had followed Newton's text to the letter, with no other resource at his disposal. Hence his insinuation that he had relied only on Newton's *publicly available* accounts.[59]

59. Desaguliers (1716), pp. 433–5, 443–4, 447, 448; Lohne (1968), pp. 189–90; Guerlac (1981), pp. 118–8; Heilbron (1983), p. 91. Newton's drafts for Desaguliers are in Newton (1714).

These experiments now had to be deployed in public. After a dry run at his house in Westminster, Desaguliers showed them to the Society. In early 1715, they were displayed to visiting natural philosophers from Holland, Italy and France. The repertoire of reformulated experiments soon became the prize exemplar of Newtonian optics. The visit of the French in 1715 was swiftly followed by successful replications in Paris and elsewhere.[60] Two aspects of this work are of importance for the career of the optical instruments: first, Newton and Desaguliers worked hard to make Mariotte's result depend on his bad glassware. Desaguliers announced in the *Philosophical Transactions* that he had proven this in 1714 and 'still shews it to those who desire to see it', presumably during his courses of experimental philosophy in the capital. News of these lectures, together with the view about the insufficiency of Mariotte's experiment, were then reproduced in Desaguliers's publications and by his colleague Pierre Coste in a French version of the *Opticks* in 1720.[61] Second, they also asserted that any experimenter must use the prisms which were available to the London natural philosophers. Personal visits to London were significant means by which natural philosophers could be won to this new practice. Once local agreement had been accomplished at the Society and among its audience in France, Italy and Holland on the doctrine of colour immutability, it was possible to define good prisms as those which displayed this result. The local and tenuous nature of this agreement was demonstrated by subsequent exchanges with European experimenters.[61]

CONTESTING 'TRANSPARENCY': THE VIEW FROM ITALY (1720–40)

In Italy Newton found several important followers and critics. His key trials had been replicated in Bologna, but the situation in Venice was less happy. As we have seen, Venetian glass had provided the standard of virtue through the seventeenth century and the English work directly challenged this status. In 1719, after reading the *Opticks*, the Venetian natural philosopher Giovanni Rizzetti began to perform

60. Desaguliers (1716), p. 435; Newton (1959–77), 6, pp. 144–5 and 7, pp. 113–14, 116–17; Guerlac (1981), pp. 128–43.
61. Desaguliers (1719), pp. 187–91; Schofield (1970), pp. 80–7. For the reaction of a French experimenter to replication of these results see Newton (1959–77), 7, pp. 114–16, 117 n. 6.

trials which denied Newton's reports and his doctrine.[62] Rizzetti's views were communicated to Newton and Desaguliers and simultaneously given publicity in the *Acta eruditorum*. Claiming that the authority of great men caused experimenters to err, Rizzetti anounced that

I have taken care to repeat all the experiments, and (is it not right that I should speak of these things?) I have found some of them false and all the rest equivocal and by no means conclusive, because of the omission of some circumstances.[63]

Newton made his own copies of Rizzetti's letters and drafted a number of possible replies. In his draft Newton demanded that his Italian critic take the *experimentum crucis* as his premise: other trials 'might indeed inform us of something new concerning light, but they could not over-turn what Sr Isaac has already established upon reasonings as free from all paralogisms as the demonstrations of Euclid'. Desaguliers obliged with a new show, first at his own house, then at the Society. Yet again, it emerged that several important techniques needed to be spelt out in detail to supplement the account published by Newton.[64] Rizzetti reacted with enthusiastic hostility. He sent a new report of his work to the Royal Society and began preparing a lengthy book on light and colour, eventually published in 1727. Rizzetti aimed to raise the standing of his trials by naming his witnesses, James Stirling and Nikolaus Ber-noulli. The choice was unfortunate: Stirling was a Jacobite mathemati-cian in exile, soon to be nicknamed 'the Venetian', while Bernoulli's uncle was the leading Leibnizian opponent of Newton. Newton soon saw Rizzetti's attack as one further attempt 'of the friends of Mr. Leibniz to embroil me'.[65]

This gave point to the London response to the Venetian experiments. Rizzetti's trials increased in number and significance, for he now turned his attention to the *experimentum crucis*. Rizzetti asked the English why Mariotte had managed to change the colour of violet rays after a second refraction. He added his own results: rays displaying pure yellow before the second refraction showed red, green and indigo after it, while the yellow colour vanished. He gave instructions for this trial: 'cate is to

62. Newton (1704), pp. 13–14; Rizzetti (1722); Rizzetti (1741), p. 91; Hall (1982), pp. 18–19, 20–21; Heilbron (1983), p. 92. The Bolognese project was inaugurated by the natural philosopher Francesco Bianchini by 1707. Bianchini came to London in 1713, witnessed Hauksbee's experiments, met Newton and was given five copies of the Latin *Opticks* to distribute in Italy, including one for his colleague at Bologna, the astronomer Eustachio Manfredi.
63. Rizzetti (1722); Desaguliers (1728), pp. 596–7.
64. Desaguliers (1722); Newton (1959–77), 7, p. 255. Newton's drafts are in Newton (1722).
65. Rizzetti (1724); Newton (1959–77), 7, pp. xli, 53–5, Westfall (1980), pp. 799, 811.

be taken that the second prism is not too distant from the first, nor the slit, through which light of one colour is transmitted from refraction at the first prism to the second, is too narrow'. These were, of course. just the opposite of the conditions Desaguliers and Newton had stipulated for a good separation of an 'uncompounded' ray. But Rizzetti denied that his results were due to imperfect separation: he asked why red-making rays in his trials did not then split into adjacent colours and why he could make yellow-making rays disappear. Nor did he allow the important claim that Mariotte's results, or his own, were due to bad prisms. He had not used 'imperfect prisms, but exact care and suitable instruments'.[66]

These new criticisms reached London at the time of Newton's death, but Rizzetti was swiftly answered there by Desaguliers. In summer 1728 the Royal Society learned that replications of the key trials were planned by the loyal Newtonian group under Eustachio Manfredi at Bologna. The English reacted by electing the promoter of these plans into their fellowship. The decisive issue was the Newtonian experimenters' faith in the virtue of English instruments. Manfredi said that

nearly all the experiments which can be read in the Optics of Newton, as well as in the little work of Desaguliers, have been done here in public displays. And when the prisms have been completely perfect, like some which we have to hand from England, the outcome has always corresponded to the doctrine.

The English claimed that Rizzetti, having made use of Prisms made at Venice, which are not of so pure a Cristall as ours, has been led into the many mistakes he has asserted for convincing proofs'. Rizzetti's bad instruments 'have rendered him ridiculous for ever'.[67] Desaguliers aimed to show the same thing in London. In August he showed a series of trials at his house, spelling out the heresies of Rizzetti and suggesting that even if the Italian had used common prisms and colours he should still have gotten better results. Desaguliers's witnesses included the President, other Fellows, and an invited group of noble Italians, who were specially named in reports published in London and sent to Italy. Using this technique to establish decisive authority, Desaguliers took the chance to add new evidence for the natural philosophy of attractive and repellent forces Newton had introduced in the *Opticks*. The campaign against Rizzetti had now made visible the scale of commitment to the novel ontology which loyalty to Newtonian principles demanded. Terms such as 'reflexion' took on polemical importance, for they were

66. Rizzetti (1724); Rizzetti (1727), pp. 37–8.
67. Dereham (1728); Desaguliers (1728), p. 597; Manfredi (1728); Rizzetti (1741), p. 112.

glossed by Desaguliers as references to an underlying dynamical theory of matter.[68]

Rizzetti bridled at these demands, as many others did. He rejected both the decisiveness of the London trials and the natural philosophy they were supposed to support. Rizzetti kept up his attacks well into the 1740s. As Geoffrey Cantor has suggested, the efforts to establish a consensus were fragile and indecisive. The 1740s saw important challenges to the basis of the dynamical theory and also specific criticisms of some of Newton's apparent claims. Cantor shows that the authority of the model for which Desaguliers and his allies argued was limited, even in Britain, to the spheres in which they most obviously exercised power: the Royal Society and the public lectures on experimental philosophy.[69] Popular presentations of this doctrine, such as that of Algarotti in the late 1730s, reported wide criticism of Newton's views. The work of Mariotte and of Rizzetti was commonly cited. Against this, propagandists used the double weapon of the crucial experiment, demonstrated by Desaguliers and supposedly proving the unchangeability of colour, together with the virtue of English prisms, allegedly evidently better than Venetian ones. Thus Algarotti severely chastised those who 'resolved to try Nature a thousand ways' when this one 'crucial' trial would serve. Newton had the Midas touch: 'everything Sir Isaac Newton handled became Demonstration'. The English optical experiments were demonstrative just because they were easily replicable: 'in looking upon a Paper of two colours with a Prism, it is of no sort of importance whether the Wind blows East or West'. Yet this replicability seemed to require English instruments. Algarotti reported that when he had tried the *experimentum crucis* he had failed to produce unchangeable colours, because 'our Prisms in Italy are of no other use than to amuse Children or hang up as a fine shew in some window in the country'. In contrast, the 'crucial' experiment worked well with prisms sent from England; 'these we esteemed as sacred'.[70]

The claims about the demonstrative authority of these experiments and the instruments needed to perform them aimed to make such instruments 'transparent'. Assent to Newtonian theories of colour was a precondition of seeing these instruments as untroubled objects. They were untroubled only to the extent that the changes in the design necessary to make the 'crucial' experiment work were viewed as nonpartisan. Convinced disciples, such as Desaguliers, Algarotti or Manfredi

68. Dereham (1728); Desaguliers (1728).
69. Rizzetti (1741), pp. 112–21; Cantor (1983), pp. 42–9.
70. Algarotti (1742), 2, pp. 27, 51–2, 35–6, 40, 55–6, 62–5.

all reported that they had needed 'improved' prisms to make the trial succeed. These 'improvements' had to be seen as such by all protagonists in order to achieve persuasive power. That power lay in control over the social institutions of experimental philosophy. In the 1670s, Newton had exercised no such power. After 1710 his authority among London experimenters was overwhelming. This authority allowed carefully staged trials before chosen witnesses and the distribution of influential texts and instruments stamped with the imprimatur of collective assent. Enemies were condemned, as was Rizzetti, either as incompetent or evilly disposed. However, just as in the 1670s, this authority was necessarily unstable and contested. It could not force assent. 'Cruciality' was not a universal feature of Newton's experiment because not all subscribed to the disciplinary history which Newton and his allies helped write. Newton's 'law' did not compel experimenters such as Rizzetti: 'it would be a pretty situation', the Italian exclaimed, 'that in places where experiment is in favour of the law, the prisms for doing it work well, yet in places where it is not in favour, the prisms for doing it work badly'. For such critics, Newton's prisms never became 'transparent' devices of experimental philosophy.[71]

71. Rizzetti (1741), p. 112.

References

Algarotti, F. (1742). *Sir Isaac Newton's Theory of Light and Colours*, 2nd edn. London: G. Hawkins.

Bate, J. (1654). *The Mysteryes of Nature and Art*, 3rd edn. London: Ralph Mab.

Bechler, Z. (1974). Newton's 1672 Optical Controversies. A study in the grammar of scientific dissent. In *The Interaction between Science and Philosophy*, ed. Y. Elkana, pp. 115–42. Atlantic Highlands: Humanities Press.

Bechler, Z (1975). 'A less agreeable matter.' The disagreeable case of Newton and achromatic refraction. *British Journal for the History of Science*, 8, 101–26.

Bennett, J.A. (1982). *The Mathematical Science of Christopher Wren*. Cambridge: Cambridge University Press.

Birch, T. (1756–57). *The History of the Royal Society*, 4 vol. London.

Blay, M. (1983). *La Conceptualisation newtonienne des Phénomènes de la Couleur*. Paris: Vrin.

Boyle, R. (1664). *Experiments and Considerations touching Colours*. London: Henry Herringman.

Cantor, G.N. (1983). *Optics after Newton*. Manchester: Manchester University Press.

Charleston, R.J. (1957). Glass. In *History of Technology*, vol. 3, ed. C. Singer, E.J. Holmyard, A.R. Hall & T. Williams, pp. 206–44. Oxford: Clarendon.

Collins, H.M. (1985). *Changing Order*. London: Sage.

Dear, P. (1985). *Totius in verba*. Rhetoric and authority in the early Royal Society. *Isis*, **76**, 145–61.

Della Porta, G. (1658). *Natural Magick*. London: Thomas Young and Samuel Speed.

Dereham, T. (1728). Correspondence with Sir Hans Sloane. British Library Sloane MSS 4049, ff. 86, 133, 165, 187, 198, 242, 246.

Desaguliers, J.T. (1716). An account of some experiments of light and colours, formerly made by Sir Isaac Newton [followed by] A plain and easy experiment to confirm Sir Isaac Newton's doctrine of the different refrangibility of the rays of light. *Philosophical Transactions*, **29**, 433–52.

Desaguliers, J.T. (1719). *A System of Experimental Philosophy*. London: B. Creake and J. Sackfield.

Desaguliers, J.T. (1722). An account of an optical experiment made before the Royal Society. *Philosophical Transactions*, **32**, 206–8.

Desaguliers, J.T. (1728). 'Optical experiments made . . . upon occasion of Signior Rizzett's opticks, with an account of the same Book'. *Philosophical Transactions*, **35**, 596–629.

Descartes, R. (1656). *Meteorologia* in *Opera philosophica*. 3rd edn. Amsterdam.

Digby, K. (1669). *Of Bodies and of Mans Soul*. London: John Williams. (first published 1644).

Dollond, J. (1758). An account of some experiments concerning the different refrangibility of light. *Philosophical Transactions*, **50**, 733–43.

Fairfax, N. (1674). *A Treatise of the Bulk and Selvedge of the World*. London.

Gruner, S.M. (1973). Defending Father Lucas: a consideration of the Newton–Lucas dispute on the nature of the spectrum. *Centaurus*, **17**, 315–29.

Guerlac, H. (1981). *Newton on the Continent*. Ithaca: Cornell University Press.

Hall, A.R. (1951). Two unpublished lectures of Robert Hooke. *Isis*, **42**, 219–32

Hall, A.R. (1955). Further optical experiments of Isaac Newton. *Annals of Science*, **11**, 27–43.

Hall, A.R. (1975). Newton in France: a new view. *History of Science*, **13**, 233–50.

Hall, A.R. (1980). *Philosophers at War. The Quarrel between Newton and Leibniz*. Cambridge: Cambridge University Press.

Hall, A.R. (1982). Further Newton correspondence. *Notes and Records of the Royal Society*, **37**, 7–34.

Hall, A.R. & Westfall, R.S. (1967). Did Hooke concede to Newton? *Isis*, **58**, 402–5.

Harrison, J. (1978). *The Library of Isaac Newton*. Cambridge: Cambridge University Press.

Heilbron, J.L. (1983). *Physics at the Royal Society during Newton's Presidency*. Los Angeles: Clark Memorial Library.

Holtsmark, T. (1970). Newton's experimentum crucis reconsidered. *American Journal of Physics*, **38**, 1229–35.

Hooke, R. (1665). *Micrographia*. London: Royal Society.

Hooke, R. (1666). A method by which a glass of a small plano-convex sphere may be made to refract the rays of light to a focus of a far greater distance than is usual. *Philosophical Transactions,* 1, 202–3.

Hooke, R. (1705). *Posthumous Works,* ed. Richard Waller. London: Royal Society.

Hooke, R. (1935). *The Diary,* ed. H.W. Robinson & W. Adams. London: Taylor and Francis.

Huxley, G.L. (1959). Newton's boyhood interests. *Harvard Library Bulletin,* 13, 348–54.

King, H.C. (1955). *The History of the Telescope.* High Wycombe: Charles Griffin.

Latour, B. (1986). *Science in Action.* Milton Keynes: Open University Press.

Laymon, R. (1978). Newton's experimentum crucis and the logic of idealization and theory refutation. *Studies in History and Philosophy of Science,* 9, 51–77

Leibniz, G. (1965). *Die Philosophischen Schriften,* vol. 3, ed. C.J. Gerhardt. Berlin: reprinted Hildesheim: Georg Olms.

Lohne, J. (1961). Newton's 'proof' of the sine law and his mathematical principles of colours. *Archive for History of Exact Sciences,* 1, 389–405.

Lohne, J. (1967). Regenbogen und Brechzahl. *Sudhoffs Archiv,* 49, 41–6.

Lohne, J. (1968). Experimentum Crucis. *Notes and Records of the Royal Society,* 23, 169–99.

Mamiani, M. (1976). *Isaac Newton Filosofo della Natura: le Lezioni giovanili di Ottica e la Genesis del Metodo Newtoniano.* Florence: La Nuova Italia.

Manfredi, E. (1728). Correspondence with Thomas Dereham. LBC 19, Royal Society Library.

Mariotte, E. (1740). *Oeuvres,* 2 vol. The Hague.

Mills, A.A. (1981). Newton's prisms and his experiments on the spectrum. *Notes and Records of the Royal Society,* 36, 13–36.

Mills, A.A. & Turvey, P.J. (1979). Newton's telescope. *Notes and Records of the Royal Society,* 33, 133–55.

Nakajima, H. (1984). Two kinds of modification theory of light: some new observations on the Newton–Hooke controversy of 1672. *Annals of Science,* 41, 261–78.

Newton, I. (1704). *Opticks.* London: Samuel Smith and Benjamin Walford.

Newton, I. (1714). Drafts for J.T. Desaguliers, 'An account of some experiments of light and colours'. Cambridge University Library MSS Add 3970.5, ff. 608, 609.

Newton, I. (1722). Draft comments on Rizzetti (1722). Cambridge University Library MSS Add 3970.5, ff. 587, 588, 481, 482.

Newton, I. (1936). *Catalogue of the Newton Papers.* London: Sotheby's.

Newton, I. (1952). *Opticks,* based on 4th edn., 1730, ed. I.B. Cohen & D.H.D. Roller. New York: Dover.

Newton, I. (1958). *Papers and Letters on Natural Philosophy and Related Documents,* ed. I.B. Cohen & R.E. Schofield. Cambridge: Cambridge University Press.

Newton, I. (1959-77). *Correspondence,* 7 vol., ed. H.W. Turnbull, J.F. Scott, A.R. Hall & L. Tilling. Cambridge: Cambridge University Press.

Newton, I. (1962). *Unpublished Scientific Papers*, ed. A.R. Hall & M.B. Hall. Cambridge: Cambridge University Press.

Newton, I. (1967–81). *Mathematical Papers*, 8 vol. ed. D.T. Whiteside. Cambridge: Cambridge University Press.

Newton, I. (1983). *Certain Philosophical Questions. Newton's Trinity Notebook*, ed. J.E. McGuire & M. Tamny. Cambridge: Cambridge University Press.

Newton, I. (1984–). *Optical Papers*, 1 vol. published, ed. A. Shapiro. Cambridge: Cambridge University Press.

Oldenburg, H. (1965–86). *Correspondence*, 13 vol., ed. A.R. Hall & M.B. Hall. Madison and London: Wisconsin University Press, Mansell and Taylor and Francis.

Peacham, H. (1634). *The Gentleman's Exercise*. (first published 1606). London: IM.

Peddle, C.J. (1921–22). The manufacture of optical glass. *Transactions of the Optical Society*, **23**, 103–30.

Pedersen, O. (1968). Sagredo's optical researches. *Centaurus*, **13**, 139–50

Pickering, A. (1981). The hunting of the quark. *Isis*, **72**, 216–36.

Pinch, T.J. (1981). The sun-set. The presentation of certainty in scientific life. *Social Studies of Science*, **11**, 131–58.

Pinch, T.J. (1986). *Confronting Nature: the Sociology of Solar Neutrino Detection*. Dordrecht: D. Reidel.

Rizzetti, G. (?1722). Letter to Christino Martello. King's College Cambridge. Keynes MSS 103.

Rizzetti, G. (?1724). Letter to the Royal Society. King's College Cambridge. Keynes MSS 103.

Rizzetti, G. (1727). *De Luminis Affectibus Specimen Physico-mathematicum*. Treviso.

Rizzetti, G. (1741). *Saggio dell'Antinevvtonianismo sopra le Leggi del Moto e dei Colori*. Venice.

Sabra, A.I. (1981). *Theories of Light from Descartes to Newton*. 2nd edn. Cambridge: Cambridge University Press.

Schofield, R.E. (1970). *Mechanism and Materialism. British Natural Philosophy in an Age of Reason*. Princeton: Princeton University Press.

Shapin, S. (1981). Of gods and kings. Natural philosophy and politics in the Leibniz–Clarke disputes. *Isis*, **72**, 187–215.

Shapin, S. (1984). Pump and Circumstance: Robert Boyle's literary techonology. *Social Studies of Science*, **14**, 481–520.

Shapiro, A. (1979). Newton's 'achromatic dispersion law'. Theoretical background and experimental evidence. *Archive for History of Exact Sciences*, **21**, 91–128.

Shapiro, A. (1980). The evolving structure of Newton's theory of white light and colour. *Isis*, **71**, 211–35.

Shapiro, A. (1980a). Newton and Huygens' explanation of the 22° halo. *Centaurus*, **24**, 273–87.

Van der Star, P. (1983). *Fahrenheit's Letters to Leibniz and Boerhaave*. Amsterdam: Rodopi.

Van Helden, A. (1974). The telescope in the seventeenth century. *Isis, 65,* 38–58.

Voltaire, F.M. (1738). *The Elements of Sir Isaac Newton's Philosophy.* ed. and trans. J. Hanna. London: Stephen Austen.

Westfall, R.S. (1962). The development of Newton's theory of color. *Isis, 53,* 339–58.

Westfall, R.S. (1963). Newton's reply to Hooke and the theory of Colors. *Isis, 54,* 82–96.

Westfall, R.S. (1966). Newton defends his first publication: the Newton–Lucas Correspondence. *Isis, 57,* 299–314.

Westfall, R.S. (1980). *Never at Rest: a Biography of Isaac Newton.* Cambridge: Cambridge University Press.

White, T. (1654). *An Apology for Rushworth's Dialogues.* Paris: Jacques Billain.

Worrall, J. (1982). The pressure of light: the strange case of the vacillating 'crucial experiment'. *Studies in History and Philosophy of Science, 13,* 133–71.

3

A VIOL OF WATER OR A WEDGE OF GLASS

J.A. BENNETT

This chapter examines two related problems in the career and uses of experimental instruments. First, from the seventeenth century instruments were conventionally classified into mathematical, optical and natural philosophical types. This classification offers a valuable insight into the connections between differing instruments and the theoretical and practical contexts in which they were used and their behaviour interpreted. Appropriate contemporary categories such as these should be preserved in an historical analysis of instrumentation. This claim contrasts with that of historians who would impose novel taxonomies upon differing kinds of experimental instrument. The opening sections comment upon Hackmann's distinction between active and passive instruments, presented earlier in this book. Second, the chapter addresses the problem of the replication of instrumentation when the instrument is so extravagant as to be effectively unique. The case chosen here is that of the giant reflecting telescope of the nineteenth century. Historians have recently paid attention to commonplace instruments in order to understand the troubles of making secure knowledge using these devices. Schaffer's chapter in this book uses the work with glass prisms as an example. A telescope, by contrast, may not always be seen as an experimental instrument, but the builders of new telescopes found that they had to grapple with the same difficulties in establishing scientific knowledge.

ACTIVE' INSTRUMENTS: THE BAROMETER

Is a barometer an 'active' instrument or a 'passive' one? It is instructive to apply Hackmann's distinction in chapter 1 to a few examples. The passive alternative is surely the instinctive choice for the barometer.

It sits fixed and silent, should not be moved, needs no connection with other apparatus and no source of power. Without being activated, or even consulted, it indicates one of the slighter variations in Nature. Interfering with nothing and no one, it leaves the world unchanged.

The early history of the mercury column (we shall avoid 'barometer' for the present) in England begins with the Torricellian tube – an instrument for demonstrating the weight of the air, not for detecting subtle changes in the environment. John Wallis's list of the topics discussed by the Gresham circle in the 1640s, places the instrument immediately in this (to use Hackmann's term) 'intellectual framework': 'the weight of the Air, the Possibility or Impossibility of Vacuities, and Natures abhorrence thereof, the Torricellian Experiment in Quicksilver.'[1] The understood purpose of this mercury column was to demonstrate Torricelli's theory of the weight of the air, used in turn to explain phenomena associated with pumping water. The major experimental concern was the general height of the column in relation to columns of liquids of different densities.

It was initially within this context (at least in England) that interest was transferred to small changes in the column's height. According to Descartes the tides are caused by the Moon's motion through the material medium that pervades all space. An additional pressure is transmitted to the Earth's surface and moves the water of the oceans and seas. Christopher Wren suggested that, since the mercury column depends on the weight of the atmosphere, it could be used to test this account of the tides, since its height should vary in relation to the motion of the Moon. Robert Boyle acted on Wren's suggestion, and discovered that there was indeed a small variation in the height of the column, but that it had nothing to do with the motion of the Moon.[2]

At this stage it was known that the weight of the air is subject to a small variation, but there was no connection with the tides and no theoretical framework to give this variation any special significance. This came rather from the medico-meteorological programme pursued by, among others, Wren, Boyle and Hooke in the early 1660s. Properties of the atmosphere were studied as important determinants of health, and changes as possible causes of epidemical disease. Alterations in the weight of the air were seen as especially important, being related to the concentration of the vital nitrous component, essential to fire and to life.[3]

1. McKie (1960), p. 12
2. Bennett (1976b), p. 69; Bennett (1982a), p. 83.
3. Bennett (1976b).

The mercury column became a barometer – an instrument for indi-cating changes in the atmosphere – and the slight variations took over as the focus of attention from the general height of the column. All of Hooke's ingenuity was used to magnify these variations, by constricting the tube, by adding a float connected to a pointer, and so on.[4]

Throughout this history the instrument remained essentially the same in material terms, while radical changes took place in its purpose and function, changes occasioned by shifts in 'intellectual frameworks'. The barometer's passivity is not a useful indication of its interactive role with theory.

EXPERIMENTAL AND MATHEMATICAL INSTRUMENTS

Two of the most radical innovations in sixteenth-century instrumen-tation were the variation compass and the dip circle. Presumably both are passive. They were created within the mathematical science of navigation, but had reference to questions in natural philosophy, and so bridged an important disciplinary divide. The use of instruments within natural philosophy – the instrument had previously been con-fined to mathematical science – would be crucial both to theory and to experimental practice.[5] Again, Hackmann classifies the telescope and microscope as passive. Yet, their acceptance carried with it all manner of theoretical assumptions, to legitimise the deliberate distortion of perception and interpret it as a clearer vision of the natural world. 'Intellectual frameworks' clearly influenced the introduction of both instruments to roles within natural philosophy. They were invented at around the same time – the beginning of the seventeenth century – and the telescope was given immediate relevance to the cosmological debate by Galileo. This accounts for the early popularity of the telescope, while the microscope made no impact until the mid-century. It was then that mechanical philosophy taught that all phenomena depended on the shapes and motions of tiny particles, and so gave the microscope an experimental role.

Though many passive instruments interact crucially with theory, there is, of course, an important difference between, say, a sundial and an air-pump. But the difference is accommodated by the traditional distinction between mathematical, optical and natural philosophical instruments. This was not, as Hackmann implies, devised by the eighteenth-century trade; rather the makers inherited a set of categories

4. Bennett (1980), p. 36, note 19.
5. Bennett (1986).

that had been created by developments in instrumentation in the previous century. As such, the traditional classification is likely to be a more useful analytical device than one imposed by the historian.

Before the seventeenth century, mathematical instruments were the only representatives of our much broader class of scientific instruments. The mathematical instruments offered means of applying techniques based on arithmetic and geometry to ends in astronomy, navigation, surveying and general calculating. It was no part of their function to uncover new truths about the natural world, and their only potential interaction with natural philosophy derived from the occasionally controversial relationship between natural philosophy and astronomy. It was clear that the new optical instruments were quite different. They were not mathematical, and their relevance to the natural philosophy of the heavens at least was quickly demonstrated by Galileo. In the way learning was organised in 1600, there was no reason to associate the telescope and the microscope with the astronomer's quadrant or the navigator's cross-staff. Craft organisation also distinguished their origins, since they were made in the workshops of spectacle-makers rather than mathematical instrument-makers.

The instruments of natural philosophy raised further problems of definition. They were not optical, but were certainly related to natural philosophy rather than mathematical science; indeed some, like the air-pump and the electrical generator, manipulated the natural world in an unprecedented way. Until the eighteenth century, unlike the mathematical and optical instruments, they had no clear craft location. Once this came, and the most entrepreneurial makers began to describe themselves as dealing in 'mathematical, optical and philosophical instruments', they were using a classification which had been fashioned by the developments of the previous century, which reflects important intellectual and craft distinctions, and which historians must, therefore, be careful to preserve.

'TRANSPARENCY' AND THE TELESCOPE

Schaffer's chapter has a more precise focus. He reconstructs the procedures adopted to create a 'transparent' scientific instrument. The example he chooses – the triangular glass prism – seems to have been particularly difficult to prescribe, because it was relatively familiar. There was admiration among seventeenth-century natural philosophers for theories constructed on the commonplace. Wren, for example, wrote that

Experiments for the Establishment of Natural Philosophy are seldom pompous; 'tis upon Billiards, and Tennis-Balls; upon the purling of Sticks and Tops; upon a Viol of Water, or a Wedge of Glass, that the great Des Cartes hath built the most refined and accurate Theories that human Wit ever reach'd to.[6]

Schaffer has uncovered the difficulties Newton encountered in fashioning an experimental instrument from the common prism, and achieving replicability between different trials. He shows how this process depended on closure of the natural philosophical debate, and the extent to which closure relied on 'control over the social institutions of experimental philosophy'.

It is interesting to set against this example of an early and primitive optical instrument the first instances where optical instruments moved so far from the commonplace that they were for all practical purposes unique. What strategy might be adopted to achieve 'transparency' when replicability became impossible?

The very large reflecting telescope was introduced into astronomy by William Herschel in the late eighteenth century,[7] and it was to raise special problems for assessing both instrument and operator. Before his spectacular discovery of a planet in 1781, Herschel was a self-educated astronomer, working outside the professional community. He was thus unskilled in scientific conventions, and his early papers had to be cleansed of such aberrant components as reports of lunar vegetation and speculation about accompanying inhabitants.[8] To J.-D. Cassini it seemed that the discovery of Uranus had fallen to a man who was not an astronomer at all, as he specifically says, but a musician.[9] Music had, till then, been Herschel's profession. Without Uranus it is difficult to imagine how his wildly unconventional astronomical programme, with its implausible instrumentation, could have been seriously regarded by the astronomers. Herschel proposed to study the three-dimensional structure of the heavens, which conventional astronomy used as a convenient backdrop to lunar and planetary work, and he planned even more extravagant instruments than he had already built. As Schaffer has shown elsewhere, the astronomers were inclined to think him 'fit for Bedlam'.[10]

It was difficult, however, to dismiss a man who had achieved royal patronage by adding a primary planet to the solar system − the only such discovery throughout the entire history of written astronomy. It was a curious situation for Herschel, thus plucked from obscurity to,

6. Wren (1750), p. 225. 7. Bennett (1976a). 8. Bennett (1982b).
9. Schaffer (1981), p. 21. 10. Schaffer (1980).

Figure 3.1. William Herschel's 40 foot reflecting telescope. (Royal Astronomical Society)

if not scientific respectability, at least a secure platform in the community, as the only person equipped to pursue astronomy in the unconventional direction he was enthusiastically taking it. No one had telescopes at all like Herschel's, but worse than that, no one could use such telescopes as he did – could 'see' as he did. As he explained, 'seeing is in some respect an art, which must be learnt'.[11] To ask someone to observe with a high magnifying power was, he said, like asking him to play one of Handel's fugues on the organ – it could not be done at sight or 'à livre ouverte'.[12] Herschel's credit-rating thus became a crucial issue for astronomers.

Though Herschel's platform remained secure, closure was not achieved during his lifetime. The grand theoretical problems that his telescopes were built to address were not resolved to the community's satisfaction; neither were his instruments 'transparent' in their judgment. Their heroic status was confirmed by the community when it was accepted that the general direction he had been pursuing was appropriate to serious astronomy, and it is ironic that Herschel's monu-

11. Bennett (1976a), p. 76.   12. Lubbock (1933), pp. 101, 104.

Figure 3.2. The emblem of the Royal Astronomical Society. (Royal Astronomical Society)

mental failure – the great, but in practice useless, 40 ft telescope – was enshrined on the seal of the Royal Astronomical Society. But during their working careers it had been problematic that to those outside his immediate circle Herschel's telescopes had remained objects of admiration and astonishment, but technological mysteries. He made telescopes for sale, but his financial dependence on this trade only made it more difficult for him to divulge the vital techniques used in making the speculum metal mirrors. Of several hundred instruments sent to other observers, not one was used in important astronomical work.

It is significant that when the third Earl of Rosse took up the challenge of extending the telescope's range – to penetrate further into space in pursuit of the structure of this and other worlds – he stated that a central aim was to establish publicly-known techniques. As a result, 'the art might no longer be a mystery known to but few individuals',[13] or as his friend Thomas Romney Robinson put it, 'a perilous adventure, in which each individual must grope his way'.[14] In fact the problem of credibility proved just as troublesome for his fabulous, but almost inaccessible instrument – his great 6 ft aperture reflector, completed in 1845, peering into space through the few gaps in the cloud and rain of central Ireland.

The observer's credibility was enhanced in a number of ways. Lord Rosse, of course, shared some of the advantages enjoyed by Newton; he became President of the Royal Society in 1848 and occupied a prominent and powerful position in the scientific community. Robin-

13. Parsons (1926), p. 8.   14. Parsons (1926), p. 14.

son, who was the most convinced champion of the astronomical work
at Birr Castle (Rosse's estate), frequently pointed out that the telescope-
builder was a nobleman, whose astronomical work was undertaken in
a selfless quest for truth, since career, status and worldly success were
irrelevant goals. In Ireland Rosse was presented as the great ornament
of the nation, and his telescope as an object of patriotic pride.

We have seen that Rosse himself emphasised that his methods were
to be fully available to others. He published detailed accounts of his
procedures, and he invited astronomers to come to Birr, to use his
instruments and to learn his methods of making them. One telescope—builder who benefited from this was William Lassell. Rosse also
stressed that his assistants had no previous experience of practical optics:
'All my workmen were trained in my own laboratory without the
assistance of any professional person, and none of them had previously
seen any process in the mechanical arts'.[15] These innocents, as Robinson
added, were 'persons taken from the surrounding peasantry';[16] unlike
the instruments of Herschel and the commercial opticians, Rosse's telescopes did not depend on any trade secrets.

But when visitors did come, those not already committed to the
enterprise like Robinson and James South, were not always convinced.
The telescope was beset with adverse observational reports, to set
against its spectacular success as a feat of engineering. 'They showed
me something they said was Saturn', reported one observer, 'and I
believed them'.[17] To cope with these problems the Birr astronomers
began to exploit the deficiencies of the telescope, just as Newton had
done with prisms. Rosse said that the critical nature of the polished
surface of the objective, which depended both on the success of the
latest polishing operation and the extent of subsequent deterioration,
compounded the difficulties caused by poor weather for observing.
Eventually he admitted that 'we can scarcely say that any one object
has been examined under a combination of favourable circumstances'.[18]
In the end all hope of replicability was abandoned, when Rosse's
astronomer son, the fourth Earl, made explicit what his father had
already implied in print: that the telescope was radically changed each
time the speculum was repolished. He did so in order to answer 'remarks . . . both from persons who have at some occasion or other had
an opportunity of observing with that instrument, and also from others
who have never been to Parsonstown [Birr]'.

15. Parsons (1926), p. 12.   16. Parsons (1926), p. 23.   17. Jones (1968), p. 372.
18. Parsons (1926), p. 125.

Figure 3.3. Workmen at Birr Castle, with one of the 6 foot aperture mirrors of the great telescope. The astronomer Otto Boeddicker is seated in front. (The Birr Observatory and Museum Trust)

Every time the speculum is removed from the tube and repolished, the old figure, whether it be good or bad, is lost in the process, and a new one formed, whose merits in no way depend on those of the last, and the telescope, though in mechanism the same, is optically speaking a new one.[19]

The attempt to establish 'transparency' held some ground for a time, but eventually proved inadequate to deal with the combination of speculative theorising and non-replicable experiments. What it illustrates, however, is the similarity, in terms of Schaffer's analysis, between a case involving unique and extravagant instrumentation and that of Newton's common prism.

19. Parsons (1880).

References

Bennett, J.A. (1976a). 'On the power of penetrating into space': the telescopes of William Herschel. *Journal for the History of Astronomy*, **7**, 75–108.

Bennett, J.A. (1976b). A note on theories of respiration and muscular action in England c. 1660. *Medical History,* **20,** 59–69.

Bennett, J.A. (1982a). *The Mathematical Science of Christopher Wren.* Cambridge: Cambridge University Press.

Bennett, J.A. (1982b). Herschel's scientific apprenticeship and the discovery of Uranus. In *Uranus and the Outer Planets,* ed. G. Hunt, pp. 35–53. Cambridge: Cambridge University Press.

Bennett, J.A. (1980). Robert Hooke as mechanic and natural philosopher. *Notes and Records of the Royal Society of London,* **35,** 33–48.

Bennett, J.A. (1986). The mechanics' philosophy and the mechanical philosophy. *History of Science,* **24,** 1–28.

Jones, K.G. (1968). *Messier's Nebulae and Star Clusters.* London.

Lubbock, C.A. (1933). *The Herschel Chronical.* Cambridge: Cambridge University Press.

McKie, D. (1960). The origins and foundation of the Royal Society of London. *Notes and Records of the Royal Society of London,* **15,** 1–37.

Parsons, C. (1880). Observations of nebulae. *Scientific Transactions of the Royal Dublin Society,* **2.**

Parsons, W. (1926). *The Scientific Papers of William Parsons, Third Earl of Rosse,* ed. C. Parsons. Bradford and London: Lund, Humphries & Co.

Schaffer, S. (1980). Herschel in Bedlam: natural history and stellar astronomy. *Journal for the History of Astronomy,* **13,** 211–39.

Schaffer, S. (1981). Uranus and the establishment of Herschel's astronomy. *Journal for the History of Astronomy,* **11,** 11–26.

Wren, C. (1750). *Parentalia: or, Memoirs of the Family of the Wrens.* London.

PART II

EXPERIMENT AND ARGUMENT

Figure 4.1. Ptolemy, Aristotle and Copernicus in dialogue, as represented in Galileo's *Two Great World Systems* (1632)

4

GALILEO'S EXPERIMENTAL DISCOURSE

R.H. NAYLOR

Of these two new sciences, full of propositions that will be boundlessly increased in the course of time by ingenious theorists, the outer gates are opened in this book, wherein with many demonstrated propositions the way and path is shown to an infinitude of others, as men of understanding will easily see and acknowledge.

(*The Printer's Preface to the* Two New Sciences *of 1638*)

THE BEGINNING OF A NEW DISCOURSE

No one doubts that Galileo played a major part in changing the European view of physical science. Looking at his *Two New Sciences* (1638) and comparing it with the work of his immediate predecessors and contemporaries we sense the existence of an immense gulf. This radical break becomes all the more striking when, on looking closer, we discover its presence in his own work. How it came about is one of the great unsolved mysteries of European intellectual history. There is, in particular, an almost total contrast between his view of experiment in his earliest writings and that found in the famous masterworks of his declining years.

It has been said that on reading his early Pisan essay *De Motu*, one may imagine oneself transferred to Buridan's lecture room. That is so, but nevertheless there is a distinct difference in the tempo of Galileo's discussion. It is a lively discourse in which experience is appealed to, repeatedly, even if only in the most general terms and primarily as a basis for criticising Aristotle. Experience and, on occasions, experiment are employed as part of a rhetorical repertoire aimed to refute Aristotelian physics while championing an alternative quasi-Archimedean theory. However, experiment does not by any means play a posi-

117

tive confirmatory role able to establish the claims of Galileo's theory; on the contrary, it proves incapable of corroboration, and Galileo's response to the failure is both interesting and significant.

The experiment that highlights this failure is one in which spheres of differing material are rolled down an inclined plane. According to his theory they should roll at very different speeds – bronze several times faster than marble, for instance – but such is not the case. Galileo does not contemplate obscuring this difficulty – which reveals him to be working within an episteme in which experiments on motion were not actually expected 'to work' in our modern sense. He did argue, nevertheless, that the failure was only apparent. Though an experimental failure, it was not a theoretical one. This rather relaxed attitude to empirical issues was by no means unusual and stemmed from the general response to such matters within both Aristotelian and Archimedean accounts of motion. But the text does prove him rather more concerned to defend his theory than was customary. In this he shows signs of paying more heed to observation than his academic contemporaries Girolamo Borro and Francesco Buonamici.

As it was, he felt the need for a suitable ancillary theory to account for the failure of his inclined plane experiment. With this found, it formed a protective shield around his Archimedean theory, allowing him to ward off any threats posed by the evidence. The experiment had failed, he concluded, because of the disturbing effects produced by friction and air resistance in addition to the impossibility of obtaining perfectly smooth, hard planes and spheres. As there is no way of judging what such effects might be, Galileo's defense is of course perfectly rational and being so demonstrates the implausibility of the idea that he might have regarded such a refutation as an effective guide to action.

Galileo could, in time, have concluded at most that the only role he had found for 'esperienza' in *De Motu* was one of refutation. As such, if it did have a methodological role, it told almost as much against his position as Aristotle's. It has been suggested by both Koyré and Drake among others that it was the lack of positive success in accounting for the phenomena that led Galileo to lose faith in the *De Motu* theory. It is true that Galileo at this point could have seen that his efforts had not moved the discussion of motion significantly nearer to achieving the standards of empirically successful Archimedean science. Compared to the precision of which he proved the Archimedean hydrostatic theory capable in his little essay *La Bilancetta*, written a few years before in 1586, the failure of his theory of motion would have provided a sad contrast. *La Bilancetta* revealed his conviction that a mathematical science as well formulated as Archimedes' needed to be capable of very exact confirmation. As a consequence he rejected the traditional myths concerning Archimedes' discovery and by a simple practical method showed how he could have established specific gravities with exactitude.

In reality we do not seem to be faced simply with two different theories of motion in *De Motu* and *Two New Sciences* but with two distinct epistemes. In the earlier Pisan essay, as Dijksterhuis sensed but did not articulate, the discourse is located within the episteme of medieval physics. Indeed even the sixteenth-century Archimedean discussions of motion suggest a stronger resemblance in outlook to earlier Aristotelian treatments of motion than might be expected, in that both forms of discourse are primarily concerned with logical and mathematical issues. In the fourteenth-century Aristotelian studies of motion, the discussion of complex examples, the sophismata, were not experimental. Its theories were not meant to be tested by an appeal to Nature. *De Motu* does not prove entirely different by any means. It suggests that Galileo's ambition was to develop a theory superior to Aristotle's but that all he could provide by way of empirical evidence was the refutation of Aristotle without the necessary confirmation of his own theory.

Within the episteme of *De Motu* it was evidently never anticipated that the central theory could be expected to provide any close correspondence with reality. In a very similar manner, the work of Benedetti, who was Galileo's immediate predecessor in the Archimedean tradition, leaves the impression that empirical issues are not of primary importance. Benedetti's theory, like Galileo's, proposed speeds of fall that were proportional to specific gravity, but he never appeared aware of, let alone concerned about, the observational implications of this idea. Lead does not fall at ten times the rate of wood – be the natural motion uniform or accelerated, and both ideas were, as it happens, discussed by Benedetti without any evident concern about a conflict between them or their empirical validity. Moreover he also managed to argue for the principle of the equal rate of fall of all bodies in a void irrespective of their specific gravity.

So although Benedetti made an important innovation in his mathematical and logical approach to motion, his lack of any method of distinguishing between a variety of conceptual possibilities led to an equivocation on the matter of natural motion. Having opened up a new vista of interesting possibilities with his *Resolutio* in 1553 he found it hard on mathematical grounds alone to pursue any one of them consistently. This suggests a relationship between his ambivalence on 'mathematical' issues and the 'empirical' uncertainty in his subsequent works.

In *De Motu* Galileo had moved beyond this point – but not a great deal further. He did pursue one central theory quite consistently. On the other hand, though he had not reached the point of expecting a correspondence between theory and experiment, he did show signs of concern that even a qualitative fit with the phenomena was unobtainable. He evidently felt compelled to call upon secondary explanations to 'save the appearances'.

In contrast the observation that bodies do not in actuality fall with speeds proportional to their specific gravities nor with uniform speeds did not apparently trouble Benedetti unduly. It was the novelty and mathematical interest of his refutation of Aristotle – on logical grounds – that seemed to appeal to him. He proved more concerned about the potential plagiarism of his ideas than their empirical disproof. As he published his theory on four occasions between 1553 and 1585, we can reasonably conclude he was justifiably proud of his accomplishments and that some, at least, of his contemporaries regarded his theory as an interesting achievement. Why else would Taisnier have plagiarised it in 1562 and Richard Eden translated it into English some sixteen years later?

If in *De Motu* Galileo was still working within what was essentially a pre-empirical episteme he does show signs of an erosion of confidence in the adequacy of such an approach. This is suggested by the recognition that the gross mismatch between theory and observation, even if an accepted feature of contemporary discourse on motion, was nevertheless an issue that could not be easily disregarded by a theory aspiring to the status of an Archimedean science.

The prime example of this issue was that of the lack of observation of steady motion in fall – which was the basic idea of his theory. In the first draft, written in essay form, the observed acceleration is accounted for by means of a secondary theory – the theory of a declining impetus. In the second draft, this time a dialogue, a style commonly used for scientific discussions in that period, he advanced an additional and quite novel explanation to defend his basic belief that natural motion is uniform motion.

Writing in dialogue had the great advantage of allowing Galileo the opportunity of a lively discussion, where the difficulties faced by his theory could be raised, examined and defused. There were considerable didactic advantages in using this dialectical process as it allowed him to present a defense to objections he had anticipated and others that had been made to him in discussion. In the dialogue Alessandro first explains the observed acceleration in terms of a declining impetus – but having done so Domenico responds by admitting the argument sounds fine but asks, in effect, why some sign of the decrease of acceleration anticipated by the theory isn't seen? Alessandro responds by saying one would see it for heavy bodies were it not for the effect of perspective – which gives a false impression of accleration when the body's fall is viewed from below. Shortly after this the dialogue breaks off to be left incomplete – an indication perhaps of Galileo's realisation that this defense of the theory, so appealing in terms of its mathematical simplicity, was doomed. The introduction of one secondary theory may help retain credibility but this search for ever more means of defending the theory from empirical failure leads to a steady drain of plausibility in *De Motu*.

A total contrast is presented by Galileo's response to such issues in the *Two New Sciences* which reveals his ultimate conclusion that an empirical theory had to be confirmable directly and precisely – no fudging of the issue was to be contemplated. If he started out by being relaxed about empirical issues as a young man he certainly presented himself in his old age as relentless in the pursuit of precision.

EXPERIMENT AND EXPERIENCE IN GALILEO'S MATURE DISCOURSE

Not surprisingly those who have puzzled over his mature work have concluded that the understanding of Galileo's scientific attitude can only be established by a careful study of his experimental discourse. Alexandre Koyré made an important advance in arguing that a critical approach to the experimental discourse would reveal far more than its naive acceptance. In his view there were no grounds for accepting Galileo's experimental accounts as true ones. Contrary to customary belief, Koyré's critical close reading showed a number of implausibilities in the *Two New Sciences* where the experiments are claimed to provide a near perfect confirmation of the theory of motion. To Koyré this degree of precision looked unbelievable. As perfect experimental confirmations of any mathematical theory are unobtainable there had to be some other explanation for Galileo's claims – Koyré concluded they must be thought experiments.

Oddly Koyré did not consider that such a situation might be the result either of Galileo's literary style or of his desire to persuade others of the merits of his new mathematical science. Koyré had laid particular stress on two aspects of Galileo's work, firstly that its inspiration was Copernican and that this spurred him on to create the new physics, and secondly that the *Dialogue on the Two Great World Systems* (1632) was a didactic work, one of persuasion, argument and rhetoric. There was perhaps a disinclination on Koyré's part to think through the implications of this thesis fully. For if the new physics and the new astronomy were indeed complementary aspects of a new episteme, in the support of which no argument or emphasis was to be overlooked – would not this influence the presentation and discourse of the new physics?

There does seem in fact to be a strong degree of compartmentalisation in Koyré's atttitude to Galileo's work, which is not at all uncommon in the history of science. In the *Dialogue* we all accept that Galileo is engaged in persuasion, or, as Koestler termed it, a crusade, whereas when he turns to the *Two New Sciences* his discussants are said not to 'engage in polemics and speak only in the objective interests of science and truth.'[1] Ludovico Geymonat suggests, in echoing Koyré's distinctions that we find in the *Dialogue* 'the Copernican Manifesto' whereas with the *Two New Sciences* we see 'the return to pure science'. He quotes with approval Koyré's assessment: 'Actually, the *Dialogue* is neither a book of astronomy nor a book of physics.' Geymonat concludes, 'Neither is it a treatise on physics, or in short a purely scientific work of any kind.'[2]

This is true, but the idea that somewhere we might find a 'purely scientific' work in the sense underlying this division of pure and impure science is hardly credible. Metaphysical commitment invariably influences all scientific discourse – including experimental discourse. The notion of pure science is a positivist myth in so far as it claims to be free of subjective elements. At most we are dealing with distinctions of degree rather than kind when we consider the discourse of the *Dialogue* and *Two New Sciences*. It is hard to see why Koyré, who placed such a great emphasis on the symbiotic relationship that existed between the new physics and astronomy, should have failed to consider the persuasive powers of Galileo's experimental claims in terms of argument. There is no question that as well as advancing a new scientific theory the *Two New Sciences* is *also* a didactic work determined to persuade its readers of the merits of the new physics. It too was a manifesto – this time for the new physics.

1. Brodrick (1964).   2. Geymonat (1965), p. 132.

THE EXPERIMENTAL DISCOURSE OF THE 'TWO NEW SCIENCES'

To Alexandre Koyré some of Galileo's experiments appeared so improbable that he concluded that they must be imaginary. What is puzzling is that Koyré presumably saw them as theoretical anticipations. It was the unbelievable nature of an experiment in the 'First Day' of the *Two New Sciences* that showed that wine and water did not mix which led Koyré to deny its reliability. It has however been shown to be quite real by James MacLachlan.[3] Its reality suggests a significant aspect of Galileo's development of experimental discourse. It shows that he was an acute observer of Nature who in noticing the unexpected perceived its relevance to theoretical issues. It also reveals that though the experiment lent support to a particular idea, there is, on examination, no obvious way he could have predicted the outcome by some process of reasoning. It is, like other of his earliest experiments, a simple form of experiment supporting a simple idea. It is a demonstration which either works or does not – there is no half way. It resembles the traditional use of experiment whereby a physical principle was made manifest from appearances, and it is some distance from the mathematical theory and process of deduction we find in the inclined plane experiment of the 'Third Day'. Only in such mathematical experiments do such issues of measurement and precision arise.

It was by means of simple experiments that Galileo was led to perceive their broader possibilities as means of testing and confirmation. In this he was evidently influenced by his father Vincenzio, by his patron Guidobaldo del Monte and, through the latter, by the exemplar of the Florentine study of perspective, specifically in Guidobaldo's *Perspectivae* of 1600. To this has to be added the suggestive influence of Giordano Bruno and William Gilbert. Galileo's earliest successful experimental discovery was a very simple one – the isochronism of the pendulum.

Another idea that must have drawn his attention to the role of simple experimental tests in the last decade of the sixteenth century was Bruno's interesting argument for the Earth's motion, which appeared at just the time Galileo began to take Copernicanism seriously. What is certain is that Bruno was instrumental in undermining the credibility of refutation as a means of arbitration between the Copernican and Ptolemaic theories. We know too that Galileo's work in the pre-empirical episteme of *De Motu* and the related reliance on experimental refutation ceased during his early years in Padua. Bruno had returned to Venice in 1592, the year after Galileo's arrival in Padua and several

3. MacLachlan (1973).

years after the appearance of his works *Cena de la Ceneri* and *De l'*
Infinito Universo et Mondi in England (1584). Bruno's arrest in 1594 and
the significance of his views could hardly have failed to prove a topic
of some fascination within the Venetian Republic.

Bruno's ideas were of great importance. He was, as Koyré said, the
first to conceive the philosophical importance of the new science.
Though preoccupied with this philosophical and, as he believed,
theological revolution, he was not lacking in a grasp of some important
methodological aspects of the debate. What he pointed to was the flaw
in a major Aristotelian argument against the Earth's motion. It had
been claimed that the fact that bodies fell vertically demonstrated the
Earth's immobility. Bruno argued that on dropping a stone from the
top of a ship's mast it arrived at the foot whether the ship was stationary
in harbour or moving steadily at sea. The fact that both logical analysis
and experimental tests supported this argument undoubtedly swayed
Galileo and others towards the Copernican theory.

THE RHETORICAL USE OF THOUGHT EXPERIMENT

In the *Dialogue* the discourse is primarily concerned with persuasion.
This being so, Galileo, by employing the Socratic method, aims to
convince the reader of the necessary truth of the argument irrespective
of experiment. This has the advantage of removing the reader's reliance
simply on the authority of a reported account of experiment. It
demonstrates that merely by calling on the reader's existing knowledge
of Nature the conclusion must be that the bullet always falls to the
foot of the mast – irrespective of the ship's motion. This use of thought
experiment rather than any claimed reliance on real experiment is a
very effective rhetorical technique. In a work of persuasion one would
expect this, naturally. It would nevertheless be rash to conclude that
such methods of persuasion represented Galileo's own methods of work
and discovery.

Though the evidence available is sparse, it suggests that Galileo was
not as rash in advancing experimental arguments as Koyré supposed.
Generally he appears to have considered the theoretical and practical
aspects fairly carefully, and where he could not provide an adequate
explanation he declined to publish – as with the experiments on the
parabolic trajectory – for which clear evidence of this research survives.
In the case of the ship-mast and wine–water experiments, it would
have been risky to publish experiments that could be easily tested and
proved wrong – if they were simply thought-experiments. We know

that G.B. Baliani and Pierre Gassendi did test the ship-mast experiment as soon as they could, perhaps directly as a consequence of Galileo's implication that it was only a thought-experiment.

His earliest experiments began by establishing a single feature, such as the isochronism of the pendulum, but by the the middle of the first decade of the seventeenth century he was involved with experiments of much greater complexity. Not only were several variables involved but careful measurement was required too. He discovered that he could not always obtain the necessary agreement between theory and experiment for each of these variables, as in his experiment on the projectile trajectory found by Drake on folio 116 of volume 72 of the Galilean manuscripts in Florence.[4] Galileo struggled with this sort of difficulty for years before finally allowing his *Two New Sciences* to go to press. Although his experimental researches had convinced him of the validity of his theory he was nevertheless dissatisfied with the overall relationship between his theory and his experiments. He had discovered that the new physics was not perfect and that it lacked the certainty of mathematical demonstration – but he would never concede this as he wished to advance his claim as being one of truth rather than probability.

The two most important experimental arguments in this theory of motion relate to the central element – his law of fall, which states firstly, that all bodies fall at the same rate and, secondly, that the distance travelled in fall increases as the square of the time. With the first he begins with his doubts as to the empirical basis of Aristotle's claim that speed of fall is dependent on weight – which he puts into the mouth of Salviati:[5]

SALVIATI: I seriously doubt that Aristotle ever tested whether it is true that two stones, one ten times as heavy as the other, both released at the same instant to fall from a height, say of one hundred braccia, differ so much in their speeds that upon arrival of the larger stone upon the ground, the other would be found to have descended no more than ten braccia.

SIMPLICIO: But it is seen from his words that he appears to have tested this for he says "We see the heavier . . . " Now this "We see" suggests that he had made the experiment.

SAGREDO: But I, Simplicio, who have made the test assure you that a cannonball that weighs one hundred pounds or two hundred or even more does not anticipate by even one span the arrival on the ground of a musket ball of half, both coming from a height of two hundred braccia.

4. Drake (1973); Naylor (1974a).   5. Galileo (1638), p. 66.

This is the famous tower experiment around which the mythology of the leaning tower has arisen. According to myth Galileo is supposed to have refuted Aristotelian philosophy by its means. The belief that this sort of experimental demonstration would have been enough to refute the Aristotelians or lend great support to Galileo's theory is not easy to credit. It had been employed by Aristotelians to show, for instance, that light and heavy objects fall at differing speeds, and that similar objects of wood and lead fall at different rates. It was this particular case of differing materials that was crucial to Galileo's theory, and in view of which he was bound to require considerably more than the tower experiment to convert his audience to his novel views on free fall. As he had himself managed to entertain such views on differing rates of fall in his youth it doesn't seem that these free-fall experiments were unambiguous.

Though one cannot doubt that Galileo is on far surer ground than the Aristotelians there appears to be a degree of rhetorical exaggeration quite unnecessary to his case. Being convinced he is correct he feels free to emphasise the novelty of his views and the strength of his case by enhancing the contrast between his view and Aristotle's. Rather than drop the quite adequate one and ten pound spheres referred to by Aristotle from a height of a few score feet (which is precisely what Simon Stevin did to make this point) Galileo refers to a musket ball and a 200 pound cannonball dropped from a 360 foot tower. This claim for the scope and exactitude of his theory of course looks fictious as well as irrelevant – but it is rhetorically effective.

Following this appeal to experiment Galileo provides his by now equally famous though-experiment to show that two equal spheres fall at the same speed whether falling singly or joined together. This experiment, a relic of his earlier ideas, works just as well for theories of uniform speed of fall as for accelerated motion. Rather than being an aid to Galileo it could easily have proved a hindrance, for his though-experiment could shed no light on the idea of an equal speed of fall of all bodies, nor indeed on the form of natural motion. It is clear that the Archimedean theory proved a false start for Galileo rather than being a stepping stone to a new and better theory. So both the over–confidence in the hydrostatic theory of fall and in such things as thought-experiment probably proved a distraction that had to be relinquished before progress was possible.

It is at this juncture that Galileo produces his final and, it seems, crucial argument for the equality of the rate of fall. The difficulty with free-fall experiments, as Salviati explains, is that in dropping heavy and light materials from any height the effect of air resistance always masks the equality in the rate of fall. For this reason he came to the idea of comparing many *equal* small falls of a bob of cork and a bob of lead by means of the simple pendulum. In this way, he argues, the effect of the resistance of the medium may be avoided. Two pendulums of equal length, one with a cork bob and the other with lead, pulled apart by the same amount of 30 degrees, are seen to swing together for hundreds if not thousands of oscillations. Observation simply reveals that the cork and lead released together from an angle of 30 degrees complete arcs of quite different dimensions. While the lead completes an arc of almost 60 degrees the cork completes and arc of 50 degrees or less. As a consequence the fall of the cork and the lead in the second and all successive swings must be different. It is true that Galileo's argument looks plausible; intuitively one feels that the only reason for the remarkable isochronism observed is likely to be the equality of the rate of fall. But such a vague explanation would not serve for Galileo's purpose. He required a conclusive demonstration of the equality of fall. fall. From both mathematical and empirical aspects his arguments are inadequate and do not measure up to the requirements of a mathematical science. As a result, from a perfectly real experiment Galileo felt compelled, between observation and theory, to claim the existence of phenomena that are quite unreal.

PROBABILITY AND TRUTH: GALILEO'S SEARCH FOR CONCLUSIVE ARGUMENTS

It is noticeable that there are changes both in style and emphasis between the discourse of the *Dialogue on the Two World Systems* and that of the *Two New Sciences*. In the earlier work Galileo appears unconcerned with issues of experimental precision. He does not, for example, claim that isochronism is exact – as he does in the later work. His main arguments in the *Dialogue* are not of the same form and do not rely on the precise agreement called upon in the *Two New Sciences*. The only argument in the *Dialogue* that makes any call on quantitative issues is his tidal theory – which proves a remarkably poor fit with observation. There was little prospect of Galileo substantiating any claim that his theory actually *predicted* two tides a day. Rather than rely on observation he had to argue that such observations as there were proved rather discordant and revealed no clear pattern. By contrast, with the mast–fall experiment he attempted to reveal by means of the Socratic method, that Nature had already spoken on this issue, if only in an ambiguous manner. Merely by reflection Simplicio could on his own come to

recognise the truth – even if he had never seen the experiment performed. As the *Dialogue* progresses and as the ever more complex and
deductively structured arguments are produced there is less and less
scope for the Socratic questioning. Nature speaks less and less directly
it seems, requiring ever more subtle interrogation. Even then, Nature
responds solely on the condition that carefully phrased questions are
posed by an interrogator familiar with her mathematical language.
Simplicio proves unfamiliar with this language and can only follow it
haltingly, even with the aid of an intermediary.

When in the *Dialogue* questions are put to Nature concerning the
reality of the Earth's movement she proves capable of only a somewhat
Sphinx-like response. Neither logic nor observation can elicit an unambiguous answer. It is noticeable that intellectual midwifery hardly seems
to figure when we get to the nub of the *Dialogue*. When we read the
'Third Day' we encounter first a battery of arguments based heavily
on observation, of which the most telling relate to Galileo's telescopic
discoveries, followed by the Copernican explanation of the retrogressions of the outer planets, and finally Galileo's argument concerning
sunspot movement. As this unfolds the discourse takes on more and
more the appearance of an exposition while Simplicio drops further
and further into the background. Thus he makes no response whatever
to the Copernican explanation of the planetary retrogressions before
Salviati presses on with the final argument of the day on sunspots.
When at last he breaks his silence, Simplicio, far from sharing a common
understanding with Salviati, only confesses that he has not entirely
mastered the content of the discourse and points out that logically
there is more than one possible interpretation of the evidence. Galileo
had to recognise, however disarmingly, that shared experience does
not lead to shared belief. Moreover, when Salviati has completed this
argument, Simplicio declares himself incapable of making a decision
on its merits. The reader is intended to sympathise with Salviati in the
face of this sophistical nit-picking and to sense that after all Simplicio
has just not heard the voice of Nature.

A similar dilemma is presented to Simplicio in the *Two New Sciences*
when Salviati attempts to elicit the acceptance of his interpretation of
the pendulum experiment. His reaction to the argument is bafflement:[6]

The reasoning seems to me conclusive and also it seems it isn't; my mind feels
a kind of confusion that arises from the moving of both movables now quickly,
now slowly, and again extremely slowly, so that I can't get it straight in my
head whether it is true that their speeds are always equal.

6. Galileo (1638), p. 89.

This resembles the Platonic dialogues where, by the Socratic interrogation, those seeking knowledge are reduced first to the *aporia* or impasse. The irony here is that Simplicio's perception of an impasse is perfectly justified – he has observed 'Nature' well enough – but such naive empiricism just will not do. Though they have all witnessed the same experiment they do not see the same things.

As the voice of Nature seems muffled in Simplicio's ears, Salviati is prepared to do all he can to clarify the message. There is no suggestion that *anamnesis* might be of service here – in the matter of ultimate scientific truth. On the contrary, rather than simply recollect what he has gleaned from Nature, Simplicio must listen to Sagredo rephrase and tidy up Nature's message – which he does by asking:[7]

Tell me, Simplicio, whether you grant that it may be said with absolute truth that the speed of the cork and that of the lead are equal every time they both start at the same moment from rest and, moving along the same slopes, always pass equal spaces in equal times.

Naturally Simplicio cannot deny this, but it does not and never could correspond to any actual observations. This is why, beginning from a real experiment, we have arrived at an imaginary one. Galileo has transformed the real experiment that convinced him of the truth into a though experiment, in order to convince others of the truth.

THE FINAL DISCOURSE

In composing the *Two New Sciences* Galileo had to tackle three very difficult issues of presentation and argument. As we have seen there were the considerable problems of linking observation and theory via a coherent and unambiguous discourse, the question of whether theories were true, false or probable and the issue of the empirical precision of the new science. The three are interrelated. His response to all these matters was influenced, to a very noticeable degree, by the reception of the *Dialogue*. Galileo had been forced by the church into a public admission that the best evidence for the new astronomy was inconclusive. He had always known this to be a possible construal by others. But he had believed that the arguments he provided in favour of Copernicanism were more than sufficient to convince any mind open to reason. Yet they were judged unconvincing by many, not only the Aristotelians but by the Jesuit scientists of the Roman College and the Censors of the Holy Office.

7. Galileo (1638), p. 89.

Galileo's public admission of fallibility could be construed not only as an admission of a particular error relating to a misjudgment of evidence relating to the Copernican theory, but also as one raising doubts as to scientific truth. Were scientific theories after all, merely probable? It seemed that they might be – and presumably the greater the evidence obtained from Nature in favour of a theory the higher the probability of its truth. What Galileo had always known but here had been forced to admit – to his considerable anguish – was that the voice of Nature could be taken to be saying such discordant things.

Not one of his four main arguments in the *Dialogue* met an unambiguous response. The first two, based on his telescopic discoveries and the planetary retrogressions, were given an alternative explanation by the Jesuits in terms of the Tychonic system. The second two arguments highlighted Galileo's difficulty in employing complex deductive arguments which were not easily related to the empirical evidence. They were sufficiently complex to leave many of his readers dubious and confused. Of these arguments the best – which is an argument from simplicity – is based on the observed motion of sun spots. It reveals that merely to have a good grasp of the scientific principles involved – here three dimensional kinematics – is not sufficient to produce an effective argument. In fact he handled the argument very ineptly. This was undoubtedly a significant factor in the subsequent reaction to it and helps us understand why it was so readily disregarded.

Galileo's sun spot argument reveals a great deal about how the 'form' in which an observational argument is presented, as well as its 'content', has a considerable effect on its reception. Mark Smith has recently claimed that Galileo's proof is in fact absolutely conclusive. This poses a very interesting question: what are we to make of the history of the reaction to it?[8] To argue that the study of 'Nature' could make Galileo's argument clear and that those who did not accept it were simply mistaken is to advance a naively empiricist view of science and scientific discourse. To persist with this argument as a means of explaining the subsequent reaction would be to advance an equally crude view of history.

Galileo's 'proofs' highlighted the reality that Nature appears ambiguous – capable of more than one explanation – and scientific arguments are accordingly frequently perceived to be inconclusive. In response his opponents dismissed them. Examination of the disparate responses to Galileo's arguments does naturally involve the consideration of their observational base, but that proves quite inadequate as a means of

8. Smith (1985), p. 544.

understanding the variety of reactions to them. Galileo found with the *Dialogue* that from the same observations more than one conclusion was always possible; such ambiguity had to be dispelled from the *Two New Sciences*.

The great difficulty Galileo experienced with the *Dialogue* was that his two main proofs of the Earth's motion failed to convince for a variety of reasons: they were conceptually complex, they appeared confusing if they were not actually confused, they were not empirically justified and they were not well presented. Consequently they were judged less plausible than the theory the proof of which they claimed to provide. In the case of the tides he singularly failed to make many converts on all these grounds. The most readily appreciated doubts arose over the implausibility of the implication that there was one primary daily tide – and that the second was due to additional factors – while the tide was associated with the Sun's position. None of these arguments proved credible.

His difficulties with the sun-spot argument arose not only because he handled it ineptly but because his description of the appearances was many months out of agreement with observation. What he did, apparently, was to fabricate the observations at the last minute without having made the year-round measurements of sun-spot positions essential to his argument. As he had not made the measurements he did not know the orientation or inclination of the Sun's axis. He guessed, wrongly, that the orientation of the Sun's axis was similar to the Earth's. This explains why his account of the appearances is confused and incomplete, although correct in outline. The revelation of Galileo's error followed immediately. Scheiner's very accurate description of years of painstaking observation of sun spots appeared in his *Rosa Ursina* in 1630 – even before the *Dialogue* was published.

Compared to his brilliant introduction of observational arguments into astronomy with the *Starry Messenger* (1610) the *Dialogue* proved a humiliating outcome. If evidence increased probability of truth, Galileo had not apparently increased the probability of Copernicanism beyond the point it had reached in 1610.

CONCLUSION

As Koyré discerned, the claims for the accuracy of Galileo's experiments in the *Two New Sciences* are unreal and there was a reason for this. Koyré's intuition was that it was the experiments that were unreal, but as we have seen his conclusion proved premature. If the experiments were real, could it be that the claims for accuracy were unreal? If Galileo had not performed his experiments one would expect, as Koyré did, that they would not work. Or if they were thought–experiments some rather large observational discrepancies would be unearthed – as with the sun-spot argument. But this has not been found to be the case. It is, by and large, only in the fine detail of Galileo's experiments that we discover an exaggerated claim for precision. (The explanation of this is simply that in Galileo's view exactness of confirmation is a demonstration of truth rather than probability). In contrast with Galileo's work in the pre-empirical episteme, the experiments in the *Two New Sciences*, with few notable exceptions, do work. Guesswork, prevision or thought experiment will not repeatedly produce experiments accurate even to within 30%, let alone the 15% or occasional 5% achieved by Galileo. Indeed, Koyré's rejection of Galileo's claims was formulated despite the background knowledge that 'errors' of the above order are just those found in modern physics – why should we expect Galileo to achieve better?

The reading of the *Dialogue* reveals that Galileo is not consistent in his attitude to scientific truth. He sometimes speaks of plausibility, but also frequently as though a theory has to be true *or* false. There is no consideration of an intermediate middle ground on such occasions. His response to the relationship between theory and experiment in a truly mathematical science proves rather different in the *Two New Sciences*, primarily because he is determined at all costs to establish that his new science of motion is an empirically true theory rather than merely probable. As a consequence, simple experiments that establish directly an observational law, such as that of isochronism, are claimed to be precisely true. The cork and lead pendulum of the 'First Day' are said to be exactly isochronous over hundreds, indeed thousands, of oscillations. This is presented as a direct appeal to Nature – but it is not.

Galilean science revealed a dilemma – it was a victory and yet a defeat. On the one hand, it changed the basis of scientific discourse in rejecting the teleological approach of the Aristotelians for an approach more abstract and yet more empirical. Aristotle had found the cause of motion in terms of his cosmology and in ideas of natural place. Galileo at first, in *De Motu*, rejected Aristotle's theory but regained the cosmological explanation of motion in terms of place – just a Benedetti had done before him. In *De Motu* Galileo claimed not only to have found the correct mathematical form of motion but to know its ultimate cause. However, in the *Two New Sciences* Galileo was left only with the former. He had no explanation of why bodies fell: this had become an inexplicable feature of the world. Thus there is a complete reversal of value. For the Aristotelians it was cause that was paramount while form played an entirely secondary role. Form was really beneath consideration – which is just the view expressed by Simplicio in the *Discourses*. Such incidental details were left 'to lowly technicians'. Galileo's position was the polar opposite. It was cause (in the Aristotelian sense of Final cause) that could be left aside but on the contrary the *precise* form of motion *was* of overriding significance.

Galileo has been much complimented for this freeing of his science from metaphysical dominance, for avoiding futile discussions of ultimate causes and providing a purely phenomenological kinematics. To the Aristotelians a major strength of their theory of motion was that it formed part of a cosmological theory: the weakness of Galileo's was that it lacked any comparable explanation in terms of an ultimate cause. Galileo had no counterclaim in terms of ultimate causes. A relationship looked a poor substitute for a *vera causa* of motion. He could only respond by insisting that what he had found was its true empirical description – but this placed a whole new emphasis on empirical validity and accuracy.

References

Brodrick, J. (1964). *Galileo*. London: Geoffrey Chapman.

Clavelin, M. (1974). *The Natural Philosophy of Galileo*. trans. Pomerans. Cambridge, Mass. and London: MIT Press.

Dijksterhuis, E. (1961). *The Mechanization of the World Picture*. Oxford, London and New York: Oxford University Press.

Drake, S. (1970). *Galileo Studies*. Michigan: University of Michigan Press.

Drake, S. (1973). Galileo's experimental confirmation of horizontal inertia: unpublished manuscripts. *Isis*, **64**, 291–305.

Drake, S. (1978). *Galileo at Work*. Chicago: University of Chicago Press.

Drake, S. & Drabkin, I. (1960). *On Motion and On Mechanics*. Madison: University of Wisconsin Press.

Drake, S. & Drabkin, I. (1969). *Mechanics in Sixteenth-Century Italy*. Madison, Milwaukee and London: University of Wisconsin Press.

Ferni, L. & Bernadini, G. (1961). *Galileo and the Scientific Revolution*. New York: Basic Books.

Galileo. (1632). *Dialogue Concerning the Two Chief World Systems*, trans. S. Drake (1967). Berkeley and Los Angeles: University of California Press.

Galileo. (1638). *Two New Sciences*, trans. S. Drake (1974). Madison: University of Wisconsin Press.

Galileo. (1890–1909). *Le Opere di Galileo Galilei*, ed. A. Favaro. Florence: Barbera.

Geymonat, L. (1965). *Galileo Galilei*, trans. S. Drake. New York, Toronto and London: McGraw–Hill.

Koestler, A. (1959). *The Sleepwalkers*. London: Hutchinson.

Koyré, A. (1978). *Galileo Studies*, trans. J. Mepham. Hassocks: Harvester.

Lindberg, D. (1978). *Science in the Middle Ages*. Chicago and London: University of Chicago Press.

MacLachlan, J. (1973). A test of an 'imaginary' experiment of Galileo's. *Isis*, **64**, 374–9.

McMullin, E. (ed.) (1967). *Galileo, Man of Science*. New York and London: Basic Books.

Naylor, R. (1974a). Galileo and the problem of free fall. *British Journal for the History of Science*, **7**, 105–34.

Naylor, R. (1974b). Galileo's simple pendulum. *Physis*, **16**, 23–46.

Naylor, R. (1974c). The evolution of an experiment: Guidobaldo del Monte and Galileo's *Discorsi* demonstration of the parabolic trajectory. *Physis*, **16**, 323–46.

Naylor, R. (1975). An aspect of Galileo's study of the parabolic trajectory. *Isis*, **66**, 394–6.

Naylor, R. (1976a). Galileo: the search for the parabolic trajectory. *Annals of Science*, **33**, 153–72.

Naylor, R. (1976b). Galileo: real experiment and didactic demonstrations. *Isis*, **67**, 398–419.

Naylor, R. (1977a). Galileo's need for precision: the Fourth Day pendulum experiment. *Isis*, **68**, 97–103.

Naylor, R. (1977b). Galileo's theory of motion: processes of conceptual change in Galileo's theory of motion in the period 1604–1610. *Annals of Science*, **34**, 365–92.

Naylor, R. (1979a). Mathematics and experiments in Galileo's New Sciences. *Annali dell'Istituto di Storia della Scienza*, **4**, 56–63.

Naylor, R. (1979b). Galileo's scientific activity. *Nature*, **279**, 457–8.

Naylor, R. (1980a). Galileo's theory of projectile motion. *Isis*, **71**, 550–70.

Naylor, R. (1980b). The role of experiment in Galileo's work on the law of fall. *Annals of Science*, **37**, 363–78.

Naylor, R. (1982). Galileo's law of fall; absolute truth or approximation. *Annals of Science*, **39**, 384–9.

Naylor, R. (1983). Galileo's early experiments on projectile trajectories. *Annals of Science*, **40**, 391–4.

Settle, T. (1961). An experiment in the history of science. *Science*, **133**, 19–23.

Settle, T. (1968). Ostilio Ricci, a bridge between Alberti and Galileo. *Actes du XIIe Congrès International d'Histoire des Sciences*, **3b** 121–6.

Shea, W. (1972). *Galileo's Intellectual Revolution*. London: Macmillan.

Smith, A. (1985). Galileo and the movement of sunspots. *Isis*, **76**, 543–51.

Wiener, P. & Noland, A. (ed.) (1957). *Roots of Scientific Thought*, New York: Basic Books.

5

FRESNEL, POISSON AND THE WHITE SPOT: THE ROLE OF SUCCESSFUL PREDICTIONS IN THE ACCEPTANCE OF SCIENTIFIC THEORIES

JOHN WORRALL

INTRODUCTION: A PRIZE STORY

One of the most attractive stories in the whole history and philosophy of science folklore concerns the award of the French Academy's prize of 1819. As usually told, the story could almost provide a script for 20th Century Fox. A prestigious prize was at stake. The hero of the story was a comparatively young man from the provinces, with unorthodox views. Facing him was a prize commission whose members were very much part of the Parisian scientific establishment and a majority of whom were committed to the rival, entrenched orthodoxy. Most dramatically, there was a time when all seemed lost for the hero, but then came an amazing *volte face*, and his unorthodox views finally won through. Add to this the facts that there was constant concern for the frail hero's health and · that the episode was played against the background of political turmoil – the hero himself rushing to join the royalist army which opposed the return of Napoleon from Elba – and the story looks like a sure-fire box-office winner.

Here are some of the details. The topic for the prize was the diffraction of light; the hero Augustin Jean Fresnel, in 1819 newly restored to his post as officer in the Engineering Corps, and just 30 years old. Fresnel sent to the Academy a long memoir[1] which attempted to show in detail how all diffraction phenomena could be explained on the supposition that light consists of wave-like disturbances transmitted through an all-pervading medium. This theory was, of course, not new in the early nineteenth century, but it had long been regarded as discredited. The prevailing view, invariably described as Newtonian, amounted (at any rate on a 'realistic' interpretation) to the theory that light consists of

1. Fresnel (1819).

135

material particles subject to short-range forces emanating from ordinary 'gross' matter. This emission theory formed a central plank of the Laplacian approach to physics which was dominant in France at the time. The commission of prize judges was a 'Who's Who' of contemporary French physics; it comprised Biot and Poisson, both committed Laplacians, as well as Laplace himself, plus Arago and Gay-Lussac. Arago was already known as a champion of Fresnel and his ideas. As for Gay-Lussac, while he, like the others, had been an active member of the Arceuil group around Berthollet and Laplace, he never seems to have taken a particular interest in optics and can probably be assumed to have been impartial. Verdet describes Gay-Lussac as 'little familiarised through his studies with the question at issue, but disposed by character to a wise impartiality'; Maurice Crosland asserts that 'Gay-Lussac . . . on this question could be relied upon to be impartial'.[2] But with or without Gay-Lussac, the corpuscularists held a clear majority on the commission. Indeed it is probable that the whole idea of a prize competition on diffraction had been conceived by the Laplacians, and principally Biot, as a stimulus to young scientists to provide a successful corpuscular account of this range of phenomena. Fresnel's chances, apparently quite slim at the outset, seemed to shrink to zero when, during the commission's deliberations, Poisson demonstrated that Fresnel's wave theory of diffraction has a patently absurd consequence: it implies that if a small opaque disc is held in the light emanating from a small hole, the centre of the disc's shadow will be bright, just as bright indeed as if no obstacle had been placed in the light's path. Arago tested this consequence; lo and behold, he found the white spot! The Academy's prize was promptly awarded to Fresnel. More importantly, the outcome of the competition proved a turning point in the history of science. Fresnel's success in 1819 signalled an end to the long domination of the corpuscular theory, and by the mid- to late 1820s the wave theory of light became generally accepted in the scientifically advanced countries. This successful novel prediction of a hitherto unsuspected, indeed hitherto *counter*expected, phenomenon had played a central, perhaps even crucial, role in the acceptance of the new theory from which it was drawn.

Something like this version of the story has penetrated the history and philosophy of science folklore. One of Kuhn's illustrations of a 'conversion' to a 'new paradigm' is, for example, the following:

2. Verdet (1866), p. xlii; Crosland (1967), p. 409.

In the case of the wave theory, one main source of professional conversions was even more dramatic [than any of the factors leading to the Copernican revolution]. French resistance collapsed suddenly and relatively completely when Fresnel was able to demonstrate the existence of a white spot at the center of the shadow of a circular disk. That was an effect that not even he had anticipated but that Poisson, initially one of his opponents, had shown to be a necessary if absurd consequence of Fresnel's theory.[3]

And Ronald Giere tells the story this way:

Fresnel's memoir was referred to a commission in which well-known advocates of particle models, Laplace, Poisson and Biot, held a majority. The commission was apparently not fully convinced by the evidence Fresnel had presented, and Poisson devised a further test. He applied Fresnel's model to the case of a shadow produced by a circular disk and deduced that the resulting diffraction pattern would have a bright spot at the centre of the shadow. Even from superficial accounts of this incident it seems clear that no one involved had ever seen such a spot. Moreover, it seems that Poisson and his fellow commissioners did not expect the spot to appear. It certainly was not a consequence of any current particle model that such a spot should exist. The experiment was performed by Francois Arago . . . The spot appeared as predicted and the commissioners yielded.[4]

Earlier, but essentially similar accounts are to be found in Jevons and Whewell. Most recent brief accounts of the episode refer to the following passage from Whittaker:

A champion was indeed needed when the [prize] memoir was submitted; for Laplace, Poisson and Biot, who constituted a majority of the Commission to which it was referred, were all zealous supporters of the corpuscular theory. During the examination, however, Fresnel was vindicated in a somewhat curious way. He had calculated in the memoir the diffraction-patterns of a straight edge, of a narrow opaque body bounded by parallel sides, and of a narrow opening bounded by parallel edges, and had shown that the results agreed excellently with his experimental measures. Poisson, when reading the manuscript, happened to notice that the analysis could be extended to other cases, and in particular that it would indicate the existence of a bright spot at the centre of the shadow of a circular screen. He suggested to Fresnel that this and some further consequences should be tested experimentally; this was done, and the results were found to confirm the new theory. The concordance of observation and calculation was so admirable in all cases where a comparison was possible that the prize was awarded to Fresnel without further hesitation.[5]

These accounts differ *slightly* in the exact claim made about the impact of the white spot episode: no doubt none of the authors, if pressed,

3. Kuhn (1962), p. 155.   4. Giere (1984), p. 280.   5. Whittaker (1951), p. 108.

would commit himself to the view that the episode was, in isolation, *the* crucial factor in winning the prize for Fresnel (and, more importantly, winning general acceptance for his wave theory). But several accounts come close to espousing this view, and certainly all agree in attributing some sort of special significance to the appearance of the white spot.

I shall argue in what follows that this story is historically incorrect. But my purpose is not simply to pour some factual cold water on an appealing story. The story has often been cited as an important illustration of a *general methodological thesis*: the thesis that favourable *novel* evidence – evidence first discovered only as a result of testing some already articulated theory – carries greater weight in support of that theory than does favourable, but already known, evidence.[6] The main aim of the present paper is to show that the real history of the reception of Fresnel's wave theory of diffraction supports, *not* this 'novel facts count more' view, but a rather different account of empirical support.

A MORE ACCURATE VIEW OF THE WHITE SPOT EPISODE

It is not usually easy to provide convincing evidence of what did and did not impress a particular group of scientists in their assessment of a particular scientific work. The case of Fresnel's prize memoir of 1819 is unusual in that there exist seemingly reliable public sources of exactly this information. These sources have been notably underexploited by those who have commented on the episode.

First and foremost there is the report of the prize commission itself.[7] This was read at the French Academy as part of the process of revealing the result of the competition, and was subsequently published, along with some additional notes, in the *Annales de chimie et de physique* in May 1819. I shall rely principally on this report in what follows. Some of the history of that report itself can, however, also be traced. Fresnel had submitted his first memoir on diffraction[8] to the Academy some three years before the prize competition. This earlier memoir contains several particular claims which Fresnel soon abandoned and replaced. None the less the overall theory presented there has a good deal in common with the 1819 theory. As was the custom, a commission was appointed (consisting in this instance just of Arago and Poinsot) to report on the merits of this 1816 memoir to a full meeting of the Academy, who then decided whether or not the paper should be pub-

6. For a discussion of, and references to, this debate see Worrall (1978).
7. Arago (1819).   8. Fresnel (1816).

lished in its *Recueil des Savants Étrangers*. Arago and Poinsot's report[9] was published in Fresnel's *Complete Works* and so is readily available today. More remarkably, so is a record of the reactions expressed at the meeting which heard Arago and Poinsot's report by various Academy members, including Biot, Poisson and Laplace. This record was taken by the secretary, kept by the Academy, and published in the early twentieth century as part of a *Procès-verbaux*. The final prize commission report of 1819 seems to represent a compromise between the 1816 report on Fresnel's first memoir and the rather less enthusiastic reactions of the Laplacians, recorded in volume 6 of the *Procès-verbaux* and, no doubt, expressed again during the prize commission's deliberations. These records, together with other papers and letters, permit a fairly complete reconstruction of the reaction of the French scientific elite to Fresnel's theory of diffraction and in particular to the white spot episode.

One common illusion about this episode can be immediately dispelled in the light of the prize commission's report. It is easy to form the impression from recent accounts that the prize was a highly competitive affair with a long list of entrants, that it was always unlikely that Fresnel would win and that he therefore needed something as dramatic as the unexpected success of the white spot prediction derived by Poisson. In fact there were just *two* competitors. There seems to be no record of who Fresnel's rival was – the candidates were officially anonymous, each memoir being identified by an epigram and being referred to by the commissioners as 'number one' and 'number two'. 'Number one's' memoir was not published. However, some judgment can be made of the scientific value of his paper from the fact that of the whole report of the commission, which in its published version covers some 18 printed pages, just one short paragraph at the end is devoted to poor old 'number one':

The author of the Memoir marked number one is certainly a trained physicist; but, the means of observation which he employed not being sufficiently precise, some of the phenomena exhibited by light in passing through small apertures, or merely in the neighbourhood of opaque bodies have escaped his attention. The author seems to know neither the work for which we are indebted to Dr Young nor the memoir which M. Fresnel had published in 1816 in the *Annales de chimie et de physique*: also the part of his work which concerns the influences which rays of light exercise on one another in mixing, far from adding anything to what was already known, includes several obvious errors . . . [10]

9. Arago & Poinsot (1816).   10. Arago (1819), p. 236.

No wonder then, that the paragraph continues (and concludes): 'accordingly the commission has decided to award the prize to the memoir marked number two, and carrying as epigram: *Natura simplex et fecunda.*' As Verdet remarked '[Number one's] work was not for an instant put in the balance with that of Fresnel.'[11] The competition facing Fresnel could hardly have been less stiff.

A second common illusion about this episode is that the appearance of the white spot brought about not only the decision to award the prize to Fresnel, but also an immediate recognition on the part of the commissioners of the superiority of the wave theory championed by Fresnel over its previously entrenched rival. In fact, the report does its best to ignore basic questions concerning the nature of light: despite Fresnel's own strong emphasis on general theory, and his wholehearted commitment to the wave theory, the report manages to avoid any discussion of this and indeed the word 'wave' occurs nowhere in it. Fresnel's theory is given a rather severe positivistic reinterpretation. It is taken as consisting essentially of its mathematical expression – the famous 'Fresnel integrals'. The various different cases of diffraction (straightedge, single slit, etc.) are characterised by different limits of integration, and the resulting definite integrals yield expressions for the varying light intensities. Fresnel himself, had, of course, presented these integrals as *consequences* of the underlying wave nature of light. But this general theoretical aspect of his views is played down in the commission's report, being replaced only by a rather bloodless version of the Huygens' principle: a version which talks simply of the 'rays' of light (an *apparently* theory-neutral term), and makes no reference to any medium, let alone to any disturbances within a medium.

As for conversions to the wave account, here is Poisson, the very man who drew out the allegedly crucial prediction, writing to Fresnel in March 1823, four years after the prize competition:

The theory of emission and that of waves both encounter great difficulties; time and the future work of physicists and mathematicians will perhaps end by settling all these doubts and entirely clarifying the question; but I believe that one can be sure at present that *if* the second theory [i.e., the wave theory] is true, it is certainly not for the reasons that have been given up to now . . . [12]

Poisson continued long after 1819 to point to difficulties (mostly of a basic mathematical kind) in Fresnel's theory; and, although one sometimes reads of Poisson being converted to waves sometime in the 1830s, convincing evidence of any full conversion is not easy to find.

11. Verdet (1866), p. xlii.
12. Poisson's letter was published in Fresnel (1866–70), II, p. 189.

Biot, another of the prize commissioners, also continued to resist the wave theory long after 1819. He remained a staunch Newtonian until *at least* the 1830s, attributing defects in the Newtonian theory largely to the fact that relatively few talented people were working on it. As for Laplace, he was already 70 years old in 1819 and published nothing on optics afterwards, but having been such a strong advocate of the Newtonian theory, any conversion on his part to the wave theory would surely have been reported somewhere. No such report seems to exist.

So there was no instant conversion to Fresnel's wave theory even among the prize commissioners, let alone of the scientific communities in Britain and France more generally. Still, even if it was much more gradual and less universal than the usual over-dramatised accounts imply, there was undoubtedly a general shift in scientific opinion away from a Newtonian theory of light (which was dominant in the eighteenth century) towards the wave theory (which became dominant by the 1830s). Moreover, focusing just on the prize competition, something must have convinced the commissioners of the high scientific merit of Fresnel's work – in some sense of high scientific merit compatible with an uncommitted attitude towards the truth of the underlying theory. (After all, the commissioners could always have decided to make no award at all.) Doesn't a slightly watered-down, more realistic, version of the story still stand? If the success of the white spot prediction was not *the crucial* factor (on reflection, we should no doubt not expect there to be any such thing), was not that predictive success at least the most important single factor in winning the prize for Fresnel, and in gradually winning his theory, if not always outright belief, at least a more and more sympathetic reception? This is something like the position adopted by Emile Verdet. Being much closer to the primary historical material, Verdet knew well enough that the white spot episode had *not* changed 'the basis of [the commissioners'] convictions'; but he none the less described it as a 'remarkable incident' which 'made a great impression on the minds of the judges' and 'probably determined the unanimity' of the decision.[13]

The question of what influenced the British and French scientific communities generally is, of course, a large one; but if we focus just on the prize commissioners then it is, I believe, clear from their report that even the watered-down claim about the white spot's impact is false.

First, so far as just the prize itself goes, it seems likely that Fresnel would have won even had his memoir contained no theory whatsoever. Fresnel's memoir after all records his invention of a new method of

13. Verdet (1866), p. xlii.

observing and measuring diffraction fringes. Earlier investigators had observed the fringes indirectly, either by casting them on a white screen or by viewing them from behind a plate of unpolished glass. Fresnel discovered that this was unnecessary and that the fringes could be viewed directly, in mid-air so to speak, using a simple magnifying glass.[14] This direct method led immediately to greatly enhanced visibility of the fringes and it allowed them to be observed much closer to the diffracting object itself than had hitherto been possible. Moreover, using an instrument of his own construction, which allowed the position of the lens to be nicely adjusted *via* a micrometer gauge, Fresnel had measured the distances between fringes with greatly increased precision. Almost one half of the prize commission's report is taken up with Fresnel's new observational method and its advantages – before any mention at all is made of any account which might codify the observational results achieved *via* the method.

The commissioners did, however, eventually come to 'the hypothesis on which the integral is founded which [Fresnel] presents as the general expression of all diffraction phenomena'[15] and to the question of what support this hypothesis receives from various experimental results. One interesting point, already mentioned, is that even though now concerned with theory, the commissioners present a bloodless, positivistically reduced version of Fresnel's views. They produce a brief (and in fact rather garbled) statement of the 'Huygens' principle' in terms, *not* of real mechanical disturbances travelling through a real mechanical medium, but of 'rays' spreading out from each point and interfering with one another. The central issue from the present point of view, however, is not how they interpreted the theory, but how they appraised the strength of various pieces of evidential support. Here there is no doubt that the commission placed the emphasis firmly on the *already known* straightedge diffraction cases, rather than on the white spot prediction.

The history of this central feature of the report can easily be traced. As I already remarked, Arago and Poinsot were charged with reporting on Fresnel's *first* memoir on diffraction, published in 1816. The feature of Fresnel's theory that chiefly impressed Arago and Poinsot was that it yields precise details of the diffraction fringes observable both inside and outside the geometrical shadow of a narrow opaque object, and that it yields these details in a straightforward way, one which needed to borrow only one quantity from observation. Here, for example, is

14. Fresnel (1819), pp. 262–3.   15. Arago (1819), p. 235.

what Arago and Poinsot said about Fresnel's theory and the interior fringes:

The simplest circumstances of the formation of the interior bands are inexplicable or at least unexplained on the usual [corpuscular] theory; that of M. Fresnel shows at the same time how they are propagated, what sizes they must have when the screen is placed at different distances, how these sizes are, for a given position of the body, in inverse proportion to its diameter; it explains also why the position of these fringes is independent of the distance of the point source . . . It determines finally when and how each interior fringe must leave the shadow to take its place among the visible exterior fringes . . . If we add that, in the same theory, the sizes and positions of the bands of different shades are deducible from the general formula, by replacing only [the parameter] *d* by corresponding values furnished by the observation of the coloured rings of the first order; that the fixity of the deviation, whatever the density or the refringent force of the body which produces the shadow, is not only an intelligible, but even a necessary consequence; that there is finally no diffraction experiment, known at present, which cannot, I shall not say be explained, but even calculated [from the theory]; one will not be able to avoid avowing, whatever opinion one holds on the basic question, that M. Fresnel's hypothesis deserves to be followed and to win the attention of physicists and geometers.[16]

When this early report was presented to the Academy, it met with a certain amount of criticism. This was mainly to the effect that many difficulties still faced the wave theory, that the corpuscular theory remained preferable in at least some respects, and that, therefore, more emphasis should be placed on the experimental laws discovered by Fresnel and less on the general theory which, he alleged, explained them.[17] This criticism, which emanated from the Laplacians, was no doubt repeated during the deliberations of the prize commission and is certainly reflected in their eventual report, as we saw earlier.

If the story about the white spot is at all correct, another important difference between the 1816 Arago–Poinsot report and the 1819 prize report is surely to be expected; for it was only during the 1819 commission's deliberations that the white spot episode occurred. If this predictive success really had the impact usually attributed to it, a major difference between the two reports would lie in the section dedicated to the empirical evidence for Fresnel's 'hypothesis': the white spot evidence should take pride of place in the 1819 report. In fact the 1816 and 1819 reports differ hardly at all in this respect: pride of place is definitely retained in 1819 by the straightedge cases – cases which had long been the subject of experimental scrutiny (beginning in the seventeenth century with Grimaldi, and then Newton).

16. Arago & Poinsot (1816), p. 86. 17. *Procès-verbaux*, 6, pp. 317–21.

The prize report emphasised, as before, that in these straightedge cases Fresnel's theory yields fringe sizes for all values of the distances between the source and diffracting object and between that object and the point of observation of the fringes. And it is again explicitly emphasised that the theory deals with these cases in a particularly straightforward way. The integral that Fresnel's theory yields as their solution is completely determined by that theory itself, except for one quantity which can be fixed independently of the straightedge diffraction experiments (for example, by doubling the numbers given by Newton for his 'intervals of fit'). These numbers had in turn been derived by Newton from his famous measurements of the diameters of 'Newton's Rings' – a phenomenon which does not at all involve diffraction as usually understood. The report recorded that Fresnel had made a series of 125 experimental measurements of the external fringes outside the shadow of a straightedge, and that in this whole series the difference between observation and the value provided by Fresnel's integral was only *once* as much as 5/100 mm, only *three* times 3/100 mm and *six* times 2/100 mm. In all the other 115 cases disagreement between theory and observation did not exceed 1/100 mm. Similar success was achieved by Fresnel in the case of a narrow slit and that of the fringes inside the geometrical shadow of a narrow object; again, both cases whose qualitative features had long been known, though admittedly Fresnel was now treating them with unprecedented precision.

It was only after this relatively long discussion of these already known cases that the prize commission's report turned to the white spot. The report devoted precisely *two sentences* to this phenomenon – two sentences in the penultimate paragraph of a report whose main text amounts to some nine printed pages. These two sentences read:

One of your commissioners, M. Poisson, had deduced from the integrals reported by the author the singular result that the centre of the shadow of an opaque circular screen must, when the rays penetrate there at incidences which are only a little oblique, be just as illuminated as if the screen did not exist. This consequence has been submitted to the test of a direct experiment, and observation has perfectly confirmed the calculation.[18]

The consequence deduced by Poisson is indeed described as 'singular' and there is no doubt that its experimental confirmation was another impressive feather in Fresnel's cap. But there is no indication of any *exceptional* importance being attributed to this result, let alone any indi-

18. Arago (1819), p. 236.

cation of everything hanging on it. This new piece of evidence in favour of Fresnel's theory simply takes its place alongside other favourable evidence concerning already well-investigated cases.

So much for the main text of the commission's findings which contains the report actually made to the French Academy. The version published shortly afterwards in the *Annales de chimie et de physique* contained some extra material in the form of five appended notes. The last, and briefest of these (Note (E)) contains further reference to Poisson's deductions from Fresnel's theory. It records that Poisson had remarked to Fresnel, *some time after the commission had reported,* that his intensity integral could 'easily' be evaluated for two cases: the centre of the shadow of a small circular screen and points on the axis of a small circular aperture. The screen case having already been dealt with in the main text, nothing was added here. As for the circular aperture case, the note records that Fresnel himself had subsequently supplied the detailed calculation and found intensities varying between four times that expected in the case of an unobstructed wave and zero, depending on the distance of the point of observation from the centre of the aperture. Fresnel had gone on to confirm this prediction empirically. This is surely just as 'singular' a result as the screen case. Fresnel had found that his theory entails that if light from a point source is shone on an opaque screen in which there is a small circular *opening*, then at certain points beyond the screen but along the axis of the aperture (that is, the extension of the line from the source to the centre of the aperture) there is total darkness.

This 'black spot' prediction is 'complementary' to the white spot case – it is just as counterintuitive as the latter and its empirical success just as unexpected and remarkable. But again no special fuss was made of it and again no indication at all was given that any more weight was attributed to this new success of the theory than to the successes – both 'novel' and 'old' – already reported in the main text. In fact the most interesting aspect of this part of the story is that the consideration of this new prediction, and its empirical testing, were left until *after* the award of the prize. Poisson had seen, during the commission's deliberations, that the Fresnel integral could 'easily' be evaluated both for the small circular screen and the small circular aperture. He had performed the detailed calculation for the first case, and the results predicted had been empirically tested. But the second case was left uninvestigated until *after* the commission's decision. If everything hinged on predictive successes of this kind, or even if special weight was being given them, then this seems very strange behaviour. Under such circumstances one would surely expect the commissioners to urge Poisson to supply the detailed calculation and Arago to test the resulting prediction as soon as possible. One would also expect the commissioners to postpone their decision until they learned the result. Clearly they felt no such urgent need. To sum up: the original published documents provide no historical evidence that any special weight was given by the prize commission to Poisson's prediction of the white spot and its empirical confirmation. The emphasis in these documents is very much on Fresnel's success in dealing with the already long-known cases of straightedge diffraction.

DOES SUCCESSFUL PREDICTION CARRY ANY SPECIAL EPISTEMIC WEIGHT?

There is a long tradition in the *philosophy* of science which tries to articulate a logic of empirical support. It shows how theories *ought* to be appraised 'objectively' in the light of the empirical evidence and, in particular, when one theory is to be preferred to another in the light of the available empirical evidence. Mill and Whewell were distinguished nineteenth-century contributors to this tradition. In this century, Reichenbach, Carnap and other inductive logicians have contributed, as have Popper, Lakatos and others. Unshaken by Kuhn and the sociologists of science, the adherents of this tradition hold that there is an important element of rationality in the development of science which has not been present (or has not been present to anything like the same extent) in the development of say, art, or philosophy. Scientists generally have switched theories because the new theory was 'objectively superior' to the old according to clearly articulable, general principles. They claim that means that the new theory was *better empirically supported* than the old. 'Switching to' or 'preferring' a theory need not – in my view – involve believing it to be true. Einstein, for example, clearly thought that the quantum theory was overwhelmingly the best-supported theory in its field – but he also believed that it could not be true and that it would sooner or later be replaced by a theory which would be still more strongly supported, and also deterministic. A major problem for this 'logic of support' tradition is that of exactly how this *normative* enterprise is meant to mesh with the *descriptive* details of the history of science. This problem is the focus of a famous (in some circles, notorious) paper by Lakatos. I have tried to improve on Lakatos's solution.[19] Without going into details, it is clear that, just as a normative theory of goodness would be in bad trouble

19. See Lakatos (1971) and section 5 of Worrall (1976).

if a large number of generally recognised saints turned out to be evil according to its criterion, so a normative theory of science would be in bad trouble if such notables as Fresnel, Arago and Poisson turned out to be judging theories unscientifically. The aim then is to construct a 'logic of empirical support' which *both* seems *a priori* plausible *and* captures the judgments of most prestigious scientists. Or, if this logic fails to capture some particular judgment of that kind, it should provide a convincing and historically well-supported account of why the judgment went awry. As we shall see, one of the consequences of the account of support to be defended in this section is that Fresnel and his contemporaries *as a matter of fact* judged the theoretical import of the straightedge and circular screen diffraction results exactly as they *ought* to have done, according to that account.

Most recent contributors have recognised that the support an empirical result lends to a theory is not merely a question of whether or not the theory entails the result, nor merely a question of the logical strength of the result itself (conjunctions generally giving more support than disjunctions, etc.). These contributors have recognised that even among single, 'atomic' experimental results (each deductively implied by some theory plus initial conditions), some results support the theory more strongly than others. The white spot episode, as I indicated earlier, has often been cited as evidence for the claim that what characterises those results which yield greater support is their *temporal novelty*. Results which were first discovered only as a consequence of some theory's predicting their existence weigh especially heavily in that theory's favour, more heavily than any fact for which the theory successfully accounts but which was already known before the theory was articulated.

This claim of the special role of temporally novel predictions has a long and distinguished history; but it is false. Those who have argued for it have made a rather natural mistake. They have noted that there are outstanding cases in the history of science where a theory logically entails a certain empirical result but the theory was intuitively accorded little or no support from that result. They have correctly concluded that there is some extra factor involved in empirical support beyond mere deducibility. They have then remarked that in all such cases the empirical result concerned was already known at the time the theory concerned was first articulated, and conversely that there are no cases where a theory logically entailed some hitherto unknown result and was *not* regarded as impressively supported by it. They have therefore concluded – this time incorrectly – that the extra factor involved in genuine support is *novelty of the evidence*.

Zahar and I have argued[20] that the important extra factor in empirical support – beyond mere deducibility of the evidence from the theory – is concerned with the way in which the theory was developed or constructed. In all cases where scientists have *not* recognised evidence *e* as fully supporting theory *T*, despite *e*'s following from it, *T* had been modified or tinkered with or otherwise developed precisely so as to yield *e*. Nothing hangs on whether we say in such cases that the theory is *not at all* supported by the evidence *e*. The important judgment to capture is that any such support is *less than* that accruing to a theory which entails the result without any tinkering or modification. Suppose that there are two rival theories, *T* and *T'*, and that *T* entails *e* directly and without artifice. *T'*, on the other hand, was, say, developed out of a more general theory which contained a free parameter, a parameter whose value had been fixed using exactly the evidence *e*. In such a case scientists will generally regard *T* as receiving *more* support from *e* than *T'*, even though both theories now entail *e*. Philosophers of science have often dealt with this phenomenon as a problem concerning scientific *explanation*: *T'* will not generally be regarded as explaining *e*, because it was introduced *ad hoc*, relative to *e*. The point can, therefore, be stated in a second way: it is wrong to regard the downgrading of *ad hoc* explanations and the apparent upgrading of genuine predictions as two *separate* methodological phenomena – they are at root the *same* phenomenon.

The approach to empirical support developed by Zahar and myself has been called the 'heuristic account'.[21] One advantage it has over the temporal account is that it comes equipped with a rationale. If the time-order of theory and evidence *was* in itself significant for scientists then we should, I think, be reduced merely to recording this as a brute fact. For why on earth *should* it matter whether some evidence was discovered before or after the articulation of some theory? (This was John Stuart Mill's main point in this part of his argument with Whewell.) On the other hand, there is a clear reason why it matters whether or not some evidence was involved in the construction of a theory. If theory *T* straightforwardly entails some observationally decidable statement *e*, then checking on the truth of *e* will, in general, constitute a test of *T* – *T might* have got this observable state of affairs wrong and hence have been refuted. And this is quite independent of whether or not *e* was already known. However, if some particular feature of *T* was in fact tied down on the basis of *e*, so that *T* had been engineered to entail *e*, then checking *e* clearly constitutes no real test

20. See, in particular, Worrall (1985).   21. See Musgrave (1974).

of *T*. It is obviously no test of *T* to ask it to get right some result which had been explicitly incorporated into it in the first place. In such a case even though *e* follows from *T* and hence not-*e* is, in Popper's terminology, a potential falsifier of *T* – it wasn't *really* a potential falsifier of *T*, since *T* was, because of its method of construction, never at any risk from the facts described by *e*. This shows, I think, that the intuitions behind the notion of a genuine test cannot be captured in purely logical terms but must involve consideration of how the theory concerned was constructed. (It is interesting that, while certain informal remarks by Popper show that he is aware of the dependence of falsifiability on heuristic factors, his formal account explicitly excludes such factors.) This rationale for the heuristic theory of support no doubt raises further problems and needs careful handling, but at least it provides *a* rationale for the heuristic account, while the temporal account seems to have no rationale at all.

A second argument for the heuristic account is that, unlike the temporal account, it seems to capture judgments made by working scientists in particular cases.[22] The Fresnel–Poisson case provides further evidence. The heuristic account makes a clear-cut pronouncement about this case: the novel prediction about the circular screen case and the implications about the already investigated straightedge cases count equally in support of Fresnel's theory, *provided* these latter facts were not used in the construction of the theory. This proviso was *clearly* met. Fresnel produced a unified theory of diffraction, covering all cases in a general way. No special conditions needed to be made for special cases. The whole precise theory followed, in a natural way, from general and rather abstract theoretical considerations. Fresnel in effect argued as follows. Consider a series of waves emanating from a point source and spreading out through some medium. The waves necessarily arrive at points in the medium more distant from the source *via* points which are less distant. We should therefore be able to take the points on *any* wave front and reconstitute the disturbance at any point more distant from the source as a certain composite – as, in fact, a composite of the separate motions which *would* emanate from each point of the chosen wave front if the effects of the disturbance of that point could be isolated from all others. Every point on the chosen wave front would, in other words, be regarded as a 'secondary source' of disturbances. The process of recomposition of the 'primary' wave must take into account the differences of phase of the disturbances from the secondary sources as they arrive at the more distant point. This consideration was, of course,

22. For examples, see Worrall (1978).

dictated by the principle of interference. The interference principle itself was derived by Fresnel from the assumption that light is a wave motion in an elastic medium, together with various simplicity assumptions. Fresnel's knowledge of the history of his subject was not great. He named the whole assumption about the recomposablity of a light wave out of secondary waves, the 'Huygens Principle', although in fact its connection with any doctrine advanced by Huygens is remote. This is how Fresnel himself introduced his theory of diffraction in his prize memoir:

I am now going to show that one can give a satisfactory explanation and a general theory [of diffraction] within the system of waves *without the aid of any secondary hypothesis,* by depending only on the Huygens principle and that of interferences, which are the one and the other consequences of the fundamental hypothesis.[23]

Here Fresnel referred to the Huygens principle as a straight consequence of the fundamental idea of light as waves in a medium. This is not, however, correct. Some aspects of his derivation of the principle from the 'fundamental hypothesis' – notably concerning the 'obliquity factor' – never really satisfied those with sensitive mathematical consciences. Fresnel does seem to have been aware of some difficulties on this score – in one place, for example, he referred to the Huygens principle as only 'an *almost obvious* consequence of the fundamental hypothesis'.[24] Nonetheless Fresnel did show that the principle is a 'natural' consequence of the fundamental hypothesis in the sense that, while not following just from the hypothesis alone, it does follow from it, together with some plausible, and *general,* considerations. Fresnel's reasoning amounts to an attempt to reconstitute the initial or 'primary' wave in the unobstructed case out of the mutual interference of the 'secondary' waves relying only on general mechanical considerations: *no special case was made and hence no observational result about the propagation of light in any particular circumstances was involved at any stage.*

Given the Huygens principle, the basic idea behind any case of 'Fresnel diffraction' is in principle straightforward. A wave front is taken tangent to, or intersected by, the diffracting object. The resultant disturbance at any point beyond the object is then found by integrating the effects at that point, not from *every* point of the wave front (as in the case of the unobstructed wave) but instead just from every point on the non-occluded part of the wave front. The theory is that the diffract-

23. Fresnel (1819), pp. 282–3; emphasis added.
24. Fresnel (1819), pp. 294; emphasis added.

ing object in fact performs no positive action on the light and plays no role beyond that of simply absorbing the portion of the incident light which happens to fall on it. Apart from details, it makes no difference what the shape of the diffracting object happens to be – straightedge, narrow strip, circular screen or circular aperture – the method is always the same: just omit from the integration those 'secondary sources' blocked by the object. The shape of the object simply supplies the limits of the integral which is otherwise completely determined by the *general* theory (aside, that is, from one parameter corresponding to the wavelength of the light employed and hence not adjustable for the different cases of diffraction).

Given all this, it is obvious that the heuristic account is correct. No objective weight should be carried by the time at which the diffraction patterns concerned were actually discovered. So far as Fresnel's theory is concerned it is *nothing more than an historical accident* that the straightedge cases happened to be known before the theory was articulated while the circular screen case was investigated only later. Prior knowledge of the straightedge fringe patterns did not help, indeed *could not have helped*, Fresnel; the particular theory of these phenomena was generated by his general wave-theoretical approach or research programme in a way which left no room for adjustment to detailed observation. Hence these straightedge results were just as much a test of the theory as was the novel case of the circular screen. Although as a matter of fact it didn't, Fresnel's theory *could have* predicted the straightedge cases in advance. It is clear that Fresnel himself saw the situation in precisely this way, more than once emphasising that the important feature of his theory of diffraction was that it yielded the right empirical results 'without the aid of any auxiliary [or sometimes 'secondary'] hypothesis'.

The equal theoretical impact of the straightedge and circular screen results not only seems reasonable, it accords with the reaction of the French Academy's commissioners to Fresnel's work. It also agrees perfectly with the explicit reactions of some of the later scientists who worked on the wave theory. Here, for example, is Stokes reviewing the case of Fresnel's theory and the straightedge diffraction results some 50 years after the event:

The theoretical distances of the several fringes from the geometrical shadow were a matter of pure prediction; for the only unknown quantity involved in the theoretical expression, the length of a wave, had been determined by Fresnel by independent methods . . . so that not a single arbitrary constant was left to be determined by some measurement of a fringe in some one particular case, whereby an at least partial accordance between theory and observation might have been brought about.,[25]

Stokes talks about the straightedge results as matters of 'pure prediction', even though they were already known when Fresnel developed his theory. And the reason he does so is exactly the one emphasised by the heuristic account, namely, the absence of prior empirical tinkering with theory, indeed the absence of any room for such tinkering in respect of the empirical results concerned: 'not a single arbitrary constant was left to be determined by some measurement of a fringe'. Many scientists follow Stokes in intuitively characterising 'prediction' in this way. By emphasising the special role of prediction they can appear to be supporting the temporal, when in fact they are supporting the heuristic account. Thus I do not, of course, deny the importance of predictions – re-emphasised by Allan Franklin in his chapter – so long as it is understood that already known phenomena can be 'predicted'.

There are two complicating factors in this historical episode which I have not so far mentioned. Neither in fact endangers my case, although each might appear to. I shall end this section by considering them. The first complicating factor is that Fresnel added a great deal of quantitative detail to the already known straightedge diffraction results. The straightedge cases had certainly been extensively studied before Fresnel, but never in anything like such minute detail. A defender of the temporal account might therefore accept my argument that the straightedge cases supported Fresnel's theory just as well as the circular screen case but attribute this to the *novelty* of some of the quantitative evidence about straightedge diffraction. But the question of quantitative evidence is not at issue between me and the temporalist: everyone agrees that quantitative, detailed agreement between theory and evidence is especially striking support for the theory. Where we disagree is over the impact of the success of even the qualitative predictions in the straightedge and circular screen cases. History supports my view of equality of impact and thereby refutes the opposing view of the temporalist: the fact that straightedge diffraction cases had long been known about just does not seem to have signified at all for the commissioners.

A second complicating factor is the *possibility* that the white spot might have been observed before 1819. Whittaker asserts, in an enigmatic footnote to his treatement of the Poisson prediction, that the 'bright spot in the centre of the shadow had been noticed in the early part of

25. Stokes (1884), p. 63.

the eighteenth century by J.N. Delisle'.[26] He gives no reference, no further detail, and no consideration of how this earlier discovery – if indeed it occurred – affects his main claims about the episode.

I have not been able to verify Whittaker's claim. It might appear to be vital to my methodological case that I show that Delisle did *not* make this discovery, since otherwise the case does not provide a 'crucial experiment' between myself and the temporalist. If the white spot was *not* unknown before Poisson's prediction, then the temporalist too accords the success of that prediction no special significance. This consideration exposes an ambiguity in the temporalist's position. In order to count as a 'novel fact', must that fact have been entirely unknown to *anyone* before its prediction by theory? Is it enough if the fact was not known to any of the scientists in the field at the time the theory was first articulated? It is easy to imagine an investigator making an experimental discovery, even publishing it in a distant corner of some out of the way periodical, and that discovery not becoming widely known. Some years later no one in the field might have any knowledge at all of the result. Indeed this may have happened in Delisle's case. Even if Delisle *did* discover the white spot, it seems that no one on the commission, nor so far as I can tell anyone in the wider scientific community in the early nineteenth century, knew anything of it. Thus the white spot was certainly *new* to the scientists of the time, even if it had in fact been observed before. It seems clear that the temporalist has no choice but to adopt the latter characterisation of 'novel fact' – that is, he must count as 'novel' any fact which was not consciously known at the time, which was not part of 'background knowledge' at the time. The other, more 'objective' characterisation might at first seem more appealing for the temporalist: a fact counts more if it was predicted by the theory and was hitherto totally unknown to any human being. But how could scientists actually *apply* (even implicitly) the temporalist's criterion in this form? If Arago, Poisson and the rest really were applying this form of the temporalist's criterion, they would have had to have exhibited some mysterious sixth sense which allowed them to intuit that someone had earlier observed the white spot even though they were consciously unaware of this. So the temporalist cannot use Delisle as an escape route. The temporalist's position has to be that an empirical result is 'novel' at time *t*, if that result was not part of background knowledge at *t*, and *not* that it is novel at *t*, if unknown to *anyone* before *t*. Even if Delisle did observe the white spot before 1819, the result was certainly not part of 'background knowledge' in

26. Whittaker (1951), p. 108.

1819. Hence the temporalist *is* committed, Delisle or no, to according special significance to the white spot – contrary, as we saw, to the historical record. This discussion surely underlines the complete untenability of the temporalist view: the objective support which a theory receives from a piece of evidence cannot rest on such purely accidental features of history as whether or not a particular empirical result is discovered, but fails to attract sufficient attention and is subsequently forgotten.

CLARIFICATION AND DEFENCE OF THE 'ZAHAR–WORRALL CRITERION'

The white spot case, so often cited in favour of the view that there is something special about successful (novel) predictions, in fact supports the heuristic account. In this section I shall try to use the historical case to clarify some further aspects of this latter account and to defend it against some recent criticisms.

One perennial criticism is that the account makes theory-confirmation unacceptably vague and unacceptably subjectivist. If we really did need to know how a theory was constructed in order to decide on its empirical support, we would seem to need access to the psyche of its inventor. But who really knows what went on in Fresnel's mind, or in Newton's or in Einstein's, when arriving at their great theories? Why should such subjective considerations matter at all once the theory is 'on the table'?[27]

The Fresnel case indicates that these fears of vagueness and subjectivism are groundless. We need know nothing about Fresnel's psyche and need attend only to the development of his theory of diffraction as set out in great detail and clarity in his prize memoir.

Although it is his integrals which do the work of yielding the observationally checkable details, Fresnel did not simply lay these integrals down. Instead he painstakingly derived them from the Huygens principle – a principle which he derived in turn from general mechanical considerations. It is an important, though underemphasised, aspect of science that scientists do not simply propose their theories as mere conjectures to be evaluated, only afterwards, in terms of their empirical consequences. Instead they *argue* for their specific, detailed theories – indicating how they can be arrived at from more general (and perhaps widely accepted) considerations. These arguments, as accounts of the construction of detailed theories, are just as much in the public domain

27. See, for example, Musgrave (1974, 1978).

as is the theory itself. It is this fact which saves the heuristic account of empirical confirmation from hopeless vagueness and subjectivism. In order to decide whether or not the straightedge diffraction results, say, were involved in the construction of Fresnel's theory, one need not look at his psyche but only at his published papers.

As for the question of *why* it should matter, once a theory has been produced, *how* it was produced, my answer in outline is this. Whether or not it can be given some further rationale, we *do* seem to regard a striking empirical success for a theory as telling us something about the theory's overall – what? Truth? Verisimilitude? Probable truth? General empirical adequacy? Closeness to a natural classification? Take your pick. The reasoning appears to be that it is unlikely that the theory would have got this phenomenon precisely right just 'by chance', without, that is, the theory's somehow or other 'reflecting' the blueprint of the Universe. The choice between the 'chance' explanation and the 'reflecting the blueprint' explanation of the theory's success is, however, exhaustive only if a third possibility has been ruled out – namely that the theory was engineered or 'cooked up' to entail the phenomenon in question. In this latter case, the 'success' of the theory clearly tells us nothing about the theory's likely fit with Nature, but only about its adaptability *and* the ingenuity of its proponents. There is, of course, no suggestion that it is scientifically unacceptable to adapt a theory so that it incorporates an originally refuting (or independent) phenomenon. Indeed from the point of view of empirical applicability it is clearly vital so to amend the theory. Classical physics, as amended to include the Lorentz–Fitzgerald contraction hypothesis, is clearly scientifically superior to the unamended form in terms of empirical adequacy – since the amended form, unlike its predecessor, entails the correct result of the Michelson–Morley experiment. The point is only that – assuming the Lorentz–Fitzgerald hypothesis *was* merely an *ad hoc* device aimed at accommodating the result – the Michelson–Morley result gives no support to the *general framework* of classical physics. The straightedge diffraction results, on the contrary, support not just Fresnel's *specific theory* of diffraction, but also the general framework of the wave theory of light. This is because of the intimate connection, not dependent on any experimental results, between the general framework and the specific theory.

Because the specific theory of straightedge diffraction and the specific theory of small circular disc diffraction both follow 'naturally' from the 'fundamental hypothesis' of waves, because Fresnel *could not have* used the already known straightedge diffraction results (owing to the logical structure of his general and more specific theoretical ideas), there is no reason to differentiate between the straightedge and the circular disc in the support they give to this theoretical ideas. As we saw from the investigation of the prize commission's deliberations, in this respect at least, *real* history and its *real* rational construction coincide.

ACKNOWLEDGEMENTS

I am indebted to Elie Zahar for discussions of the ideas underlying this paper, and to John Watkins and Peter Urbach for criticisms of earlier versions and suggestions about presentation. The initial draft of this paper was written while I was a happy and stimulated visiting fellow at the Center for Philosophy of Science, University of Pittsburgh.

Some more detailed criticisms of the 'heuristic' account of empirical support have recently appeared.[28] I decided that it would be more appropriate to respond to these detailed points in a separate (and forthcoming) journal article.

References

Arago, F. (1819). Rapport fait par M. Arago à l'Académie des Sciences, au nom de la Commission qui avait été chargée d'examiner les mémoires envoyés au concours pour le prix de la diffraction. *Annales de Chimie et de Physique*, XI; reprinted in Fresnel (1866–70), vol. I, pp. 229–46.

Arago, F. & Poinsot, L. (1816). Rapport fait à la première Classe de l'Institut, le 25 Mars 1816, sur un mémoire relatif aux phénomènes de la diffraction de la lumière par M. Fresnel. In Fresnel (1866–70), vol. I, pp. 79–87 (page references are to the reprint).

Campbell, R. & Vinci, T. (1983). Novel confirmation. *British Journal for the Philosophy of Science*, **34**, 315–41.

Crosland, M.P. (1967). *The Society of Arceuil: a View of French Science at the Time of Napoleon I*. New York: Heinemann.

Fresnel, A.J. (1816). Première mémoire sur la diffraction de la lumière. In Fresnel (1866–70), vol. I, pp. 79–87.

Fresnel, A.J. (1819). Mémoire sur la diffraction de la lumière couronné par l'Académie des Sciences; Reprinted in Fresnel (1866–70), vol. I, pp. 247–382.

Fresnel, A.J. (1866–70). *Oeuvres Complètes d'Augustin Fresnel, Publiées par MM. Henri de Senarmont, Emile Verdet and Leonor Fresnel*, 3 vols. Paris: Imprimerie Imperiale.

28. Notably Campbell & Vinci (1983); Howson (1984).

Giere, R.N. (1984). Testing theoretical hypotheses. In *Testing Scientific Theories. Minnesota Studies in the Philosophy of Science*, vol. X, ed. J. Earman, pp. 269–98. Minneapolis: University of Minnesota Press.

Howson, C. (1984). Bayesianism and support by novel facts. *British Journal for the Philosophy of Science*, **35**, 245–51.

Kuhn, T.S. (1962). *The Structure of Scientific Revolution*. Chicago: Chicago University Press.

Lakatos, I. (1971). History of science and its rational reconstructions. In *Boston Studies in the Philosophy of Science*, vol 8, ed. R.C. Buck & R.S. Cohen. Dordrecht: Reidel.

Musgrave, A. (1974). Logical versus historical theories of confirmation. *British Journal for the Philosophy of Science*, **25**, 1–23.

Musgrave, A. (1978). Evidential support, falsification, heuristics and anarchism. In *Progress and Rationality in Science*, ed. G. Radnitzky & G. Anderson, pp. 181–202. Dordrecht: Reidel.

Procès-verbaux (1910-22). *Procès-verbaux des Séances de l'Académie Tenues Depuis la Fondation de l'Institut Jusqu'au Mois d'Aôut 1935*. Hendaye.

Stokes, G.G. (1984). *Burnett Lectures. On Light*. London: Macmillan.

Verdet, E. (1866). Introduction aux oeuvres d'Augustin Fresnel. In Fresnel (1866–70), vol. I, pp. ix–xcix.

Whittaker, E.T. (1951). *A History of the Theories of Aether and Electricity*. Edinburgh: Thomas Nelson and Sons.

Worrall, J. (1976). Thomas Young and the 'refutation' of Newtonian optics: a case study in the interaction of philosophy of science and history of science. In *Method and Appraisal in the Physical Sciences*, ed. C. Howson, pp. 107–80. Cambridge: Cambridge University Press.

Worrall, J. (1978). The ways in which the methodology of scientific research programmes improves on Popper's methodology. In *Progress and Rationality in Science*, ed. G. Radnitzky & G. Andersson, pp. 45–70. Dordrecht: Reidel.

Worrall, J. (1982). The pressure of light: the strange case of the vacillating 'crucial experiment'. *Studies in the History and Philosophy of Science*, **13**, 133–71.

Worrall, J. (1985). Scientific discovery and theory-confirmation. In *Change and Progress in Modern Science*, ed. J. Pitt, pp. 301–32. Dordrecht: Reidel.

Worrall, J. (1988). Scientific revolutions and scientific rationality: the case of the 'elderly hold-out'. In *The Justification, Discovery and Evolution of Scientific Theories*, ed. C.W. Savage. Minnesota: University of Minnesota Press.

6

THE RHETORIC OF EXPERIMENT

GEOFFREY CANTOR

Most contributors to this volume adopt one of the following approaches when analysing scientific experiments: while some authors concentrate on the hardware of experiment and the techniques and laboratory procedures of the scientist, other contributors, like most historians of science, rest content with working from the published reports of experiments – their literary remains, as it were. It is generally taken that these two approaches bear a close and unproblematic relationship. The simplest assumption is that the written report which appears in, say, a scientific journal contains an accurate account of what transpired in the laboratory. In the absence of other sources the historian may be forced to adopt this assumption, but where alternative sources exist one can often gain deeper insight into laboratory practice. Among these alternative sources are extant pieces of apparatus and laboratory notebooks; thus, for example, Gooding, working with Faraday's *Diary*, and Holmes, who has studied Lavoisier's notebooks, have offered finely-textured and insightful analyses of the laboratory practices of these two scientists.[1] Such notebooks not only provide far more detailed accounts of experimental procedures but also indicate the failures, errors and false starts that are not reported in public and those numerous particulars that are deemed unnecessary in a publication.

Yet extant laboratory notebooks also sometimes indicate more interesting mismatches between laboratory practice and published reports. Holton, for example, has drawn attention to Robert Millikan's selection of acceptable results for his oil-drop experiment. During one series of experiments Millikan omitted well over half of his results, retaining data from only 58 drops out of a total of about 140. Against some runs he annotated comments such as 'Beauty. *Publish* this surely,

1. Gooding (1985); Holmes (1985).

beautiful', whereas in other cases he dismissed the run with *'Error high will not use'*, or some such remark. His reasons for accepting some runs and not others are complex; sometimes parts of his apparatus did not appear to function properly, on other occasions the result was not sufficiently close to the emergent value for e, the electronic charge. Yet I am less concerned with whether he was fully justified in his judgments than to illustrate the far more important point that a laboratory notebook and a published journal article are two very different literary forms, serving different purposes and subject to different conventions. The published report should not be viewed simply as a tidied up version of the laboratory notes, since the former contains many conventional elements that would find no place in the latter. The publication is a retrospective narrative, an impersonal, passive reconstruction which draws attention to those theories, tests and data which are considered appropriate for consumption by the scientific community. Thus, contrary to the manuscript evidence, Millikan announced in his paper: *'It is to be remarked, too, that this is not a selected group of drops but represents all of the drops experimented on during 60 consecutive days . . . '*[2] The crucial difference between the notebook and the publication is that the latter is written for an audience and constitutes part of the public transactions of science whereas the notebook is principally for the use of the researcher (and perhaps a few intimates).

I introduce these two literary modes not only to contrast them but also to illustrate the considerable 'distance' that exists between an experiment, as such, and its published report (and therefore the chasm between the two forms of historical writing with which I opened this paper). The published narrative is the outcome of a complex process whereby an extended series of experiments is translated and condensed into prose. It is difficult to characterise this process succinctly, but its outcome may depend on the branch of science, the historical period, the (presumed) audience, literary conventions, the aims and interests of the scientist, etc. It needs to be stressed that the report of an experiment appearing in, say, a scientific journal is a highly artificial product which, far from giving an unproblematic account of work in the laboratory, is a form of narrative and is therefore open to the kinds of analysis which can be applied to other literary genres.

One question which needs to be asked about any literary composition is: what was the author's intention?[3] In the present case the question

2. Holton (1978), all quotations from pp. 63–9. See also the study by Earman & Glymour (1980) of Eddington's eclipse observations and the form in which they were publicly presented.
3. This line of questioning would not, however, be encouraged by structuralists.

can be rephrased as follows: what was the author of an experimental report trying to achieve? Of the various answers that can be given the most general and important is: in writing an experimental report a scientist is trying to *persuade*. By the discourse of experiment the author tries to convince an audience – be it a specific scientist, a community of scientists, the lay public, the dispenser of research grants, etc. – of the validity of the author's position (and perhaps the falsity of some opposing view). The branch of language concerned with persuasion is rhetoric. Scientific narratives in general and experimental reports in particular are *rhetorical*, in the accepted sense of the term, since they aim to persuade or influence. This makes scientific discourse a discourse of power. This point is frequently overlooked. Philosophers have been concerned primarily with the truth claims of science (and thus with epistemology) and have generally neglected its rhetorical dimensions. Indeed, Naylor, like many other authors, tends to contrast truth and rhetoric: rhetoric – 'mere' rhetoric – is often considered to be the refuge of false opinions and therefore to have no legitimate place in science. However, unless we are to believe that truth is manifest we need to view rhetoric as an integral part of science. Those who oppose this kind of analysis might argue that the current convention of writing scientific reports in the impersonal, and the extensive use of the propositional form eliminates rhetoric from science. While these conventions seem to preclude the kinds of *ad hominem* strategies usually associated with poetry, the pulpit and political oratory, they are nevertheless forms of rhetoric, and very sophisticated forms at that. Far from being rhetoric-free, modern scientific prose has become the most potent instrument of persuasion in our culture.

Every successful (and unsuccessful) theory in the history of science has had to be argued, sometimes against strenuous opposition. If, as many philosophers have claimed, theories are underdetermined, then we need to understand how theories come to be accepted by scientists, both individually and as a community. A study of rhetoric may help us better understand how scientists choose between theories. A key element in modern scientific rhetoric is provided by experiment, since the appeal to experiment is generally the most effective persuasive strategy when arguing either for or against some theory or doctrine. The rhetoric of experiment is endowed with considerable power.[4]

However, in this paper I will not be concerned primarily with the fascinating but greatly under-researched subject of rhetoric in present-day science; instead, my main aim is to suggest how the rhetoric of experiment provides a challenging site for historical research. The

4. Gilbert & Mulkay (1984).

approach advocated here should be conducive to the historian of science, since it offers a highly contextualised view of science in which language and its functions are firmly located in particular communities at specific times. Commensurate with the prevalent anti-Whiggism, this approach rejects the simplistic position that language is trans-historical and therefore unproblematic. Moreover, as Christie and Golinski have stressed, the study of scientific rhetoric transcends the outmoded division between internal and external history since language relates both to scientific ideas and also to the practice of science and thus to its social environment.[5] Experimental reports are therefore particularly sensitive areas for analysis.

RHETORICS, DISCIPLINES AND MYTHS

In this section I will briefly suggest several problems deserving further research, but my list is far from complete and reflects only a very limited part of the primary and secondary material on this topic.

The creation of new rhetorical forms

For many historians experiment is a natural part, even the defining characteristic, of science. Yet the question of how experiment was introduced into discourse about the natural world is a major topic for research. Such research would have to be sensitive to the differences between experience and experiment.[6] Only since the seventeenth century has there been a clear sense that Nature should be subjected to active interrogation – 'torturing the sunbeams', as Goethe later referred to Newton's prism experiments – and that this interrogation could uncover truth and help adjudicate on questions of theory. Moreover, we should be careful not to limit our analysis of the discourse of experiment to research-orientated texts, but also include the important role of experiment in didactic works.

One insightful way of articulating the scientific revolution is in terms of the new language forms used to discuss natural phenomena. Recently a few historians have begun to explore the new narrative forms of the period and from their work it is apparent that there were a number of very different rhetorical strategies employed. In his exciting book *The Chemists and the Word*, Hannaway has documented the rise of chemistry, not in terms of specific chemical theories, but as a new didactic practice.

5. Christie & Golinski (1982), pp. 235–7.
6. See, for example, note in Hannaway (1975), p. 26.

The key figure was the German polemicist Andreas Libavius who sought to create in his *Alchemia* of 1597 a discipline of chemistry which was distinct from medicine or from any other sciences and turned its back on the irrational elements of Paracelsianism. As a didactic tool Libavius's book not only described the operations of the artisan but also reduced the diversity of manipulations to a tidy system. Libavius thus emerges as the reformer of chemistry, the man who brought order to chemical discourse.[7]

The subject of experiment is more directly confronted in Shapin's analysis of Robert Boyle's 'literary technology'. Shapin's contention is that Boyle created a literary form in which scientific facts were given prominence. Sceptical of the theories and speculations of his predecessors, Boyle sought to raise the status of facts and to make them the currency among natural philosophers. Facts had to form the foundations of the new science and therefore had to be expressed in an appropriate rhetorical form. He thus described apparatus and experimental procedures in great detail so as to make them manifestly clear to the reader. Hence his prolix style. The detailed descriptions were meant to allow other scientists to repeat them. Yet perhaps the most interesting aspect of Boyle's narrative concerns the role of witnesses. Seventeenth-century scientists often listed the witnesses present at the experimental trial but Boyle sought through his text to create in the reader the impression of 'virtual witnessing', as if the reader had been present in the laboratory. Boyle's literary techniques were aimed at persuading the reader of the experiment's adequacy and thus of the truth of the reported facts.[8]

However it should not be assumed that Boyle's literary technology was the only rhetoric of experiment developed in the seventeenth century. The comparison with Libavius or with Galileo, Descartes or even Newton points to the lack of consensus on this matter and the problem of creating a literary form which could convey observational and experimental reports, often integrated with other types of material. This is clearly a topic deserving further historical study.

Changes in rhetorical form

Moving beyond the mid-seventeenth century, a number of questions arise concerning subsequent changes in experimental discourse. Most obviously, the highly personal tone of Boyle's work was replaced by the impersonal. The persuasive, authoritative power of a scientific fact

7. Hannaway (1975).   8. Shapin (1984); Shapin & Schaffer (1985); Golinski (1987).

was no longer underwritten by the social standing of the experimenter and the witnesses. Instead the fact stood on its own as an impersonal truth which, at least for the non-scientist, bore the imprimatur of the scientific community. Nature, as it were, spoke directly to the reader. However, there is no canonical way of reading a text, and an experimental report which generally brought about conviction might, for some readers, be highly problematic. Thus in scientific controversies much ink has been spent on trying to attack – if not deconstruct – the experimental reports of the opposition. Again, descriptions of experiments and visual representations became less realistic and more condensed and schematic. Particularly in the physical sciences, experimental discourse became progressively more numerical and subsequently more concerned with errors and the precise fit between theory and experiment.

The above characterisation is obviously very brief and partial, nevertheless it is meant to draw attention to some ways in which the discourse of experiment has changed. Obviously not all sciences have changed in the same ways and at the same times, and there are some interesting questions to be asked about both the general trends and the differences between subject areas. One major change has been the growth of specialist scientific languages and their progressive differentiation from the language of the non-expert. Hence one topic for research is the relation between such languages and the professionalisation and social differentiation of science.

Disciplinary relationships

The modes of discourse adopted in one branch of science may differ greatly from those used in another – compare, say, contemporary ornithology with solid state physics. Surveying the sciences at any given time the historian could draw a 'map' – certainly a very complex map – of the different discourse forms in use. Changes over time also present interesting problems concerning the development of different modes of discourse. One might expect such development to occur both within a particular discipline and also between disciplines, and in the latter case the interaction might involve not only another branch of science but also inputs from other aspects of culture.

Two brief examples will suffice to suggest some of these kinds of interaction. The great successes in physics have often led the practitioners of other subjects to model their discourse on that of physics –

'physics envy'. Most obviously, the social sciences have sought to copy from physics its impersonal style, its theoretical structures and even its deployment of mathematical relationships. Texts in the social sciences have adapted the physicists' experimental discourse in a number of ways, ranging from fieldwork to computer simulations and from historical studies to laboratory situations. Such developments are not to be dismissed as trivial; instead they provide insight into the changing discourse patterns in the social sciences. My second example concerns the role of metaphor, since in the articulation of experiments terms are often drawn metaphorically from other areas. In the eighteenth century it was generally accepted that light consisted of small particles of matter projected from a luminous body. Scientists therefore expected a beam of light to possess momentum and a series of experiments were carried out to detect this momentum (which was not predicted according to the rival, wave theory). The experimental apparatus was described as a 'steelyard', a 'lever' or a 'balance'; the 'impulse' of the incident light on one such apparatus set it into vibration 'as if it had been struck by a stick'. As these examples indicate, the discourse of optics had been drawn into the discourse of mechanics.[9]

Didactic functions

So far I have been concerned almost exclusively with the language of communication between scientists. Yet as Kuhn noted in the oft-quoted opening paragraph of *The Structure of Scientific Revolutions*, 'the aim of such books [textbooks] is persuasive and pedagogic'.[10] The rhetorical forms used not only in textbooks, but also in public lectures, inaugural addresses and the media deserve considerably more historical analysis than they have so far received; although a few historians have refused to compartmentalise research and teaching and instead insist that the language of science is shaped by its didactic context.[11] To quote Kuhn again, it is from textbooks that 'each new scientific generation learns to practice its trade'.[12] Thus textbooks play an important role in defining a subject, its organisation and its division; they offer models of how to write about the subject, and provide the student with terminology and metaphors necessary for any further exploration. The teaching context, now generally undervalued, is the crucial site for understanding much

9. Cantor (1987). 10. Kuhn (1962), p. 1.
11. See, for example, Hannaway (1975); Christie & Golinski (1982); Golinski (1984).
12. Kuhn (1962), p. 1.

about the language of science and especially how scientists perceive experiment. This approach undermines the highly unsatisfactory but often stated view that in facing the wider public the populariser merely simplifies and dilutes (and may even misuse) the language of the research scientist.

Mythology

One important function performed by textbooks (and not only textbooks) is to convey the values of the scientific enterprise. Very often these values are encoded in the accounts of experiments; indeed, descriptions of experiments are particularly potent sources for understanding the mythology of science. Such accounts of experiments are deceptive since they appear to deal with reality – both historical reality and the real structure of the physical world. Yet, like all myths and even dreams they are very condensed, invariably glossing over the numerous difficulties (often the immense difficulties) which arose during the construction of the experiment (except to evoke the reader's awe). Likewise, controversy over the experiment and its interpretation are usually suppressed. In the resulting discourse experiments emerge as very persuasive devices. They tell the reader the way things are and inculcate the kind of empiricism which philosophers of science have been at pains to undermine. For their part historians of science have been kept in business 'demythologising' science but have paid little attention to the function of myth. As with the Millikan oil-drop experiment, cited above, there is no shortage of case studies showing that canonical experiments are not what they appear. The mythic dimension of experimental discourse thus becomes apparent.

There are mythic components in all reports of experiments. They are all invested with a symbolic meaning that goes beyond the results, as such. Not all experiments possess identical mythological content and function. At one end of the spectrum stand those experiments which dramatise empiricism and truth, such as Newton's imputed discovery of gravitation from his observation of a falling apple. Equally dramatic is the story of Galileo's dismissal of Aristotelianism by dropping two balls of different weight from the Leaning Tower of Pisa.[13] The fact that *we* know that these involve historical inaccuracies has done nothing to abate their popularity or their symbolic significance (which deserve the attention of anthropologists). But why stop there? We should equate

13. On Newton's apple see Westfall (1980), pp. 154–5; on the Galileo example see Cooper (1935).

mythology not with falsity but with a text's symbolic meaning. In that case the experimental reports in the latest issues of the *Journal of Atmospheric and Terrestrial Physics* and *Molecular and Cellular Biochemistry* can also be analysed as myths.

EXPERIMENTS IN DIALOGUE

As Naylor rightly points out in chapter 4 of this volume, Galileo remained ambivalent about the demonstrative power of experiment. While in many cases Nature could unambiguously confirm or refute a theory, there were other instances, such as his discussion of a stone falling inside a tower, where the evidence could be used to confirm both Salviati's views and Simplicio's.[14] In the *Dialogue* in particular Galileo did not deploy the dialogue form simply to articulate his own theoretical system. Instead he exploited some of the literary possibilities of the work. Thus even Simplicio often emerges as a realistic character with some strong arguments and not simply as a foil to Salviati.

Although dialogue had been used extensively both before and after the early seventeenth century, and had served a variety of different purposes, it possessed several clear advantages over a treatise. The relative merits of the dialogue were considered at the outset of Hume's *Dialogues Concerning Natural Religion* (1779). One of the interlocutors, Pamphilus, noted that while some 'order, brevity, and precision' may be lost in adopting dialogue, there are some subjects to which it is particularly well suited. Dialogue can provide the best means to emphasise to the reader an obvious yet crucial point. Again, it enables the writer to present a topic effectively in a variety of different lights. Turning to issues closer to hand Pamphilus noted that a philosophical question 'which is so *obscure* and *uncertain* that human reason can reach no fixed determination with regard to it – if it should be treated at all – seems to lead us naturally into the style of dialogue and conversation.'[15] Natural religion was thus well suited to the dialogue form since it contained both unassailable truths and points of contention.

Hume's comments may assist us in understanding Galileo's use of dialogue. Did Galileo adopt the dialogue form in the *Two New Sciences* because his subject matter was '*obscure* or *uncertain*'? In one sense his purpose was the very opposite. He sought to dispel obscurity by articulating his theory of how bodies move. However, experiment did not provide the unambiguous proof he sought. Moreover, the theory he proposed was novel, contentious and internally imperfect. Like

14. See also Tamny (1988).   15. Hume (1948), p. 3.

Hume, Galileo constructed a discourse which incorporated opposing viewpoints but which permitted him to articulate a new theory and at the same time contradict the views of his opponents. The dialogue allowed him to achieve these ends. Compared with the *Two New Sciences*, the *Dialogue* is the more highly-charged agonistic work and the more successful dramatically, with Simplicio suffering defeat on numerous occasions at Salviati's hand. By contrast the *Two New Sciences* trades far less explicitly on controversy, and there are only a few references to Galileo's antagonists.[16]

The role of Simplicio differs markedly between these two works; in the *Dialogue* he is a far stronger advocate of Aristotelianism and to the end remains unconvinced of Galileo's arguments, whereas in the later work he largely concedes Salviati's position. Whatever political reasons lie behind these two differing strategies, both works employ similar models for scientific and, more particularly, experimental discourse. Towards the end of 'Day Four' of the *Dialogue* the interlocutors compliment one another on their civil, gentlemanly behaviour. Salviati asks Simplicio's pardon 'if I have offended him sometimes with my too heated and opinionated speech', and Simplicio responds by assuring his companion that he is used to public debates which tend to be far more abusive.[17] Throughout these dialogues the interlocutors behave calmly and reasonably (in contrast to the disputes in which Galileo was involved). Although Simplicio remains unconvinced by many of Salviati's arguments the circle of agreement is considerably increased through rational dialogue, particularly in the *Two New Sciences*. By his use of the vernacular Galileo indicated that rational, partly-consensual discourse could flourish outside the accepted institutions of learning.

A further relevant point can be drawn from Hume's *Dialogues*. By employing the dialogue form Hume effectively hid his views from the reader. This aim could not have been achieved had he written a treatise on the subject. The game is played out between 'the accurate philosophical turn of Cleanthes', the 'careless scepticism of Philo' and the 'rigid inflexible orthodoxy of Demea'.[18] But, many commentators have asked, who speaks for Hume? We all know that Hume was a sceptic; so Philo, of course, represents Hume's position. But wait a minute, doesn't the work end with Cleanthes being proclaimed closest to the truth? Surely Cleanthes, then, must stand for Hume. However, in the first half of the work, Demea scores many devastating points against Cleanthes. So perhaps Demea is Hume? The game can go on indefinitely, and scholars

---

16. See, for example, Galileo (1974), pp. 34, 86, 89. 17. Galileo (1962), pp. 463–4.
18. Hume (1948), p. 4.

have indeed argued all three positions.[19] Standing back from the fray it seems impossible to decide which interlocutor speaks for Hume. Before declaring that Hume has deceived us we should go back to the passage quoted earlier and remind ourselves that Pamphilus recommended the use of dialogue in discussing questions where 'human reason can reach no fixed determination'. Perhaps, then, we have no warrant to find a single victor and proclaim him to speak for Hume. To press this line of argument is to render the question 'who is Hume' redundant. Moreover, Hume may have been playing a very different game, deploying the dialogue form to interactively articulate the three positions, to probe their strengths and weaknesses, and to avoid identifying himself with any one theological position.

I'm not sure how far one can legitimately carry this point over to Galileo's works. The roles of the interlocutors are generally taken as unproblematic in the *Dialogue*, but are far less easy to define in *Two New Sciences*; indeed in respect to this work Drake has suggested that all three interlocutors represent Galileo – Sagredo and Simplicio espousing positions he had held earlier in his life, while Salviati 'remains the spokesman for Galileo'.[20] The problem of identifying which actor plays the part of Galileo is further complicated by the numerous references to 'our Academician', 'the Author' or, more rarely, 'our friend'. The net result is that the author and his opinions are effectively kept off stage, as it were, while he is omnipresent throughout the proceedings. Such concerns defuse the question, 'which actor is Galileo?' and, as in the case of Hume's *Dialogues*, suggest that we are likely to miss much if we fail to read Galileo's works from the standpoint of rhetoric and drama. Finally, both Galileo and Hume had reasons for casting veils over their personal views – Galileo having been forbidden to hold the Copernican theory while Hume was generally viewed as a sceptic (albeit an amiable sceptic).

NATURE AS INTERLOCUTOR

As mentioned earlier there was not, at the turn of the seventeenth century, a settled, conventional form of language in which experiments could be expressed. This issue has recently been explored by Dear who argues that contemporary Jesuit mathematicians were striving 'towards a formulation of techniques designed to incorporate singular experiences, discrete events, into properly accredited knowledge about the

19. For example, Hurlbutt (1956); Pearl (1970); Carabelli (1972).
20. Stillman Drake's introduction to Galileo (1974), pp. xii–xiii. Cf. Finocchiaro (1980), pp. 46–7.

natural world'.[21] Their problem takes us back to the classical distinction between mathematics and physical sciences, and the need for the latter to be based on evident, general principles. Galileo was part of this movement, but he also strayed into deeper, uncharted waters in pushing the notion of experiment far beyond the realm of common experience so as to include specific experiments. From this perspective the *Dialogue* and the *Two New Sciences* emerge as highly tentative works in which Galileo deployed new forms of narrative to convey the behaviour of Nature and in which he sought to incorporate experimental reports within his analytical theoretical framework.

Even the criteria of what constitutes an experiment appear to be far from settled in the early seventeenth century. While some of Galileo's experiments clearly could have been carried out (and in some cases definitely were performed), others have been classed as 'thought experiments'. As Kuhn has emphasised, while 'thought experiments' cover a large and unruly class, they have often been rhetorically effective in drawing attention to confusions in one theory and in emphasising the clarity of a rival.[22] Such experiments can be very pleasing because they deal with situations that cannot (or cannot easily) be obtained in practice and thus allow for a precision and perfection not readily attainable in the laboratory, such as pendulums swinging in a void or a ball falling in a hole passing through the centre of the earth. Yet the term 'thought experiments' is somewhat misleading since it implies that these are variants of laboratory experiments and also sets them apart as thoroughly different from what we understand as real experiments. Hence the concern among scholars to identify which experiments Galileo actually performed and which he did not or could not. Even Naylor seeks to police the boundary between them and to label 'thought experiments' as mere rhetorical devices. However, my perspective downgrades but does not eliminate this distinction. Instead I draw attention to the common rhetorical aim and similarity in form of both types of experiment in Galileo's works.

The *Two New Sciences* is a complex text, combining two traditional but antithetical forms of discourse: the antithesis is most apparent in the 'Third' and 'Fourth Days' which consist of a prose treatise embedded in a dialogue. The treatise is in latin and is constructed round some 50 propositions, whereas the dialogue is in italian. The contrast is not simply between academic and vernacular languages but also between two different but long-established forms of discourse. Mathematical works and, to a lesser extent, scientific and philosophical ones have

21. Dear (1987), p. 173.   22. Koyré (1968); Kuhn (1977).

often been cast in the form of treatises employing deductive, analytical arguments, usually constructed from a set of propositions. Such works convey the formality, certainty and authority of knowledge. By contrast, the dialogue form was deployed in academic disputations and non-academic settings. During a disputation a student would have to defend a proposition against the questions and objections of the examiners. Far from being vehicles of manifest truth, disputations depended on the skill of the participants. Despite the differences between these two forms of discourse, their propinquity in the *Two New Sciences* suggests that Galileo was seeking a more dogmatic form of narrative than he could achieve using dialogue alone.

The problematic style of the *Two New Sciences* reflects Galileo's difficulty in finding a suitable narrative form to articulate the complex relationship between theory and experiment. Thus particularly in the 'Third' and 'Fourth Days' the Latin treatise bore most of the mathematical material and demonstrative arguments while all references to experience and experiment appear in the surrounding dialogue. The relative novelty of Galileo's undertaking has also been stressed by Dear, who argues that Galileo was freely intermixing two subject areas (associated with the above two different forms of discourse) that had traditionally been kept separate. While Jesuit mathematicians cautiously sought to develop ways of incorporating mathematical arguments into areas of physics, such as optics, so as to create 'mixed mathematics', Galileo's shot-gun marriage between mathematics and physics rode roughshod over the niceties of such philosophical distinctions.[23] Yet Galileo's innovativeness was not confined to redefining the relationship between mathematics and physics, his literary technology was also innovative.

Although both Galileo and Boyle challenged authority and sought to reform not only science but the existing forms of scientific narrative, the contrast between their literary technologies is instructive. While Boyle was primarily concerned with articulating and promoting facts Galileo's aims were far more ambitious since he sought to construct a narrative which interrelated mathematical theory and experiment. However, his experiments are portrayed as subservient to his theoretical concerns and they are described very briefly and tersely with the minimum of detail. Consider, for example, the passage in the 'Third Day' of *Two New Sciences* where Galileo first set out (as part of his Latin

---

23. Dear (1987), pp. 142–3. Galileo was also undeterred by philosophical discriminations when he conflated his belief that Nature is written in philosophical characters with the view that experiment reveals this reality in mathematical terms. See Butts (1978), p. 61.

treatise) the proposition that the distance travelled by a falling body is proportional to the square of the time. In the ensuing dialogue Simplicio expresses admiration of the 'Author's' reasoning, adding, 'But I am still doubtful whether this is the acceleration employed by nature in the motion of her falling heavy bodies.' The 'Author', assures Salviati, really performed the appropriate experiments, and even 'I have often made the test in the following manner in his company':

In a wooden beam or rafter about twelve braccia long, half a braccio wide, and three inches thick, a channel was rabbeted in along the narrowest dimension, a little over an inch wide and made very straight; so that this would be clean and smooth, there was glued into it a piece of vellum, as much smooth and clean as possible. In this there was made to descend a very hard bronze ball, well rounded and polished, the beam having been tilted by elevating one end of it above the horizontal plane from one or two braccia, at will.

Although his ingenious method of measuring time (by weighing the water from a water clock) receives some discussion, the whole experiment is accorded but three short paragraphs. The experimenter (the 'Author') and witness (Salviati) play prominent roles in this narrative, guaranteeing the truth of the procedure. Moreover, the perfect fit with the 'Author's' theory was invariably found, the experiment being repeated 'a full hundred times'. This section of the dialogue ends with Simplicio saying:

It would have given me great satisfaction to have been present at these experiments. But being certain of your diligence in making them and your fidelity in relating them, I am content to assume them as most certain and true.[24]

Simplicio is here portrayed as accepting Salviati's account of this experiment, Galileo thereby offering a literary road to consensus. Unlike Boyle, who sought consensus by appeal to prolix detail and related failed or irrelevant experiments, Galileo's stark appeals to experimental truth increase rather than diminish the authority and persuasiveness of his narrative. The reader is simply *told* how Nature behaves under certain circumstances.

This rhetorical, persuasive strategy also fully satisfies Salviati's doubt 'whether this is the acceleration employed by nature'. Throughout the *Dialogue* and the *Two New Sciences* there are numerous appeals to Nature. It is as if there were a fourth (or fifth, if we include the 'Author') interlocutor – NATURE. Thus when Simplicio asks whether bodies accelerate in the way proposed by our 'Author', Nature answers 'Yes' in a clear, unwavering and incisive voice. Or, to take an example from the

24. Galileo (1974), pp. 169–70.

'First Day' of the *Dialogue*, Salviati tries to persuade Simplicio that a rough surface looks brighter than a smooth one. 'Now please take that mirror which is hanging on the wall', he instructs Simplicio,

and let us go out into the court; come with us Sagredo. Hang that mirror on the wall, there, where the sun strikes it. Now let us withdraw into the shade. Now there you see two surfaces struck by the sun, the wall and the mirror. Which looks brighter to you; the wall or the mirror? What, no answer?[25]

The point, of course, is that Simplicio does not need to answer. Nature has answered the question in complete agreement with Salviati's prediction.

However, as Naylor argues, Nature's answer was often open to a variety of discordant interpretations. If Nature answers with one voice, not all auditors hear the same message. If the metaphor appears rather strained, another metaphor seems more natural and more appropriate in that Galileo employed it in the dedicatory epistle prefacing his *Dialogue*. Here he refers to the 'great book of nature': what 'we read in that book is the creation of the omnipotent Craftsman'.[26] God is the author, we are the readers. It is for us to read the signs, but there may be many interpretations – a point that applies also to the reading of historical texts.

EXPERIMENT AND DRAMA

Worrall opens his impressive analysis of the 'white spot' experiment in chapter 5 with a dramatisation of the traditional interpretation of this incident. For Worrall the heroic adventures of Fresnel and the wave theory serve the purpose of introducing the rather more sober interpretations offered by Kuhn, Giere, Whittaker, and others. Moreover, in descending to parody, Worrall effectively contrasts the flimsy mythology with his own solid, rational reconstruction in which he shows that the diffraction disc experiment played only a very minor role in the deliberations of the 1819 prize commission and in the rise to dominance of the wave theory of light. Yet Worrall's parody also performs a disservice since it conveys the impression that the diffraction disc experiment and its novel prediction have been recurrent and constant elements in the history of science for the last century and a half. From a rather limited survey of optics texts (primarily textbooks) I have been surprised by how few mention the incident. I also have found very few texts that support Worrall's contention that the diffraction disc experiment is

25. Galileo (1962), p. 72. 26. Galileo (1962), p. 3.

generally assumed to have played 'a central, perhaps even a crucial, role in the acceptance of the new [wave] theory from which it was drawn'. There were certainly some nineteenth-century writers who portrayed the experiment as a novel prediction, but the 'myth' which Worrall is attacking appears to have only taken root during the past few decades and then as grist for the mills of certain historians and philosophers of science.

Far from wanting to dismiss Worrall's opening melodrama I would like to savour it since it points to a series of historical problems that fall outside the scope of his paper. These questions concern the role of rhetorical devices in accounts of the diffraction disc experiment, primarily those found in nineteenth and early twentieth-century texts. Even before concentrating on the accounts which portray the diffraction disc experiment as a novel prediction, I want to draw attention to certain texts, dating from early this century, in which the very opposite was implied. When writers such as Arthur Schuster attributed the discovery of the white spot not to Poisson or even to Fresnel but to the early eighteenth-century writer Joseph-Nicolas Delisle, they robbed Poisson's prediction of much of its novelty.[27] Another particularly instructive example is provided by Ernst Mach who, in his *Principles of Physical Optics*, contrasted the 'at first sight surprising deduction' by Poisson with the 'scarcely surprising' reaction of anyone 'well acquainted' with the observations by Grimaldi and Young of the bright central diffraction fringe produced by a straight wire. For the informed positivist there was no room for surprising, and thus trivial, novelty; instead, Poisson's deduction contained a manifest truth derivable from existing physical laws.[28]

The example of Mach illustrates points with which Worrall would doubtless concur, that the portrayal of an experiment reflects the writer's interests and that those interests may arise from any of a number of sources, one of which is the writer's philosophy of science. Turning to texts dating from the mid-nineteenth century we find that the diffraction disc experiment was portrayed as a novel prediction by some of the early proponents of the wave theory. The context is important, for it suggests that the episode formed part of the rhetoric of the 'optical revolution' in which the wave theory of light gained dominance over its particulate rival. In this context optical writers were not concerned with developing a general logic for novel predictions; rather their appeal to Poisson's subsequently-confirmed prediction functioned as a dramatic device to persuade readers in favour of the wave theory.

27. Schuster (1904), p. 104.   28. Mach (1913), p. 285.

There are several standard ingredients in this drama. The actors play clearly defined roles. It is significant that Fresnel, the author of the paper and a partisan of the wave theory, did not initially apply his analysis to the case of an illuminated disc. Only when the paper was in the commission's hands did Poisson, an avowed opponent of the theory, articulate the inference that a bright spot was predicted at the centre of the shadow. Whatever the logical implication of the theory or the supposed irrelevance of who makes a prediction, the roles of the actors do have dramatic significance. Had the prediction been made by Fresnel or Arago (the only member of the commission clearly favouring Fresnel's theory) the account would have a lesser appeal. Fresnel or Arago might be in league with Nature, might know the result beforehand and be suppressing it. Poisson on the other hand was the doubting Thomas whose expectations could not influence affairs. *Nature* showed him to be wrong.

Most accounts dwell on the result being 'unanticipated', 'unexpected' or a 'startling result'. Appeal to novelty – the unexpected, the sudden and (apparently) arbitrary change of fortune – is a recurrent dramatic strategy. According to Aristotle, in tragedy the 'change of fortune will be, not from misery to prosperity, but from prosperity to misery, and it will be due, not to depravity, but to some great error.'[29] One interesting aspect of scientific drama is that when Nature (the oracle?) speaks out against a theory, it is difficult to decide where the error lies. While experimental refutation can assist the demise of a theory, it would be inappropriate to speak of theories suffering tragedies. Their proponents, on the other hand, are the subjects of tragedy. Although theory-centred philosophies of science do not recognise the fate of scientists who are offered up as sacrifices to Truth, personal tragedies often accompany the failure of a theory which is found wanting at a trial by experiment. In this sense the diffraction disc experiment was a tragedy for Poisson and, indirectly, for the fortunes of the particle theory of light.[30]

Accounts of the diffraction disc experiment generally pay far less attention to its tragic component than to its supposed fillip for the wave theory. The dramatic strategy involves the wave theory's reversal of fortune; that is 'a change from one state of affairs to its opposite, one which conforms . . . to probability or necessity.'[31] An experiment may thus bring about such a reversal in the fortunes of two competing theories. Although nineteenth-century optical writers rarely portrayed

29. Aristotle (1965), p. 48.
30. On the role of tragedy in the history of science see Ravetz (1973); on the decline of the Laplacian programme see Fox (1974).
31. Aristotle (1965), p. 46.

Geoffrey Cantor

the diffraction disc experiment as the sole cause producing a change in theory, it was often viewed as having enhanced the wave theory as against its rival. But perhaps the most important dramatic element in the diffraction disc experiment corresponds to what Aristotle calls appropriately 'discovery', for discovery 'is a change from ignorance to knowledge, and it leads either to love or to hatred between persons destined for good or ill fortune.'[32] Thus the discovery by Oedipus that he had slain his father and that Jocasta was his mother led to the reversal of his fortune. What brought about this discovery was the tracing of the shepherd who was generally supposed to have killed Oedipus as a baby. For Aristotle there are several possible causes which precipitate discovery (and thus reversal), the 'least artistic' of these being 'visible signs or tokens'.[33] In a sense the result of the diffraction disc experiment functions as the 'visible sign' while the discovery is the spectacular confirmation of the wave theory. Indeed the episode discussed in Worrall's paper contains all the key elements of a recognition scene – Worrall uses the word 'recognition' on p.140 – in classical drama.

THE CRUCIAL EXPERIMENT

While experiment in general provides the scientist with 'tokens' which may lead to discovery – the recognition of things as they are – and possibly even to reversal, one particular class of experiments seems to place greater emphasis on reversal, particularly when the experiment manifestly favours the theory which had previously been the pauper. This is the so-called *experimentum crucis*. Such experiments, or 'instances of the fingerposts' as Bacon called them, 'afford very great light, and are of high authority, the course of interpretation sometimes ending in them and being completed'.[34] Philosophers of science now generally downplay the significance of such experiments, denying them any major role. This is partly because they fail to recognise them as rhetorical, dramatic devices. For those seeking excitement in science, the historical record is packed with these dramatic episodes, the history of optics being no exception. During the optical revolution of the early nineteenth century, when the wave and particle theories of light were seen as menacing rivals, *experimenta crucis* abound. Thus in his *Prelimi-*

32. Aristotle (1965), p. 46.   33. Aristotle (1965), p. 53.
34. Quoted in Hacking (1983), p. 250. See pp. 246–61 for Hacking's reappraisal of crucial experiments.

nary Discourse to the Study of Natural Philosophy (1830) John Herschel portrayed the two contestants as almost equally matched:

When two theories run parallel to each other, and each explains a great many facts in common with the other, any experiment which provides a crucial instance to decide between them or by which one or other must fall, is of great importance.[35]

The experiment which Herschel cited at this point lacks the expected rhetorical impact. It concerns the interference fringes produced between two glass plates, one of which is slightly convex, and viewed through a prism. Although both theories predict the appearance of bright fringes, Fresnel had predicted that, according to the wave theory, the dark bands would be *'absolutely black'*, whereas, by the particle theory, they would be *'half bright'*. Fresnel had found the result 'to be decisive in favour' of the wave theory.[36]

For many writers after 1850 a conceptually simple but technologically sophisticated experiment offered *the* crucial test (although, by that time, the choice between the wave and particle theories was no longer a contentious issue for most scientists). Even a novice could readily appreciate that in refraction the particle theory required the velocity of light to increase on entering an optically denser medium (e.g., in travelling from air into glass or water), while the wave theory required a decrease in velocity. The two rival theories offered manifestly incompatible predictions concerning this commonplace phenomenon. Although there were earlier attempts to carry out this experiment, the independently-obtained results of Foucault and Fizeau were hailed as definitive. In showing that the velocity of light was less in water than in air 'the Newtonian explanation of refraction, the last remnant of the Emission Theory, was proved false'.[37] Humphrey Lloyd proclaimed this research as providing the *'experimentum crucis* between the two theories'.[38] Although the techniques of the Foucault–Fizeau experiments were considerably more complex and refined than the interference experiment that Herschel had so enthusiastically proclaimed as the crucial test between the two theories, the conceptual simplicity of the former contributed greatly to their rhetorical appeal.

35. Herschel (1851), p. 206.
36. Herschel (1851), p. 207. In Herschel (1845), which was written c.1827, he refers to this experiment as an *'Experimentum Crucis'*, p. 675.
37. Whewell (1857), 3, p. 483.   38. Lloyd (1873), pp. vi, 33.

Compared with numerous references to the Foucault–Fizeau result as providing the *experimentum crucis,* very few of the writers I have encountered made a similar claim about the 1819 diffraction disc experiment. One example is provided by A.V. Guillemin's book, *The Forces of Nature: a Popular Introduction to the Study of Physical Phenomena* (1877) which was clearly not intended for an audience of scientists. After dwelling on the early nineteenth-century optical controversies Guillemin described the disc experiment, adding,

This experiment affords one of the most beautiful triumphs of the theory, – a decisive proof of the undulatory theory of light and of the existence of the ether[!][39]

If Guillemin's rhetoric arose in the context of a popular physics text, the context of Ronchi's *The Nature of Light* (1970) is less easy to determine since his book is one of the few recent book-length studies of the history of optics. According to Ronchi the diffraction disc episode

meant the triumph of Fresnel and the final downfall of the corpuscular theory. At the open meeting of the Académie des Sciences of 1819, the prize was presented to Fresnel and this date marked the collapse of Newtonian ideas.[40]

Whatever rhetorical effect was intended, Ronchi's historical knowledge and judgment were patently at fault.

39. Guillemin (1877), p. 364. 40. Ronchi (1970), p. 251.

References

Aristotle, (1965). On the art of poetry. In *Classical Literary Criticism*, ed. T.S. Dorsch. Harmondsworth: Penguin.

Butts, R.E. (1978). Some tactics in Galileo's propaganda for the mathematization of scientific experience. In *New Perspectives on Galileo*, ed. R.E. Butts & J.C. Pitt, pp. 59–86. Dordrecht: Reidel.

Cantor, G.N. (1987). Weighing light: the role of metaphor in eighteenth-century optical discourse. In *The Figural and the Literal: Problems of Language in the History of Science and Philosophy, 1630–1800*, ed. A.E. Benjamin, G.N. Cantor & J.R.R. Christie, pp. 124–46. Manchester: Manchester University Press.

Carabelli, G. (1972). *Hume e la Retorico dell' Ideologia: Uno Studio dei 'Dialoghi sulla Religione Naturale'.* Firenze: La Nuova Italia.

Christie, J.R.R. & Golinski, J.V. (1982). The spreading of the word: new directions in the historiography of chemistry 1600–1800. *History of Science,* **20**, 235–66.

Cooper, L. (1935). *Aristotle, Galileo and the Tower of Pisa.* Ithaca: Cornell University Press.

Dear, P. (1987). Jesuit mathematical science and the reconstitution of experience in the early seventeenth century. *Studies in the History and Philosophy of Science*, 18, 133–75.

Earman, J. & Glymour, C. (1980). Relativity and eclipses: the British expeditions of 1919 and their predecessors. *Historical Studies in the Physical Sciences*, 11, 49–85.

Finocchiaro, M.A. (1980). *Galileo and the Art of Reasoning. Rhetorical Foundations of Logic and Scientific Method*. Dordrecht, Boston and London: Reidel.

Fox, R. (1974). The rise and fall of Laplacian physics. *Historical Studies in the Physical Sciences*, 4, 89–136.

Galileo, G. (1962). *Dialogue Concerning the Two Chief World Systems – Ptolemaic & Copernican*, trans. S. Drake. Berkeley and Los Angeles: California University Press.

Galileo, G. (1974). *Two New Sciences*, trans. S. Drake. Madison: University of Wisconsin Press.

Gilbert, N.G. & Mulkay, M. (1984). Experiments are the key: participants' histories and historians' history of science. *Isis*, 75, 105–25.

Golinski, J.V. (1984). 'Language, Method and Theory in British Chemical Discourse, c. 1660–1770', Ph.D. dissertation, University of Leeds.

Golinski, J.V. (1987). Robert Boyle: scepticism and authority in seventeenth-century chemical discourse. In *The Figural and the Literal: Problems of Language in the History of Science and Philosophy, 1630–1800*, ed. A.E. Benjamin, G.N. Cantor & J.R.R. Christie, pp. 58–82. Manchester: Manchester University Press.

Gooding, D. (1985). 'In nature's school': Faraday as an experimentalist. *Faraday Rediscovered: Essays on the Life and Work of Michael Faraday, 1791–1867*, ed. D. Gooding & F.A.J.L. James, pp. 105–36. Basingstoke and New York: Macmillan.

Guillemin, A.V. (1877). *The Forces of Nature: A Popular Introduction to the Study of Physical Phenomena*, trans. N. Lockyer. London: Macmillan.

Hacking, I. (1983). *Representing and Intervening*. Cambridge: Cambridge University Press.

Hannaway, O. (1975). *The Chemists and the Word. The Didactic Origins of Chemistry*. Baltimore and London: Johns Hopkins University Press.

Herschel, J.F.W. (1845). 'Light'. *Encyclopaedia Metropolitana*, 4, 341–586.

Herschel, J.F.W. (1851). *A Preliminary Discourse on the Study of Natural Philosophy*, new edn. London: Longman.

Holmes, F.L. (1985). *Lavoisier and the Chemistry of Life: An Exploration of Scientific Creativity*. Madison and London: University of Wisconsin Press.

Holton, G. (1978). *The Scientific Imagination: Case Studies*. Cambridge: Cambridge University Press.

Hume, D. (1948). *Dialogues Concerning Natural Religion*, ed. H.D. Aiken. New York: Hafner.

Hurlbutt, R.H. III (1956). David Hume and scientific theism. *Journal of the History of Ideas*, 17, 486–97.

Koyré, A. (1968). *Metaphysics and Measurement: Essays in the Scientific Revolution*. London: Chapman and Hall.

Kuhn, T.S. (1962). *The Structure of Scientific Revolutions*. Chicago and London: Chicago University Press.

Kuhn, T.S. (1977). A function for thought experiments. In T.S. Kuhn. *The Essential Tension*, Chicago and London: Chicago University Press.

Lloyd, H. (1873). *Elementary Treatise on the Wave Theory of Light*, 3rd edn. London: Longman.

Mach, E. (1913). *The Principles of Physical Optics. An Historical and Philosophical Treatment*, trans. J.S. Anderson & A.F.A. Young. New York: Dover.

Pearl, L. (1970). Hume's criticism of the argument from design. *The Monist*, **54**, 270–84.

Ravetz, J.R. (1973). Tragedy in the history of science. In *Changing Perspectives in the History of Science: Essays in Honour of Joseph Needham*, ed. M. Teich & R.M. Young, pp. 204–22. London: Heinemann.

Ronchi, V. (1970). *The Nature of Light: An Historical Survey*, trans. V. Barocas. London: Heinemann.

Schuster, A. (1904). *An Introduction to the Theory of Optics*. London: Arnold.

Shapin, S. (1984). Pump and circumstance: Robert Boyle's literary technology. *Social Studies of Science*, **14**, 481–520.

Shapin, S. & Schaffer, S. (1985). *Leviathan and the Air-pump: Hobbes, Boyle, and the Experimental Life*. Princeton and Guildford: Princeton University Press.

Tamny, M. (1988). Atomism and the mechanical philosophy. In *The Companion to the History of Modern Science*, ed. G.N. Cantor, J.R.R. Christie, M.J.S. Hodge & R.C. Olby. London: Croom Helm. (in press).

Westfall, R.S. (1980). *Never at Rest: A Biography of Isaac Newton*. Cambridge: Cambridge University Press.

Whewell, W. (1857). *History of the Inductive Sciences, From the Earliest to the Present Times*, 3rd edn. London: Parker.

PART III

REPRESENTING AND REALISING

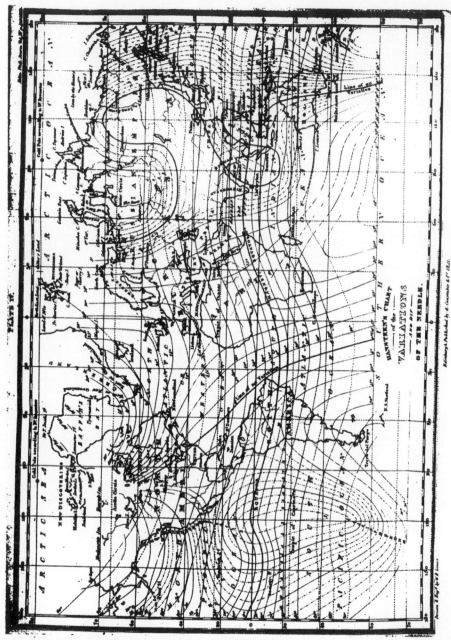

Figure 7.1. 'Christian Hansteen's 'Chart of the variations and dip of the needle'. (Hansteen, 1821b).

7

'MAGNETIC CURVES' AND THE MAGNETIC FIELD: EXPERIMENTATION AND REPRESENTATION IN THE HISTORY OF A THEORY

DAVID GOODING

CONTEXTS OF DISCOVERY

Natural knowledge is produced by people, interacting sometimes with Nature and always with each other. This chapter is about the role of experimentation in these two types of interaction. I shall look for a relationship between what experimentalists did to open up a new observational field and what they did to convey the outcomes of their action upon Nature to each other and to a wider lay audience. This draws attention to a much wider range of activities and interests than is usually associated with the 'context of discovery' as most philosophers and many historians understand it. The development of field theory was shaped by work done in places such as workshops, lecture theatres, laboratories and demonstration rooms, and with ideas and images expressed in the literary contexts more familiar to historians, such as the pages of journals, an encyclopaedia and manuscripts.

The work done and recorded in all these contexts shared at least three related objectives: to produce new electromagnetic phenomena, to communicate them to a scientific audience and to interest a wider public in them. The interest in diffusion through demonstration was particularly important. Its influence on early developments shows the inadequacy of those views which locate discovery entirely within the laboratory. Successful experimentalists move easily between the world

183

and the laboratory, drawing elements of the 'outside' world into the laboratory.[1]

Historians have approached the early development of field theory as if theory developed as a purely intellectual, disinterested and disembodied form of construction.[2] Intellectual factors were of course important, as I have argued elsewhere,[3] but not exclusively so. Here I argue the influence of a network of shared practices which embodied a new way of thinking about electric and magnetic phenomena. Where practices succeeded in realising pragmatic objectives, for example, enabling scientists to produce and disseminate phenomena, this success encouraged and perpetuated them.[4] These practices were amongst the resources on which Michael Faraday drew during the three decades in which he articulated the field description of electromagnetism that he defended as an explanatory theory from 1850. To understand the origins of the language of the field theory we need to examine the seamless web of practical, intellectual and social interactions that made up the scientific culture in which Faraday thrived. Although the analysis necessarily distinguishes between intellectual, practical, material, individual, social and other strands, I want to emphasise the 'seamlessness' of the web. To distinguish between 'intellectual' and 'procedural' or 'individual' or 'social' is not to imply that these activities and categories are separate: on the contrary, such 'factors' are sometimes so interdependent as to suggest that new ways of thinking and imaging phenomena have their roots in ways of producing and representing those phenomena. New meanings emerge from new practices. The following sketch of the context from which Faraday's radical new representation of electricity and magnetism emerged will illustrate the point.

In 1820 Oersted showed that a wire carrying a current will affect a magnetised needle. This discovery created considerable excitement. Magnetic effects of a chemically-produced current promised the unification of three natural powers, moreover electromagnetism presented something new. The magnetised needle set perpendicular to the wire, indicating that the forces at work were somewhat unusual. The early responses to electromagnetism show that observers with quite different

1. On translating phenomena into and out from the laboratory see Latour (1983) on Pasteur, Gooding (1985b) on Faraday, and Asmuss & Galison [Chapter 8 this volume] on Wilson. On the context of experimental places see S. Shapin, (1988).
2. See, e.g., Williams (1966); Heimann (1970, 1971).
3. Gooding (1978, 1980a, 1980b, 1982a, 1989).
4. For the interaction of representational style and scientific style, see Fleck (1979); pragmatic success is discussed in Gooding (1986).

theoretical and methodological beliefs agreed about what the phenomena were, though their precepts led them to interpret these differently. Their agreement was negotiated and expressed in terms of newly invented visual and verbal representations. Where these constructs succeeded – and they often did not – they enabled observers to make sense of the novel part of their experience, and to communicate that sense. During the 1820s Peter Barlow and others developed a visual, geometrical and very practical mode of representing magnetism. Christian Hansteen used lines to map the terrestrial magnetism at different places on the surface of the Earth (see Fig. 7.1.).[5] Alexander von Humboldt used this technique to represent geothermal properties of the globe and advocated the study of natural phenomena on a global scale and, as Cannon points out, argued the superiority of field studies over experimentation which confined Nature to the minute and human scale of the laboratory.[6] Humboldt sought to avoid the artificiality of man-made laboratory phenomena. Alexi Assmus and Peter Galison argue in chapter 8 of this volume that the cloud chamber physics developed by C.T.R. Wilson later in the nineteenth century is a legacy of this field-based observation. Although scientists abandoned the Humboldtian insistence on observation of phenomena as they 'really occurred' in Nature they held to the imperative to observe by imitation rather than analysis. To study atmospheric phenomena such as clouds, Nature had to be brought into the laboratory, so Wilson learned, first to produce clouds and later how to observe tracks formed in them.[7] Barlow is an earlier example of a scientist who moved from the *global* domain to a *local* one, although he did not share the morphologist's concern for the riskiness of such a translation. Barlow adapted contemporary methods of global mapping of the geothermal and geomagnetic fields which he introduced into the local context of the workshop of the Royal Arsenal at Woolwich.

A verbal form of representation also appeared early in the 1820s.[8] The term 'curves' was used to denote wires capable of carrying current. George Birkbeck used it in a lecture–demonstration of analogies between terrestrial magnetism and electromagnetism. Barlow discussed 'magnetic curves' in an 1824 encyclopaedia article on magnetism, where he sought to show that the patterns of magnetism mapped by iron filings could be constructed using well-known geometrical methods

5. Hansteen (1821a,b, 1827); von Humboldt (1820–21); von Humboldt & Biot (1805).
6. Cannon (1978), chapter 3.
7. Galison & Assmus (chapter 8 this volume).
8. Faraday (1821), and (1932–36), 1, p. 52.

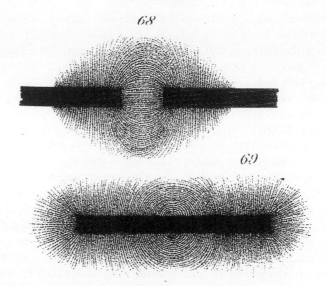

Figure 7.2. Engraving of iron filing patterns, from Barlow's 1824 article on magnetism, in the *Encyclopaedia Metropolitana.*

(see Figs. 7.2 and 7.3). Each of these contexts of introduction can be identified with certain interests which helped promote the development and acceptance of methods of experimentation and representation. Scientists' interest in the dissemination of the new phenomena is important because they had to invent methods of representing or imaging unfamiliar phenomena for scientific acquaintances and co-practitioners. They then developed these technologies of observation into technologies of display which made the new phenomena accessible to a larger audience. Since they were based on the same visual and verbal language of lines or curves of force as the observational practices, the technologies of display helped to establish that language while also reinforcing the scientific values it embodied.

A second stage began when Michael Faraday used the curves to state the new law of electromagnetic induction in 1831. He went on to theorise a well-known practical method of intensifying magnetic effects, embodied in the electromagnet. Faraday's discovery of electromagnetic rotations in 1821 and of electromagnetic induction in 1831 made him an authority. However, the acceptance of lines and eventually of a field way of thinking about forces cannot be attributed solely to his influence or that of his major discoveries. After all, most electromagnetic

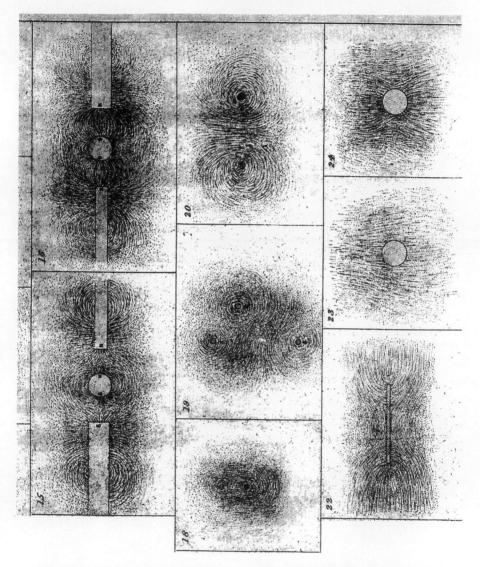

Figure 7.3. Engraving of iron filing patterns actually formed by magnetism and fixed in shellac, from Faraday's 1852 paper 'On the employment of the Induced Magneto-electric Current as a test and measure of Magnetic Forces'.

phenomena (including Faraday's results) could be explained by other theories which made no use of curves of force.[9] Part of the explanation of the acceptance of Faraday's lines lies in the earlier dissemination of the observational practices from which they were drawn. Much of Faraday's work on lines of force during the 1830s and 1840s articulated possibilities that were implicit in practices he had learned from others or developed in his own studies of electromagnetism during the 1820s. The ground for the eventual overthrow of the Newtonian–Laplacian conception of pondermotive forces had been cultivated by practical as well as intellectual means.

Between 1836 and 1850 he developed the analytical and explanatory power of the curves in order to model the interaction of electric and magnetic forces manipulated in the laboratory. Eventually, the lines represented the expenditure of natural powers as processes obeying the intellectual principles of economy and conservation. With them Faraday could draw a picture of Nature that was consistent with more widely accepted intellectual and moral principles.[10] By 1850, when he published his field theory, Faraday had articulated the practical understanding of laboratory procedures developed during the 1820s into a

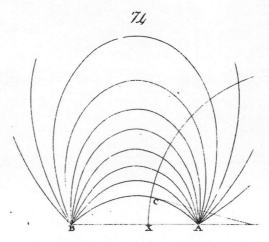

Figure 7.4. Curves as geometrical constructs, from Barlow's 1824 article on magnetism.

9. These derived from a theory constructed by Ampère shortly after he learned of Oersted's discovery. This reduced magnetism to molecular currents and so reduced electromagnetic effects to electrical 'elements' interacting by means of conventional (i.e., non–skew) forces. On Ampère see Caneva (1981); Williams (1983); Hofmann (1987).
10. See Gooding (1982a), esp. pp. 61–7.

theory of considerable explanatory power. For example, he showed how the behaviour of lines (such as those mapped by iron filings shown in Figs. 7.2 and 7.3, represented as constructs in Figs. 7.4 and 7.5) can explain the motions of matter in the field.

From 1850 Faraday concentrated more on extending and defending a theory of lines. He moved the lines out from the laboratory into the global context of terrestrial and atmospheric magnetism from which the imagery had been drawn three decades earlier (see Fig. 7.6, and compare Fig. 7.1). This was a conscious attempt to win support for the theory by showing that it could now portray the whole of Nature as what he called one 'great field'. The theory realised his aspirations for an observationally well-founded theory of great generality and explanatory power. It expressed both natural facts and values by attributing Victorian virtues such as economy to the operations of 'the great field of nature'.[11] Faraday was able to argue of the magnetic curves

Fig. 1.

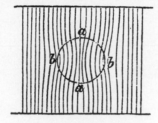

Fig. 2.

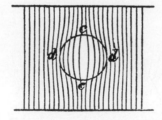

Figure 7.5. Faraday's representation of magnetic matter with respect to 'mere space', published in 1851.

11. Faraday (1851b), § 2974. Faraday's theory of atmospheric magnetism, published in 1851, has been overlooked by biographers and historians. Over 100 printed pages in length, it was an ambitious attempt to comprehend a large amount of data on the dip, intensity and diurnal variation of magnetism accumulated during the 'magnetic crusades' of the first half of the nineteenth century (Cawood, 1979). Faraday's aspirations for this theory were as 'global' as those that geologists such as Murchison had for their systems (Secord, 1982).

that – whether representing the local, laboratory domain of Figures 7.4 and 7.5 or the macroscopic domain of Figures 7.1 and 7.6 – they obey or 'respect' teleological principles of economy and conservation of force. His reference (in Fig. 7.6) to the atmosphere as a lense acting on the magnetic lines, and another manuscript reference to lines as analogous to lines of latitude and longitude, point to influential analogies to problems and techniques that had been important during the 1820s. His successful application to both local and global domains confirmed the usefulness of the method and implied the reality of the physical conceptions it suggested. Moreover, Faraday was finally able to argue the *empirical* superiority of his theory by pointing to the strong connection between his methods of observation, the facticity of his results, and his interpretation of them. The unification of methods of producing,

Fig. 14.

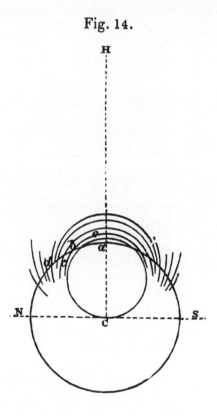

Figure 7.6. Faraday's 'illustration' of the atmosphere as a magnetic lense.

representing and explaining phenomena gave his approach a coherence that encouraged Faraday to argue for an ambitious metaphysical reversal. The inversion of the physical priority of matter and force would make electromagnetic energy a property of immaterial fields and an inherent property of discrete and mundane objects such as electromagnets. Two decades of work had produced a theory that realised far more than new ways of thinking about electromagnetic phenomena: it also realised the values and the scientific aspirations he had learned during his apprenticeship.[12]

EXPERIMENTATION AND REPRESENTATION

People act upon Nature to generate new possibilities for observation. However, these remain mere possibilities (and private ones at that) until they are successfully communicated to lay observers, whose witness confirms the existence of the phenomena. Witnessing requires a shared means of experiencing the outcomes of interventions, and so discoverers must invent new representations that present the phenomena in a way that novices can be shown what they are to witness. Lay observers are taught to see.[13] Where a phenomenon is really novel, *all* intending observers are lay observers. Some must first teach themselves to see. If representations are necessary to observation then experimentation without representation is blind. This does not mean that experimentation without theory is blind![14]

Experimenting and representing are closely connected because of the pragmatic and social constraints on the construction of representations. For example, whether an observer's tentative rendering of a phenomenon communicates successfully to other observers will immediately affect what he *does*. At the frontiers of experience, possible

12. Faraday (1851a, b, 1852a).
13. I introduced 'lay observers' in Gooding (1985b). The need to interpret new and unexpected phenomena does not arise often in the history of science, but where it has the new phenomena are usually assimilated quickly into existing theoretical frameworks. For the importance of 'witness' see Shapin & Schaffer (1985) and Shapin (1988).
14. For examples of the independence of experiment from theory see Hacking (1983); Gooding (1986). Exploratory work can create phenomena for which no acceptable theoretical explanations yet exist. Even when experiment is guided by theoretical hypotheses, the latter often fail to specify the relevant parameters and observational conditions necessary to obtain a result. These are usually learned during the conduct of an experiment. See Gooding (1985a) and my 'Technologies from texts?' (1987, unpublished). In the language of the framework for comparing case studies developed by Trevor Pinch (1985), such observations have *low externality*, that is, they are not predicted by, or otherwise dependent upon, the acceptance of a particular theory.

interpretations must make sense both of phenomena and of the experimental practices generating them. In this respect empirical access to Nature is both a cognitive and a social process.[15] It is cognitive, in that experimenters learn through engaging a real (and often recalcitrant) natural world; it is social in so far as what they learn owes its significance and even its formulation to engagements with other (often recalcitrant) observers.

Thought in action

If representations are necessary to the possibility of agreement about the experience of Nature, then what experimenters actually *did* in order to construct observable, publicly accessible phenomena will sometimes influence how they *thought* about the outcomes of their experiments. Although this assertion appears to invert the priority of thought and action (for are not actions always informed by thoughts?), I want to emphasise that there is usually a two-way traffic of influence. Intellectual considerations, apparently remote from the laboratory workbench, influence the selection of images and techniques, which themselves imply or embody images and concepts.[16] It is taken for granted that linguistic modes of representation influence the thoughts formulated in them and even that we cannot think without such a system of representation. I am making an analogous claim about non-verbal modes of representing sensory and mental experience. Renderings of new phenomena are sometimes articulated procedurally before they can be expressed in verbal or visual imagery. These representations can be articulated as *instrumentally* useful concepts before they are incorporated into a theoretical framework, so it is plausible to suppose that they shape the theories developed to interpret and explain the phenomena they describe.[17]

The techniques I describe were developed to observe, represent and demonstrate aspects of a wholly new set of electromagnetic phenomena. Some were adapted from existing methods of representing

15. This is because whether a representation communicates successfully affects what experimenters do next. What counts as a successful solution will be evaluated (at least implicitly) in terms of immediate objectives and long-term interests and values. So, by looking behind practitioners' *rhetoric* to their experimental *style*, we can get a sense of how deeply rooted that rhetoric is. Models of the process are discussed in Rescher (1979); Barnes (1983); Gooding (1986, 1989).

16. These are assumptions, convictions (theoretical, methodological, metaphysical, theological) and values typical of the wider culture. Hesse calls them *coherence conditions* (Hesse, 1980, pp. 131–4).

17. For instrumental facts and observations as the rudiments of concepts see Gooding (1982a), pp. 50–5 and (1989).

terrestrial magnetism while others, such as the use of the electromagnet to enhance the magnetic effects of currents, owed their existence to the new phenomenon of electromagnetism. They belonged to a network of practices, a repertoire of skills which typified the culture of a number of experimental natural philosophers.[18] All experimented on electromagnetism during the 1820s in London.

A NETWORK OF PRACTITIONERS

Some of the people I'll consider thought it possible that electromagnetism involved a new type of force in Nature — one which did not simply attract or repel like other (Newtonian) forces. Thus Humphry Davy claimed that it was self-evident that needles position themselves around a vertical wire as tangents to a circle drawn in a horizontal plane having the wire at its centre. Such observations were far from self-evident.[19] Others sought to assimilate the novel, 'skew', aspect of electromagnetism to conventional forces.[20]

We inhabit a world so shaped by electro-mechanical and electronic devices that it is difficult to imagine a time when electromagnetic effects were insignificant to everyone but a handful of devotees. This insignificance is illustrated by comparing the place of early representations of electromagnetism in scientific journals of the day with those that appeared within a year or two. Figure 7.7 shows a plate from *Brewster's Journal* which pictures phenomena (such as waterspouts) and wonderful contraptions (including a new means of propelling boats, devices for walking on water and for ventilating the lungs). An experiment on a phenomenon derived from Oersted's effect has been squeezed onto this plate as fig. 16.[21] This plate is typical in showing that electromagnetism vied for space with every other phenomenon and contraption thought to be important or possibly of interest. An experiment by Jack Tatum — one of Faraday's first teachers — appears in the corner of drawings of water-saving locks for the Regent's Canal. Drawings of Charles Hatchett's experiments were squeezed onto a plate concerned with the boring of a cannon. Plates were expensive to engrave and these early representations were squeezed onto them most probably

18. For more detailed treatments of what follows and for references see Ross (1965); Heilbron (1981); Williams (1983, 1985); Gooding (1985b, c). For a case study of the dependence of phenomena on practices see Pickering (1984a).
19. Davy (1821a, b). On the constructed-ness of self-evidence, see Gooding (1986); Lynch (1985).
20. These included Peter Barlow and P.M. Roget in England, and — for quite different reasons — A.M. Ampère and others in France.
21. Brewster (1821).

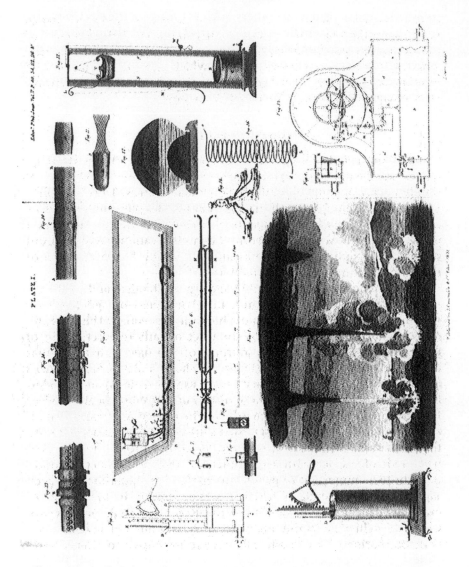

Figure 7.7. A plate from Brewster's *Edinburgh Philosophical Journal* for 1821, showing as 'fig. 16' an experiment on Oersted's effect.

to save money.[22] Yet within a year, full-page illustrations of electromagnetic images and devices began to appear. Figure 7.8 is a full page of illustrations for the uninitiated observer from Faraday's review of the experimental response to Oersted's discovery.[23] Figure 7.9 shows a page devoted to demonstration apparatus used by William Sturgeon.

The spread of these images onto the page shows the growing importance of what they depict. Why were they drawn? The images highlight the pragmatic and social aspects of observation that I mentioned earlier: the practical problems of interpreting, representing and communicating new findings and the social character of that problem-solving. We can situate the pragmatic activity in the social by looking briefly at some connections in a loose network of practitioners who had an active interest in electromagnetism during the decade after Oersted's announcement of electromagnetism in 1820.

The electromagnetic 'network'

Oersted's discovery promised to be a doorway to the unification of electric and magnetic science and possibly chemistry as well. This promise brought into contact a number of individuals whose backgrounds, class and financial circumstances did not otherwise give them much in common. These included self-educated devotees of science (such as Michael Faraday, William Sturgeon and Peter Barlow); instrument-makers and technicians (Francis Watkins and James Marsh); lecturers and publicists (Jack Tatum, George Birkbeck), and other, established scientific practitioners (S.H. Christie, W.H. Wollaston, W.T. Brande, W.H. Pepys, C. Hatchett and Sir Humphry Davy).

Their common interest in electromagnetism was never institutionalised in a way that led all of them to meet regularly or work together. I refer to them collectively as a 'network' to indicate that most were acquainted through associations with other institutions or interest groups and that all knew of each others' work. Although at least two dozen people were involved in some way with electromagnetism, surviving evidence of personal contacts and collaborations is incomplete.[24] Correspondence, references in correspondence and

22. See the plates referred to in Hatchett (1821) and Tatum (1821).
23. Faraday (1821–22).
24. There are no records of the sort that we have for the London Electrical Society (established by William Sturgeon in 1837 to promote the application of electromagnetism). Records of the London Institution appear to have been dispersed or lost and little archival material relating to Woolwich in this period survives. I have drawn mainly on collections at the Royal Society and the Royal Institution.

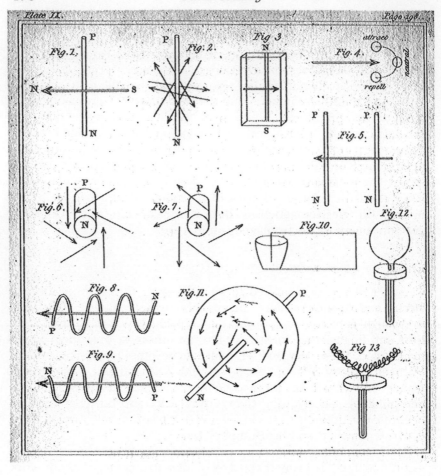

Figure 7.8. Drawings of possible interpretations of electromagnetic experiments, from Faraday's review of 1821–22.

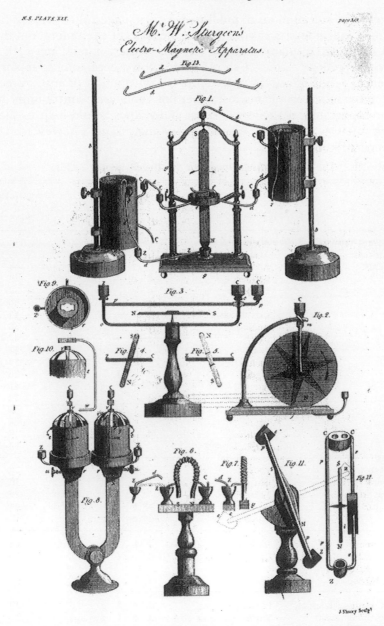

Figure 7.9. A set of devices used by Sturgeon to demonstrate electromagnetic phenomena around 1826. The new electromagnet is shown as 'figure 6'.

David Gooding

notebooks, and citations in published papers suggest that perhaps half a dozen met at least occasionally to share ideas or equipment, but they did not meet regularly and did not constitute a formal network or 'invisible college'. Table 7.1 shows their known occupations, apparent interest, period of activity and nature of contribution (if any). A more extensive search would disclose other links and other contributors and a prosopographical analysis of these individuals would help us place their interests in context. A preliminary analysis does however reveal several striking features:

Table 7.1. *The London Electromagnetic 'Network' in the 1820s.*

	Names	a	b	c	d	e	f	g	h	ñ	j	Age in 1820
A[b] Short-term or restricted interest	Allen						f				j	--
	Andersen				d	e				ñ		25
	Birkbeck						f		h	i	j	44
	Bollaert				d	e				i		13
	Brande				d	e		g		i	j	32
	Colby				d					i	j	36
	Cummings	a			d	e		g				45
	Daniell				(d)					i	j	30
	Davy				d	e	f			i	j	42
	Fox	a				e						31
	Hatchett		b	c		e			h			55
	Marsh	a	b	c	d	e	f			ñ		26
	Roget							g	h			41
	Sabine				d					i		32
	Tatum			c		e	f					nk
	Wallis			c			f					nk
	Watkins	a	b	c	d			g				18
	Wollaston					e				i	j	54
B[c] Active in several key areas during the 1820s	Babbage	a				e				i	j	28
	Christie	a	b		d	e				ñ		36
	Herschel					e				ñ	j	28
	Pepys	a	b	c		e				ñ	j	45
	Sturgeon	a	b	c	d	e	f	g		ñ		37
C[c] Active in most or all areas during the 1820s	Barlow	a			d	e		g		ñ	(j)	44
	Faraday	a			d	e	f	g		ñ	j	29
D[d] Continued work on electromagnetism into 1840s	Faraday	a			d	e	f	g		ñ		29
	Sturgeon	a	b	c	d	e	f	g		ñ		37
Totals	All: 25									16		

(1)   Over a third were associated with the Royal Institution or the London Institution and several taught, demonstrated or were technicians at military and naval establishments such as The Royal Observatory at Greenwich and the Royal Military Academy at Woolwich. Samuel Hunter Christie and Peter Barlow taught at Greenwich and Barlow conducted research in the workshop of Woolwich arsenal, assisted by James Marsh. The workshop of the Royal Arsenal at Woolwich was a particularly important 'nursery' of applied science.[25]

(2)   Several people fall under more than one of the categories used in the table. (This is unsurprising since English science had not yet become specialised or professionalised.) For example, Faraday, Sturgeon, and Watkins all depended to some extent on income earned from lectures and demonstrations. Instrument-making was for a time the livelihood of Watkins, Sturgeon and Marsh. Pepys, who was active in the London Institution, the Royal Institution, and the Royal Society, had built up his father's cutlery business into an instrument-making firm. Marsh, who assisted at Faraday's lectures at the Royal Military Academy from 1829, subsequently established himself as an electrician and chemist. Sergeant Anderson, who became Faraday's laboratory assistant at the Royal Institution in 1827, had worked in the Woolwich armoury for many years, probably with William Sturgeon. Sturgeon had recently bought himself out of the army and lived nearby.[26]

Notes:
[a] *Key to profiles:*
   a made magnetic instruments or electromagnetic devices
   b partly or wholly self-supporting as instrument-maker or inventor
   c mainly self-supporting during 1820s (business, lecturing, chemist, other)
   d salaried employment at a scientific institution, university or other
   e made/assisted with experimental investigations 1820–30 (other than demonstrations)
   f known as lecturer/demonstrator
   g published important review, treatise or text on electromagnetism
   h M.D.
   i known affiliation with or activity at, Royal Institution, London Institution, Military or Naval establishment (Greenwich, Woolwich, Addiscombe or other)
   ii 8 out of 16 were active at two or more institutions in London
   j Present at Herschel's 'contact' at the London Institution, March 1823
[b] A: 18 represents 72% of the known 'network'
[c] B+C: 7 represents 28% of the known 'network'
[d] D: 2 represents 8% of the known 'network'

25. Royal Society, Barlow Correspondence. In 1837 Joseph Henry visited most of the institutions mentioned here and met most of those I have named as members of the 'network'. His European diary contains many useful observations about London science and its practitioners. See Reingold (1979), p. 172 ff.
26. DNB, DSB; see also Thompson (1891).

(3)   Very few pursued electromagnetism on a broad front. Although the diversity of their backgrounds and livelihoods suggests that electromagnetism could be investigated by anyone having the interest and resources to take it up, only four of the two dozen people I have identified were active in more than one of the six main contexts of activity identified in the search.[27] Here the most striking finding is that only two people – Faraday and Barlow – worked on *every* aspect of electromagnetism during the first decade of its discovery.

(4)   The dates show that few experimented on electromagnetic phenomena for very long. Again only two people – Faraday and Sturgeon – pursued electromagnetism beyond the first decade of its discovery.

(5)   This means that *only one person in London continued to pursue every aspect of electromagnetism beyond the 1820s*. This was Michael Faraday.

Was there at any time during the 1820s a 'core' group of keen investigators? Of the two-dozen people identified in Table 7.1, only seven had the range or tenacity of interest that qualify them as principal investigators. They are listed in Table 7.2. Here the suggestion that they were a 'core' group of experts must be used with care, for two reasons.[28] The first is that although some practitioners clearly collaborated or assisted each other (see Table 7.3), we cannot be as sure of this as, say,

Table 7.2. *Probable membership of a 'core' group, 1820–25.*

	Names	Age in 1820
Likely members between	Babbage	28
1820–30	Christie	36
	Faraday	29
	Herschel	28
Possible additional members	Barlow	44
	Pepys	45
Active but outside	Marsh	26
'core' group	Sturgeon	37

27. These include: instrument-making, exploratory experimentation (magneto-optical effects, rotations), work on magnetic curves, and mathematical treatments. See Gooding (1985c). It is worth noting that the movement and sharing of laboratory staff between institutions in the network would ensure the exchange of many unrecorded skills and techniques.

28. Thus I refer to them as a 'core-group' rather than use the term 'core-set' proposed by Collins to explain the closure of controversies. See Collins (1985), p. 142ff.

Table 7.3.   Some known collaborations (including use of important instruments)

Collaboration	Main dates:	Location:
Davy–Pepys–Faraday	1821	London Institution
Babbage–Herschel	1823–26	at Herschel's
Barlow–Marsh (as assistant)	1820–23	Woolwich
Birkbeck–Barlow	1824	London Institution
Christie–Faraday	1831	at Christie's and Royal Society
Davy–Faraday (as assistant)	1820–21	Royal Institution
Davy–Wollaston	1820–21	Royal Institution
Faraday–Anderson (as assistant)	182?–27	Woolwich
Faraday–Anderson (as assistant)	1827-	Royal Institution
Pepys–Herschel	1823	London Institution

contemporary sociological investigators of 'core sets' can be. For example, while it is clear that Faraday and Sturgeon knew each other's work and techniques, there is no evidence of actual collaboration and even some circumstantial evidence of Sturgeon having been on poor terms first with Davy and later with Faraday, especially after Faraday published his discovery of electromagnetic induction in 1832.[29] The second reason is that my concerns differ from those for which core-set analysis was developed. I am interested in the construction of a local consensus about observations *within* a particular network of natural philosophers rather than the role of experimental expertise in the closure of controversies between rival groups. This network was not a clearly defined group of expert practitioners to whom others appealed to end controversies.[30] By contrast to most science in the twentieth century (and by comparison to French science even in the early nineteenth century) British science remained an individual enterprise at this time. 'Expertise' was thought to be a property of individuals rather than groups or schools.[31] A full account of some of the main interests and problems occupying members of this group is needed to show how the environment of technical and material resources, interests and allegiances shaped their work and thought. Here I offer only a brief sketch, concentrating on three of the most active practitioners: Peter Barlow, William Sturgeon and Michael Faraday.

29. Reingold (1979), p. 250.
30. Collins (1981).
31 See Morrell (1971); Morrell & Thackray (1981), esp. chapter 3.

TECHNOLOGIES OF OBSERVATION AND DEMONSTRATION

Oersted's discovery contained both a challenge and a promise. The apparent non-centrality of the new force challenged the received understanding of magnetic induction (and might therefore have implications for practices such as correcting the readings of ships compasses). Oersted's discovery also offered a new source of public entertainment and edification, in which magnetism and electricity could be combined. Historians emphasise theoretical motivations which epitomise the pursuit of truth or the search for important applications. They have overlooked dissemination and display, especially where these have no obvious relationship to technological advance.[32] Yet demonstration – the witnessing of a phenomenon – is essential to its acceptance into the body of natural knowledge.[33] The influence of this particular interest on scientists' development and use of experiment remains largely unexplored.

Many of the scientists identified in the previous section were concerned to win wider interest in – and public support for – scientific work. This is illustrated in the following episode. In March of 1823 a grand electromagnetic display was held at the recently opened London Institution. In 1845 John Herschel recalled this display as the occasion of his failure to detect the magneto-optic effect which Faraday had discovered. Herschel wrote to William Pepys (who had been secretary of the London Institution in 1823) asking him to recall who had been present at his 'contact' at the institution twenty-two years earlier. The occasion was an opportunity to show off the powers of Pepys' quantity battery – then the most powerful source of current available. Two things are striking about the magnetic display: Pepys' reply to Herschel suggests that most of the scientific worthies of London had attended (see column *j* in Table 7.1). Yet Herschel was allowed to connect his apparatus only at the end of a long session of public displays. By that time, according to Herschel, the battery was exhausted.[34] The demonstration of visually impressive effects must have seemed more important than a search for new ones.

Many members of the network were keen to promote science, because of utilitarian or diffusionist ideals, an interest in expanding a research programme, or to make a living. Magnetic induction and terrestrial magnetism were of obvious importance to navigation and were

32. On the importance of 'display' see Hays (1983).
33. See Gooding (1985b); Shapin & Schaffer (1985).
34. Gooding (1985a), p. 232.

thought to be important to mineralogists, for locating ores. The electromagnet – an instrument at the core of electrical technology – was first welcomed as a contribution to effective public displays of magnetic phenomena. Developed during 1823 and 1824, William Sturgeon's electromagnet received the silver medal of the Royal Society of Arts because it made electromagnetic demonstrations easier, cheaper and more portable.[35] Figure 7.9 shows Sturgeon's horseshoe electromagnet in one of its first and *most important* contexts, a set of demonstration devices dating from 1826. We shall encounter it again in contexts more familiar to historians of science and technology, as a laboratory instrument producing compact, intense magnetic fields for other purposes.

Peter Barlow's work also illustrates the convergence of experimental style with a concern to demonstrate useful results. Barlow investigated terrestrial magnetism in terms of a three-dimensional structuring of space within the laboratory. There he subjected the terrestrial field to a minute examination in order to analyse the effect of iron masses on compasses. By 1820 he had devised a method for correcting for the magnetic effect of iron used in the construction and arming of naval vessels. For this work he received the substantial sum of £500 from the Board of Longitude and the gold medal of the Royal Society of Arts in 1821.[36] Barlow wanted to make the existing theory 'practical', which meant that although mathematical in form it should yield results which would be understood as readily by a naval cadet as by a Lord Commissioner of the Admiralty.[37] This concern was typical of the Woolwich approach to applied mathematics. Faraday (who also taught at Woolwich) had similar views about the connection between the likely truth of a theory, its usefulness as a practical guide to experiment and its conduciveness to public demonstration.

Barlow's practical approach encouraged a style of experimentation which had an unintended consequence: although he applied the existing *theory* of magnetism, Barlow's *method of observation* and representation implied a new way of thinking about magnetism. His early investigations involved comparing compass bearings at a given set of points in the region of an iron cannon ball to the bearings obtained when it

35. Hudson & Luckhurst (1954). See Sturgeon (1826).
36. DNB; Barlow (1820), pp. 58–83; Wood & Oldham (1954), pp. 303–16. Barlow naturally sought to minimise the implications of Oersted's discovery for his new correcting-plate method. In 1821 he showed that the interaction of currents and needles could be described in terms of ordinary magnetic forces, construed as the resultants of a new 'tangential force' (Barlow, 1822). The new effect therefore had few implications for the application of the existing theory of magnetism to navigation.
37. Barlow (1820), p. 109.

was absent. He defined the set of points at which bearings were taken operationally, moving the sensor around in circles corresponding to lines of latitude or longitude of a cannon shell (see Fig. 7.10). To identify the effect of the iron on the Earth's field he needed a three-dimensional reference frame or 'plan' to structure the space (shown in Fig 7.11).

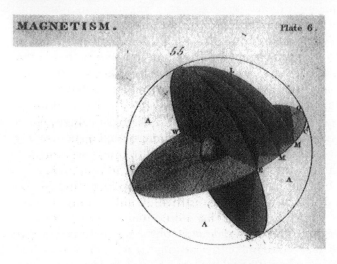

Figure 7.10. Barlow's analysis of the region of an iron cannon shell, showing plane of no magnetic action.

Barlow's plan for mapping properties and their controlled variations required no special new terms or concepts. However, his procedures implied a view of magnetism as a spatial array of dispositional properties. Barlow's method of producing phenomena embodied an implicit concept of a magnetic field, just as the methods of Canton, Beccaria and others had implied an electrostatic field concept in the eighteenth century.[38] Faraday's later renderings of electrostatic, magnetic and electromagnetic phenomena made explicit what had been implicit in these methods of observation and representation. Moreover Barlow's method places the experimenter in the same space as the magnetic properties under investigation. This introduces a global conception of the mapping of geothermal and geomagnetic properties into the local, small-scale context of the laboratory. In both cases the observer is immersed in the magnetic sea whose properties he is investigating.

38. Heilbron (1981) argues that techniques for mapping and representing electrostatic phenomena implied an electrostatic field concept long before this was articulated by Faraday.

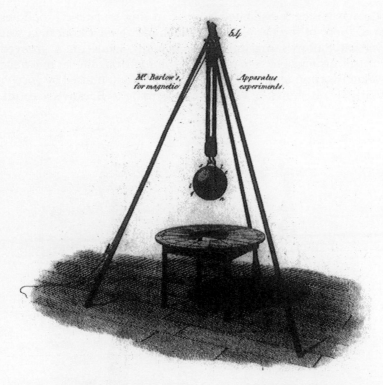

Figure 7.11. One of Barlow's set-ups for mapping local effects of masses of iron on the terrestrial field.

MAGNETIC CURVES

Observers must be located in the experimenter's space, thus bearings are needed in laboratories and demonstration rooms, just as they are at sea.[39] Technologies of display rely upon the prior development of technologies of observation, that is, methods of representing or imaging new phenomena. Representations can, of course, be visual, verbal or symbolic.[40] One of the very first visual representations of terrestrial magnetism was introduced in a context in which the verbal language of 'magnetic curves' was also used.

Faraday used the term 'curves' in 1832, to denote current-carrying wires in one of his earliest writings on electromagnetism. At that time the term had no other significance. In May 1824, at the London Institu-

---

39. For other examples see Lynch (1985); Gooding (1989), chapter 2.
40. The interaction of verbal and visual representations in Faraday's work on diamagnetism is described in Gooding (1981), p. 242ff.

tion, George Birkbeck exhibited a globe having magnetic properties similar to those of the Earth. His imitation of the terrestrial field was achieved, at Barlow's suggestion, by winding wires on a grooved wooden ball, arranged so as to produce a symmetrical, dipolar magnetic field. A later version used by William Sturgeon (and praised by Joseph Henry in 1837) is shown in Fig. 7.12.[41] According to Barlow's account,

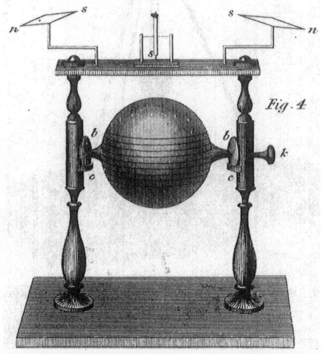

Figure 7.12. An electromagnetic globe used by Sturgeon to illustrate the analogy between terrestrial and electro-magnetism.

Birkbeck wanted to 'illustrate experimentally' that Ampère's hypothesis – that all magnetism is due to electric current – might be used to explain the terrestrial field. This meant showing by analogy how both could be caused by systems of currents flowing in 'curves'. This demonstration required that a property of the whole Earth be imitated within the walls of the London Institution.

For Barlow, Birkbeck and Faraday the term 'curve' signified *wires* carefully positioned to produce a facsimile of the terrestrial field. The term also implied something that became important later. A curve is a

41. Reingold (1979), p. 307.

definite thing, whether it is the path followed by a geometrical point or the physical course of the wire carrying the current that produces the magnetic field.[42] This context of use is important for another reason: Birkbeck and Barlow could hardly have chosen a better platform from which to connect words, images, models and interests by introducing an analogy between currents (in the 'curves' or wires) and lines of magnetic action (the terrestrial field, mapped in the style of Christian Hansteen and Alexander von Humboldt).[43]

This use of 'magnetic curves' did not signify a new theory of magnetic lines, any more than Barlow's 'plan' had introduced a field theory of terrestrial magnetism. (In both cases the favoured theory was Ampère's.) In his discussion of the geometrical representation of ordinary magnetism, written for the new *Encyclopaedia Metropolitana* in 1824, Barlow used the term 'magnetic curves' both for the well-known patterns formed in iron filings and to denote geometrical constructs (as in Fig. 7.2). He went on to argue that the curvilinear patterns are 'necessary consequences' of the inverse-square forces given in the mathematical theory of magnetism, not physical explanations of the iron filing patterns. By showing that geometrical techniques could be applied to the problem of constructing representations of magnetic phenomena, Barlow forged a link between observation and rigorous analysis. This helped to make empirical curves respectable and interesting to others with mathematical interests, such as Charles Babbage, Samuel Hunter Christie and John Herschel.

The next important development for these curves was Faraday's discovery of electromagnetic induction at the end of 1831. This made new demands on the communicative skills of its discoverer and his colleagues: Christie and Faraday both argued that lines were useful 'helps

42. During the 1830s and 1840s Faraday used the term 'curve' interchangeably with 'line'. He understood the 'curves' both practically (i.e., experimentally) and as constructs which, following Barlow, Christie and others, he called 'resultants'. This signified that the curves were produced by geometrical construction or experimental manipulations. Aware that the lines suggested a physical interpretation Faraday emphasised until 1852 that the lines were 'mere representations' and that no theory was implied by his use of them.

43. These displays attracted people having a number of overlapping interests, for example: a practical interest in both the local and the global aspects of the phenomena (shared by mineralogists with interests in the implications for geological surveys and the Admiralty whose concern lay in the use of compasses for navigation); lecturers would be interested in the potential for new demonstrations; several would have theoretical concerns, say, to incorporate terrestrial magnetism into a general theory of magnetism (Faraday, Barlow) or to explain the effects of terrestrial currents on the formation of mineral veins (R.W. Fox), and finally, public interest would be enlisted by demonstrating analogies, e.g., between ordinary; terrestrial; and electromagnetism and between minute or local phenomena and global ones.

to the memory' of 'the relative positions of the electric current and the magnet' in Faraday's new law of electromagnetic induction. Christie had refereed Faraday's paper on induction for the *Philosophical Transactions*.[44] In his report he endorsed the geometrical representation Faraday used in his statement of the law of induction. Faraday introduced the term 'magnetic curves' in a footnote added to his first paper on electromagnetic induction.[45] The perceived need for a new mode of representation illustrates the close interaction of experimentation, its reportage, and practical and intellectual demonstration. Faraday introduced the lines to help others grasp the experimental conditions and the procedures necessary for the induction of currents by magnetism. Their usefulness to material and mental manipulation of phenomena persuaded Christie and Peter Mark Roget to work on the geometrical representation of magnetic curves.[46]

Faraday was interested in the heuristic potential of the curves. They suggested – at least implicitly – a way of assimilating two quite distinct types of activity (and two types of knowledge): the procedural, practical understanding of electromagnetism and the theoretical, abstract understanding of natural forces, based on their presumed identity. The examples considered here link the construction and use of electromagnets as a research tool to Faraday's search for a unified theoretical interpretation of the forces of Nature.

TECHNIQUES AS SOURCES FOR CONCEPTS: FARADAY'S LINES OF FORCE

Faraday is usually credited with inventing the lines of force as a bold new way of thinking about electric and magnetic properties. He did not invent them at all. As we have seen, he was not even the first to analyse ordinary magnetism in terms of 'curves' and he did not publish drawings of his lines until 1851, twenty-seven years after Barlow's had appeared.[47] The lines were drawn from other representations of Nature (mapping terrestrial fields, geometrical construction) mainly to provide a means of visualising and analysing magnetic action. Though magnetic curves were not 'given' in Nature, natural phenomena seemed to require them. Why was this? Their usefulness in other contexts of established importance would have made the curves an obvious and generally acceptable resource. Faraday followed Davy, Barlow and other contemporaries in discerning pattern or structure in iron filings spread

44. Royal Society MSS
45. Faraday (1839), §§ 114–16.
46. Roget (1831); Christie (1833).
47. Although they conformed to an accepted representational style Faraday was reluctant to 'publish' the lines before 1850. His reasons are discussed in Gooding (1980b, 1982b).

near magnets and wires. Lines emerged from these patterns, partly because lines were already available and a proven means of structuring sensory experience, and partly because of Faraday's particular interest in the explanatory potential of a link between vibrations and 'striations' (suggested by his work in the late 1820s on acoustical figures). Christie's suggestion brings out these aspects of the network of experimental practices and the interests motivating them.

The 'curves' were constructs, yet they were not merely artifacts of intellect or culture, borrowed and adapted to conform to metaphysical prejudices about action at a distance or the non-existence of 'skew' forces.[48] As for the heroic image of Faraday as the inventor of lines as a powerful new image of electricity and magnetism, it has to be said that the lines acquired this status only in the 1850s, largely through the efforts of William Thomson. By then their explanatory and theoretical potential had eclipsed their humble, procedural origins. Thus, to James Clerk Maxwell, Faraday provided a new way of thinking about phenomena whose existence was beyond doubt, rather than a way of producing or disseminating effects that might be explained as artifacts of a method of observation.

Why were the lines accepted by the London practitioners? A full answer calls for a more detailed look at the character and context of the science of this period than I can offer here. This would show how experimental practices were selected and sustained by a larger network of interests and values. It would also show how alternative representations failed (because they were impracticable) or were rejected (as unacceptable, on other grounds). Here I attempt the less ambitious task of identifying a few connections between the images of lines, the local laboratory context of use and the larger cultural context.

Where the action was: the electromagnet

Faraday recognised a connection between experimenting, representing and theorising. In a manuscript note of 1850 he remarked that magnetic lines or curves were like lines of latitude and longitude. This shows that he recognised their relationship to older techniques such as the mapping of geomagnetic and geothermal properties, even if he did not recall particular sources such as Barlow's 'curves' or his method of examining the effects of iron bodies on the terrestrial field.

---

48. Compare Ampère's reduction of electromagnetism to the interaction of a force and a couple acting between current-elements, his static style of experimentation in which null effects were sought, and his selection of 'primitive facts' different from Davy's and Faraday's.

Evidence that the mapping analogy was embedded in laboratory practices of the London electromagnetic network of the 1820s can be found in their work with electromagnets. Imagery suggested by their description of the effect of electromagnets reappears in concepts used later in the construction of Faraday's magnetic field theory. Faraday and his contemporaries were interested in analogies between electricity, magnetism, light, heat and other 'forces': this made analogies between lines and rays important to them. Ampère had shown that parallel currents interact magnetically. Sturgeon's investigations of electromagnets in the early 1820's gave prominence to the phenomenon (discovered by Arago) that soft iron increased the strength of the magnetism produced by a current passing through a coil.[49] This could be interpreted by analogy to the lateral interaction of currents discovered by Ampère. The lifting power of Sturgeon's electromagnet – a wire coil with a soft-iron core – could be expressed in two ways; in terms of the weight lifted by a magnet of a specified number of turns around a given core, or qualitatively and impressionistically. The device shown in Fig. 7.13, suspended from the tripod, is one of Sturgeon's electromagnetic lifting devices. The iron core was described as having a 'concentrating' effect on the magnetism produced by the current. This language evokes the imagery of lines being made to converge like rays of light. The effect of introducing an iron core between the poles of the horseshoe electromagnet could be described qualitatively, as a 'concentrating' action of the iron on the lines. Faraday later expressed this increased power in terms of an increase in the density of lines in the region of the poles (and an apparent increase in the number of lines in the surrounding space).

'Concentration' was a very rudimentary interpretation of a common effect whose significance so far remained instrumental rather than theoretical. The interpretation was expressed with imagery borrowed from other contexts (such as geomagnetism) and interpreted by analogy to other sciences (such as the wave theory of optics and heat). It was favoured, so to speak, by the importance of those contexts of exploration and of analogical links between lines and rays of light and heat. Wave theories were favoured throughout the first half of the nineteenth century because they suggested a way of unifying theories of electricity, magnetism, heat, light, and molecular forces.[50] Figure 7.5 is taken from a major step towards such a theory, Faraday's magnetic field theory of 1850. It shows 'concentrated' lines and those which have been dispersed

49. Williams (1983).
50. Gooding (1980a, 1981).

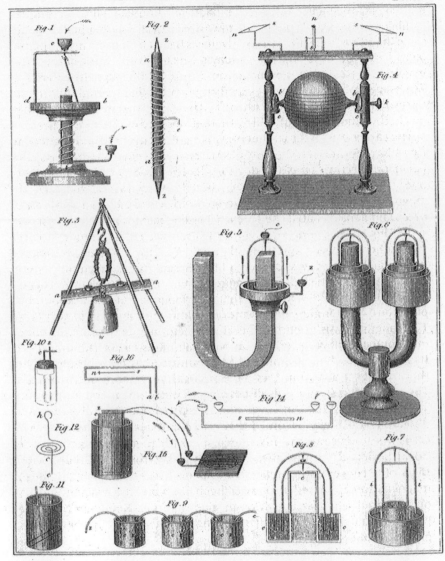

Figure 7.13. A set of demonstration apparatus including as 'fig. 3', one of Sturgeon's electromagnetic lifting devices.

or diverged. Although convergence is easily seen, the divergence of lines by diamagnets could not in fact be observed with techniques available to Faraday. This shows how 'theoretical' this fact had become since the 1820s when the effects on curves had been purely instrumental facts.[51] On the other hand, Faraday's image of the atmosphere as a magnetic lense, in his 1850 paper on atmospheric magnetism (see Fig. 7.6), suggests that he envisaged using world-wide variations of the terrestrial field as a global test of a theory of magnetism developed inside the laboratory, but which he had not been able to test there.

The earliest use of the rudimentary idea of 'concentrating' magnetism that I have found is in Christie's 1831 referee's report on Faraday's first paper on electromagnetic induction. In his report, Christie described one of Faraday's experiments in which a disc was rotated near the magnet, between the ends of 'two bars of iron which [he said] *concentrated* and approximated the poles'. Christie did not explain the expression. This suggests that the action of iron was already understood in terms of established experimental practices.[52] By 1831, electromagnets were widely used use as sources of localised, intense magnetic action. Barlow and Christie's use of 'magnetic curves' and Christie's 'concentrating' effect of iron indicate that the language of such laboratory procedures was understood in terms of lines by the end of the 1820s. This language was important because electromagnets were an important source of new experimental possibilities, as electrochemical cells had been since the beginning of the century. Methods of intensifying the field of an electromagnet were particularly important to Faraday. They were necessary to his discovery of electromagnetic induction in 1831 and to his 1845 discoveries of magneto-optic and diamagnetic phenomena. Faraday recorded that on 28 October 1831 he was 'at Mr. Christie's to-day making many expts. with the great magnet of the Royal Society'.[53] During the 1840s, Faraday wrote in his laboratory *Diary* of a 'concentrating action' of the coil and core upon the lines of magnetic force. Immediately after he had obtained a magnetic effect on polarised light passing through glass (on 13 September 1845) he observed that a 'mass of soft iron on the outside of the *heavy glass* greatly *diminished* the effect'.[54]

51. Gooding (1982a), pp. 50–5 and (1989).
52. Royal Society MSS.; cf. Faraday (1832), §§ 84.
53. Faraday (1932–36), 1, p. 380 and 4, §§ 7498, 7874–7. See also Williams (1965), p. 169ff. and Snelders (1984).
54. Faraday (1932–36), 4, § 7508.

Lines, 'the field' and economy

When Faraday used the term 'field' for the first time in 1845, he used it to describe the region of intense magnetic action between the poles of a magnet built on principles established empirically by William Sturgeon.[55] Moreover, his first use of the term 'field' as a *theoretical* concept, made in 1850, defined the field in terms of the lines of force. It is significant that on both occasions, the context was one in which the *instrumental* significance of both the 'concentrating' effect on lines and the analogy to the field of an optical instrument were of paramount importance. This shows that the practical fact did not lose its instrumental importance; rather, it acquired additional, theoretical significance as Faraday gradually constructed an evidential context capable of explaining it.[56]

The phenomenon of 'concentration' or 'convergence' is central to the general concept of magnetic conduction Faraday developed by 1850, on which his theory of the field depended: there Faraday was able to give a unified explanation of a wide range of magnetic phenomena in terms of a few basic laws of the behaviours of lines of induction. One of the most important of these is that the lines obey a principle of economy or least action. Figure 7.14 shows how lines 'converge' or are 'concentrated' in the region of good magnetic conductors (ferromagnets) and are dispersed or 'diverged' by poorer ones (diamagnetics). This is because they seek pathways of greatest conductivity or least resistance. Experiments of 1847 and 1848 had enabled Faraday to link the optical properties of certain crystals to their magnetic properties through this same concept of magnetic conductivity for lines of force.[57] The explanatory power of this theory derives from the fact that the lines permit the motions of *different* kinds of matter in the field to be seen as due to the *same law* of magnetic action. This can be read from the changing patterns of lines corresponding to those motions. The behaviour of lines (as observed with iron filing patterns or through the movements of both crystalline and amorphous 'conductors') therefore obeys or 'respects' a principle of economy in Nature. Economy was a Victorian virtue which Nature should of course express even in the the minutest detail (such as the movement of a tiny crystal in the field).

55. Ibid. § 7979.
56. Using Pinch's terminology (cited in note 14) we could say that Faraday moved observation reports having low externality into a newly constructed evidential context which gave them a higher degree of externality (Pinch 1985).
57. For the work on crystals see Gooding (1981), pp. 253–63.

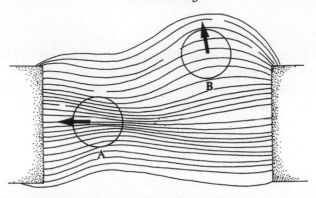

Figure 7.14. A simplified view of the relationship between matter's conductivity
for lines of force and its tendency to move in the field, according to the theory
Faraday had completed by the end of 1850. Sphere *A* is paramagnetic and *B*
is diamagnetic. Lines converge near and within *A* because it has greater
conductivity for magnetism than the surrounding medium between the poles.
They diverge near *B* because its conductivity is less than that of the medium.
(See Gooding, 1981)

Procedures and reality

Faraday's religious beliefs and allied metaphysics made forces ontolog-
ically prior to everything else. For him it was also essential to be able
to demonstrate the obedience of forces to God's will.[58] In their fully
articulate form the lines made this possible. They had the additional
advantage of being – or seeming to be – necessary to the very existence
of the phenomena his theory could now explain. We are now in a
position to interpret a statement Faraday made in 1850 about the epistemological significance of his own experimental methods. In his first
major paper on the new magnetic field theory he announced a longstanding presupposition of his experiments on magnetic and diamagnetic materials. His 'differential method of observation' of magnetic properties, so named in 1850, depended on the possibility of detecting
differences in the magnetic influence on different substances in a
magnetic field. This method of observation required *lines* because Faraday had represented the effect in terms of changes in the density of
lines. The 'converging' of magnetic lines by the 'concentrating action'
of iron made articulate a practical understanding of the experimental
techniques of the 1820s. Faraday now claimed that it was impossible
to interpret his results in any other way. He stated emphatically that

58. See Gooding (1980a); Cantor (1985).

'no other method [of representation] could be used with the differential system of observation'.[59]

After over twenty years he had articulated a radical implication of experimental practices that originated in the method of 'concentrating' magnetic action – this was the need to represent *matter* in terms of the space it occupies, that is, in terms of its effect on the distribution of lines of magnetic (or other) induction. This is more than a denial of the relevance of familiar parameters such as weight: Faraday was about to reverse the physical priority of matter over force, by arguing that matter should be construed in terms of its effect on lines of force, as a spatially distributed set of properties.

THE SIGNIFICANCE OF EXPERIMENT

I have been concerned with experiment as an influence on the imagery scientists use, first to construct experience and then to theorise that experience. The importance of experimental practices to concept-formation has been argued, for example, by operationalists. However, they treated operations merely as enabling scientists to read phenomena off Nature (rather in the way that a lamp enables us to read a printed text after dark). As Nickles points out in chapter 10, empirical studies of science make untenable the assumption that phenomena come ready made to be 'read off' Nature. Practices are sometimes constitutive of phenomena: reading the 'Book of Nature' involves having read the appropriate chapter of the book of culture or – as I have argued here – writing new skills into that book. Several historians – including three other contributors to this volume – have argued the importance of practices to the phenomena eventually disclosed as real by a particular style of science.[60] The examples discussed in this chapter draw attention to the mediating role of representations. I have argued that what scientists do when 'intervening' shapes how they represent the outcomes of their interventions. Successful representations in turn influence how they theorise those outcomes. Experimental practices are sometimes essential to the construction of ways of experiencing and theorising new phenomenal domains, although they are not the only resource for such constructs.

Since experiments enable scientists to make sense of previously unknown bits of Nature, I suggest that we identify this role as the

59. Faraday (1851a), § 2804.
60. Galison & Assmus (chapter 8, this volume); Hackmann (1980, and Chapter 1, this volume); Price (1980); Heilbron (1981); Hacking (1983); Pickering (1984b).

cognitive significance of experiment. This cognitive, constructive role needs to be distinguished from the more familiar epistemological uses of experiment since (as Nickles points out in chapter 10) it has been largely eclipsed by an emphasis on the epistemological role of experiment, i.e., its use to decide the truth or falsity of empirical claims derived from competing theories. Since the seventeenth century experiment has acquired an important epistemological role as a means of proving or deciding between theories. Studies of the context of the emergence of experiment in the sixteenth and seventeeth centuries show that its epistemologically privileged role owed a great deal to the social status of the people who conducted experiments and to the places in which experiments were performed.[61] I contend that the epistemological privilege of experiment depends *also* on a cognitive role, in which experiment is used to make practical sense of some aspect of the natural world. Whereas my examples emphasise the early stages of the process of articulating reality, Pickering (chapter 9) examines learning in the later stages of testing hypotheses. However, both chapters deal with the same sort of learning process. This process has been overlooked because scientists reconstruct their experimentation when incorporating its outcomes into the arguments which make them relevant to theories.

Recovering construction

The process of reconstruction makes natural phenomena – which have been accessible only through human construction and intervention – come to be accepted as independent of that activity. This is illustrated by the later development and use of Faraday's lines. Over a period of time his experiments made phenomena appear as objectively real, as if given in Nature. It is likely that, notwithstanding his arguments for the enabling role of his 'differential method of observation', Faraday became so accomplished with many of his experimental practices that he lost sight of them as essential to the production of the experiences he described. As the constructive and enabling practices were mastered, they could be dropped from accounts of how to produce and see natural phenomena. Most of the practice became invisible and the experimentation became transparent: only the phenomena remained. The residue of phenomena thus came to appear as objects independent of human intervention.[62]

61. This is argued in Shapin (1988).
62. See Gooding (1982a). The process of reifying entities by rendering transparent the means of producing them was described by Fleck in 1935 (see Fleck, 1979) and more recently by Latour & Woolgar (1979).

This invites two questions: why do the procedures and instruments recede into the background (or disappear entirely) rather than the phenomena?[63] For an answer we would have to determine which interests were overriding ones at any particular time. For example, Faraday had a long-term interest in developing a unified and comprehensive theory. However this was often overridden by more immediate demands of practical problem-solving. Though he devoted much of his time to developing skills, he himself emphasised – and was perceived as having left – a residue of phenomena, facts, laws and ideas rather than a residue of devices and techniques. By contrast, an inventor or engineer, whose interests were primarily in technological applications would make the outcomes secondary to techniques and the design of artifacts. The second question is, why do the instrumentalities disappear just *when* they do?[64] Here we need to recognise the emergence of two very different sorts of thing: on the one hand, a body of effective and learnable *skills*, and on the other, an *explanation* with its evidential context. In science the interest in explanation directs attention increasingly towards phenomena and laws (the main things) and away from human agency and skills (which merely enable). This is largely true of science education as well. Recognising this fact may help dissolve the paradox of transparency Nickles identifies in Chapter 10 of this volume.

Faraday began to lose sight of the means of producing phenomena because his overriding objective was to get a better correspondence between concepts, the facts they denote and the intellectual and moral precepts they express. The quality of this correspondence depended on how well central concepts (lines of force) expressed properties that his deep assumptions (economy, unity, obedience) attributed to Nature. Faraday believed that, as the author of Nature, God and not man is the source of natural phenomena. Although the correspondence of theory to Nature depended on practices described at great length in his *Experimental Researches*, the practices were not the object of the account and so they fell out of the explanation of what was produced: Faraday's interests excluded them.

63. Adapting Polanyi's distinction between focal and peripheral awareness, we can say that concern with the significance of phenomena as evidence for theoretical constructions shifts scientists' awareness from practices to the phenomena they produce, so that awareness of enabling techniques remains at best only peripheral (Polanyi 1964), pp. x, 55–7, 161–3.

64. This aspect of the micro-structure of discovery and construction is analogous to Latour's observation that the facticity of objectivity of a natural fact increases as statements about it become detatched from a particular context of discovery or proof (i.e., as its 'modalities' disappear), Latour & Woolgar (1979), pp. 151 ff., 236. In Latour (1987) this is extended to the 'black boxing' of expertise.

CONCLUSION

I have argued that practices and the experiences they produced were an important source for *and* influence upon the early development of the magnetic field concept. Procedures lost visibility during the 1830s and 1840s, as theory developed. Phenomena appeared increasingly independent of the practices which elicited them. This process led from the earlier, *cognitively* significant experiments to the later, *epistemologically* significant ones. Faraday made use of this fact in arguments for the physical reality of lines of force after 1850. These arguments reconstruct the process of discovery. They select *phenomena* as evidence for concepts and ignore *practices*.[65] Faraday drew attention to practices only where he could argue that the very existence of the phenomena must be interpreted in terms of concepts that also make sense of how we produce and experience those phenomena.[66] A well-known example is his argument that the induction of a current in a closed circuit moving across the lines of force can be explained only in terms of the prior existence of magnetism, occupying the space through which the wire is moved.[67]

I have situated theoretical concepts and arguments in a network of practices and have identified these with a group of people who actively explored electromagnetic phenomena during the 1820s. Because the results of their actions were not accepted as natural phenomena unless they succeeded in communicating the outcomes to each other, experimentation and representation were inseparable in the early stages of the discovery process. The pragmatic connection between instrumental and social objectives explains why experimentation influences representation. This influence makes experiments important to cognition and conceptualisation.

This fact should affect the way we write the history of theories. For example, if 'having a concept' requires implicit understanding based in procedures (including the use of lines in exploratory and representational practices) then a concept was certainly in use during the 1820s. However, if 'having a field concept' requires a *verbally* articulated expres-

65. The selection of the phenomenal 'wheat' from the instrumental 'chaff' must be distinguished from another form of selectivity, that of selecting aspects of a phenomenon for attention. For a general theory of the latter see Fleck (1979), esp. p. 125 ff. Selectivity of both kinds becomes evident during controversies, when particular selections and interpretations are challenged. For Faraday and Ampère's controversy about which electromagnetic phenomena were primary, see Williams (1985).
66. On the reconstruction of experiments in the context of argumentation see Nickles (1985); Cantor (chapter 6 this volume).
67. Faraday (1852a,b).

sion, then Faraday did not 'have' a field concept of magnetism until well into the 1840s.[68] Faraday's theory of the magnetic field owed much to active experimentation that is largely ignored in the histories of field theory. If the cognitive role of experimentation and representation is as important as I have claimed, then we should not be surprised that Faraday's later scientific work articulated ideas implicit in experimental procedures established during the 1820s.

ACKNOWLEDGEMENTS

This chapter is based on talks given to symposia at the Museum of the History of Science, Oxford and St. John's College, Annapolis; portions were also read at the 1985 International Congress of the History of Science at Berkeley and to the conference on *The Uses of Experiment* at Bath; later versions were read at the University of Virginia and at Princeton University in 1986. I am indebted to participants and discussants for their comments and I thank Trevor Pinch and Simon Schaffer for their suggestions. Archive work on the London 'network' was made possible by a grant from the Royal Society of London.

References

Barlow, P. (1820). *An Essay on Magnetic Attractions*, London. 2nd edn. London, 1823.
Barlow, P. (1822). Notice respecting Mr. Barlow's discovery of the mathematical laws of electro-magnetism. *Edinburgh Philosophical Journal*, 7, 281–3.
Barnes, B. (1983). Social life as bootstrapped induction. *Sociology*, 17, 524–45.
Brewster, D. (1821). Account of discoveries of Oersted, Davy, Ampère and Biot. *Edinburgh Philosophical Journal*, 4, 167–75, 406–16, 435; 5, 301, 352–5, 391–2; 6, 83–5, 178–9, 220–4, 281–3.
Caneva, K.L. (1978). From galvanism to electrodynamics. *Historical Studies in Physical Sciences*, 9, 63–159.
Caneva, K.L. (1981). What should we do with the monster?. In *Sciences and Cultures*, ed. K. Mendelson & Y. Elkana. Dortrecht: Reidel.
Cannon, S.F. (1978). *Science in Culture: the Early Victorian Period*. New York: Science History Publications; Folkestone: Dawson.
Cantor, G.N. (1985). Reading the Book of Nature: the relation between Faraday's religion and his science. In ed. D.C. Gooding & F. James, pp. 69–82.
Cawood, J. (1979). The magnetic crusade: science and politics in early Victorian Britain. *Isis*, 70, 493–518.

68. For a discussion of this issue see Nersessian (1985).

Christie, S.H. (1833). Experimental determination of the laws of magneto-electric induction. *Philosophical Transactions*, **123**, 95–142.

Collins, H.M. (1981). The place of the 'core set' in modern science. *History of Science*, **19**, 6–19.

Collins, H.M. (1985). *Changing Order: Replication and Induction in Scientific Practice*. London and Beverly Hills: Sage.

Davy, H. (1821a). On the magnetic phenomena produced by electricity. *Philosophical Transactions*, **3**, 7–19.

Davy, H. (1821b). Farther researches on the magnetic phenomena produced by electricity. *Philosophical Transactions*, **3**, 425–39.

Faraday, M. (1821). On some new electro-magnetical motions, and on the theory of magnetism. *Quarterly Journal of Science*, **12**, 74–96. Reprinted in Faraday (1844), pp. 127–47.

Faraday, M. (1821–22). Historical sketch of electromagnetism. *Annals of Philosophy*, **2**, 195–200, 274–90; **3**, 107–21.

Faraday, M. (1839, 1844, 1855). *Experimental Researches in Electricity*, vol. 1, (1839); vol. 2 (1844); vol. 3 (1855). London.

Faraday, M. (1851a). Experimental researches in electricity, twenty-sixth series, In Faraday (1855), pp. 200–73.

Faraday, M. (1851b). Experimental researches in electricity, twenty-seventh series. In Faraday (1855), pp. 274–322.

Faraday, M. (1852a). Experimental researches in electricity, twenty-eighth series. In Faraday (1855), pp. 328–70.

Faraday, M. (1852b) On the physical character of the lines of magnetic force. In Faraday (1855), pp. 407–43.

Faraday, M. (1932–36). *Faraday's Diary*, 7 vols. ed. T. Martin. London: Bell.

Fleck, L. (1979). *Genesis and Development of a Scientific Fact*, trans. T.J. Trenn & F. Bradley. Chicago and London: Chicago University Press.

Galison, P. (1988). *Image and Logic*. Chicago: Chicago University Press (in press).

Gooding, D.C. (1978). Conceptual and experimental bases of Faraday's denial of electrostatic action at a distance. *Studies in the History and Philosophy of Science*, **9**, 117–49.

Gooding, D.C. (1980a). Metaphysics versus measurement. *Annals of Science*, **37**, 1–29.

Gooding, D.C. (1980b). Faraday, Thompson and the concept of the magnetic field. *British Journal for the History of Science*, **44**, 91–120.

Gooding, D.C. (1981). Final steps to the field theory. *Historical Studies in Physical Science*, **11**, 231–75.

Gooding, D.C. (1982a). Empiricism in practice: teleology, economy and observation in Faraday's physics. *Isis*, **73**, 46–67.

Gooding, D.C. (1982b). A convergence of opinion on the divergence of lines. *Notes and Records of the Royal Society*, **36**, 243–59.

Gooding, D.C. (1985a). 'He who proves, discovers': John Herschel, William Pepys and the Faraday effect. *Notes and Records of the Royal Society*, **39**, 229–44.

Gooding, D.C. (1985b). 'In nature's school': Faraday as an experimentalist. In ed. D.C. Gooding & F. James (1985), pp. 105–35.

Gooding, D.C. (1985c). Experiment and concept formation in electromagnetic science and technology in England in the 1820s. *History and Technology*, 2, 151–76.

Gooding, D.C. (1986). How do scientists reach agreement about novel observations?. *Studies in the History and Philosophy of Science*, 17, 205–30.

Gooding, D.C. (1989). *The Making of Meaning*. Dordrecht: Martinus Nijhoff (in press).

Gooding, D.C. & James, F. (ed.) (1985). *Faraday Rediscovered: Essays on the Life and Work of Michael Faraday, 1791–1867*. London: Macmillan.

Hacking, I. (1983). *Representing and Intervening*. Cambridge: Cambridge University Press.

Hackmann, W.D. (1980). The relationship between concepts and instrument design in eighteenth century experimental science. *Annals of Science*, 36, 205–44.

Hansteen, C. (1821a). Remarks on Professor Hansteen's 'Inquiries'. *Edinburgh Philosophical Journal*, 3, 124–38.

Hansteen, C. (1821b). Account of the recent magnetical discoveries of Professor Hansteen. *Edinburgh Philosophical Journal*, 4, 114–16, 295–300.

Hansteen, C. (1827). Notice respecting Professor Hansteen's new chart of the isodynamic lines for the whole magnetic intensity, with a chart. *Edinburgh Journal of Science*, 7, 351–3.

Hatchett, C. (1821). On the electromagnetic experiments of M.M. Oersted and Ampère. *Philosophical Magazine*, 57, 40–9.

Hays, J.N. (1983). The London lecturing empire, 1800–1850. In *Metropolis and Province, Science in British Culture, 1780–1950*, ed. I. Inkster & J. Morrell. London: Hutchinson.

Heilbron, J. (1981). The electric field before Faraday. In *Conceptions of Ether*, ed. G. Cantor & J. Hodges, pp. 187–213. Cambridge: Cambridge University Press.

Heimann, P.M. (1970) Maxwell and the modes of consistent representation. *Archive for History of Exact Sciences*, 6, 171–213.

Heimann, P.M. (1971). Faraday's theories of matter and electricity. *British Journal of the History of Science*, 5, 235–57.

Hesse, M.B. (1980). *Revolutions and Reconstructions in the Philosophy of Science*. Hassocks: Harvester Press.

Hoffman, J.R. (1987). Ampère's invention of equilibrium apparatus: a response to experimental anomaly. *British Journal of the History of Science*, 20, 309–41.

Hudson, D. & Luckhurst, K. (1954). *The Royal Society of Arts, 1754–1954*. London: Murray.

Humboldt, A. von (1820–21). On isothermal lines and the distribution of heat over the globe. *Edinburgh Philosophical Journal*, 3, 1–20; 4, 23–37, 262–81; 5, 28–38.

Humboldt, A. von & Biot, J.B. (1805). On the variations of terrestrial magnetism at different latitudes. *Philosophical Magazine*, **22**, 248–57, 299–308.

Kuhn, T.S. (1977). *The Essential Tension*. Chicago: Chicago University Press.

Latour, B. (1983). Give me a laboratory and I will raise the world. In *Science Observed*, ed. K.C. Knorr & M. Mulkay, pp. 141–70. London: Sage.

Latour, B. (1987) *Science in Action*. Milton Keynes: Open University Press.

Latour, B. & Woolgar, S. (1979). *Laboratory Life. The Social Construction of Scientific Facts*. London and Beverly Hills: Sage.

Lynch, M. (1985). Discipline and the material form of scientific images: an analysis of scientific visibility. *Social Studies in Science*, **15**, 37–66.

Morrell, J.B. (1971). Individualism and the structure of British science in 1830. *Historical Studies in Physical Sciences*, **3**, 183–204.

Morrell, J.B. & Thackray, A. (1981). *Gentlemen of Science*. Chicago: Chicago University Press.

Nersessian, N. (1985). Faraday's field concept. In ed. D.C. Gooding & F. James (1985), pp. 175–87.

Nickles, T. (1985). Beyond divorce: current status of the discovery debate. *Philosophy of Science*, **52**, 177–207.

Pickering, A. (1984a). Against putting the phenomena first: the discovery of the weak neutral current. *Studies in History and Philosophy of Science*, **15**, 85–117.

Pickering, A. (1984b). *Constructing Quarks. A Sociological History of Particle Physics*. Chicago and London: Chicago University Press.

Pinch, T. (1985). Towards an analysis of scientific observation: the externality and evidential significance of observation reports in physics. *Social Studies of Science*, **15**, 3–36.

Polanyi, M. (1964). *Personal Knowledge*. New York: Harper Torchbooks,

Price, D.J. (1980). Philosophical mechanism and mechanical philosophy: some notes towards a philosophy of scientific instruments. *Anneli dell'Institute e Museo di Storia della Scienzia di Firenze*. **5**, 75–85.

Reingold, N., Pearson, S. & Molella, A. (ed.) (1979). *The Papers of Joseph Henry, vol. 3: January 1836–December 1837, The Princeton Years*. Washington: Smithsonian Institution Press.

Rescher, N. (1979).*Cognitive Systematization*. Oxford: Blackwell.

Roget, P.M. (1831). On the geometric properties of the magnetic curve. *Journal of the Royal Institution*, **1**, 311–19.

Ross, S. (1965). The search for magnetic induction. *Notes and Records of the Royal Society*, **20**, 184–219.

Secord, J. (1982). 'King of Siluria': Roderick Murchison and the imperial theme in 19th century British geology. *Victorian Studies*, **25**, 413–42.

Shapin, S. (1988). The house of experiment in seventeenth-century England. *Isis*, **79** (in press).

Shapin, S. & Schaffer, S. (1985). *Leviathan and the air-pump. Hobbes, Boyle and the Experimental Life*. Princeton: Princeton University Press.

Snelders, H.A. (1984). The electromagnetic experiments of the Utrecht physicist Gerrit Moll. *Annals of Science*, **41**, 35–57.

Sturgeon, W. (1826). Account of an improved electro-magnetic apparatus. *Annals of Philosophy*, **12**, 357–61.

Tatum, J. (1821). Electro-magnetic experiments by Mr. J. Tatum. *Philosophical Magazine*, **57,** 446–7.

Thompson, S.P. (1891). *William Sturgeon, the Electrician: a Biographical Note.* published privately.

Tweney, R. (1985). Faraday's discovery of induction: a cognitive approach. In ed. D.C. Gooding & F. James (1985), pp. 189–209.

Williams, L.P. (1965). *Michael Faraday.* London: Chapman & Hall.

Williams, L.P. (1966). *The Origins of Field Theory.* N.Y.: Harper & Row.

Williams, L.P. (1983). What were Ampère's earliest discoveries in electrodynamics? *Isis,* **74,** 292–508.

Williams, L.P. (1985). Faraday and Ampère: a critical dialogue. In ed. D.C. Gooding & F. James (1985), pp. 83–104.

Wood, A. & Oldham, F. (1954). *Thomas Young, Natural Philosopher, 1773–1829.* Cambridge: Cambridge University Press.

Figure 8.1. Ben Nevis Observatory in 1884. Wilson worked here in autumn 1894 and summer 1895, inspiring him to mimic meteorological phenomena in the laboratory (Cambridge University Library).

8

ARTIFICIAL CLOUDS,
REAL PARTICLES

PETER GALISON AND ALEXI ASSMUS

INTRODUCTION

Most twentieth century physicists would describe the cloud chamber as the earliest particle detector, a device which allowed scientists to get their first glimpse of the interactions of elementary particles. The chamber revealed the positron and the muon to Carl Anderson and allowed George Rochester and G.C. Butler to 'see' a new class of 'strange' particles. John Cockcroft and Ernest Walton used the device to demonstrate the existence of nuclear transmutation. Indeed, for generations of cosmic-ray physicists, and then briefly for accelerator physicists, the cloud chamber gave concrete meaning to the many newly discovered particles that inaugurated high-energy physics.

Cloud chambers served as the prototypes for later detectors, including the high-pressure chamber, sensitive nuclear emulsions and, most importantly, the bubble chamber. Scientists were quick to realise how essential the chamber was for modern physics – J.J. Thomson believed the chamber was 'of inestimable value to the progress of science'[1] and Lord Rutherford ranked it as 'the most original and wonderful instrument in scientific history'.[2]

But the creator of the cloud chamber, Charles Thomson·Rees Wilson (1869–1959),[3] cannot possibly be considered a particle physicist. From

1. Thomson (1937), p. 420.
2. Biography accompanying C.T.R. Wilson (1965), pp. 215–17, on page 216; see also a similar remark by Rutherford in the London *Times* Obituary for C.T.R. Wilson, November 16, 1959, p. 16.
3. There are only a few articles written about C.T.R. Wilson. Most complete is Blackett (1960); G.L'E. Turner has a short article in Gillispie (1981), pp. 420–2. Wilson himself left two retrospective articles, C.T.R. Wilson (1954) and C.T.R. Wilson (1960). His Nobel Prize Lecture, in 1927, gives a general overview of his work: C.T.R. Wilson (1965). In addition, Wilson wrote chapter seven, '1899–1902', of History (1910)

his earliest work in 1895 to his last ruminations on thunderclouds when he had turned ninety, Wilson was riveted by the phenomena of weather. Even J.J. Thomson, one of his greatest admirers, commented that Wilson's experimentation on fogs was not 'a very obvious method of approaching transcendental physics'.[4] By 'transcendental physics' Thomson meant analytical research into the basic structure of matter that shunned complex but mundane problems like fog and rain. Though the eventual use of the cloud chamber has led particle physicists to appropriate Wilson as one of their own, his life's work is incomprehensible outside the context of weather. One *must* come to terms with the dust, air, fogs, clouds, rain, thunder, lightning, and optical effects[5] that held the rapt attention of Wilson and his nineteenth-century contemporaries in order for the invention of the cloud chamber to make historical sense.

Our purpose is to resurrect mundane as well as transcendental physics and thus set Wilson's cloud-chamber work in its properly Victorian context. By doing so we will find a startling source of modern physics and exhibit a coherence in Wilson's work that is entirely lost when it is divided into 'meteorology' on the one hand and 'fundamental physics' on the other. Between, or perhaps overlaying the two domains lies an area we will identify as 'mimetic experimentation', a term that will

   Crowther discussed Wilson in two of his books, but without much detail: Crowther (1968), chapter 1, pp. 25–55, and Crowther (1974), chapters 9, 10, 11, 12, 16, pp. 126–75, 213–24. Tomas (1979), uses art-historical techniques to fix the authorship of certain chambers; he does not use Wilson's notebooks. For a modern explanation of the cloud chamber see J.G. Wilson (1951). Meteorologists recall Wilson as an important figure in their discipline, see e.g., Halliday (1970). The most important sources for the cloud chamber are Wilson's notebooks (note 38).

4. Thomson (1937), p.416. This language was not unique to Thomson, as is evident when Airy wrote Stokes in April 1867:: [The Mathematical Tripos] 'gave no adequate encouragement or assistance to men of the second degree, who might by proper direction of their studies become men highly educated as philosophers though not so transcendent as the first'. Cited in David B. Wilson (1982), p. 338.

5. Several different meteorological optical effects will be mentioned in this paper: A *corona* is a series of concentric coloured rings formed by the diffraction of light from the Sun or Moon when it hits water droplets in clouds or high fogs. A *halo* is also a series of coloured rings, produced, however, by refraction through ice crystals. These terms are defined by meteorologists, but are often used inconsistently. A *glory* consists of coloured rings seen around the shadow cast by an observer on a cloud or a fogbank. It is caused by the diffraction of light from water droplets and can occur with a corona. The entire phenomenon of the shadow and rings is sometimes called a *Brocken Spectre*. A *Bishop's Ring* is not due to water droplets, but instead is due to the interaction of light with solid particles in the atmosphere. Bishop's Rings were often seen after the Krakatoa eruption of 1883, but are very rare in normal circumstances; they depend on a large number of particles in the atmosphere and therefore are usually seen after catastrophic explosions, F. Whipple, 'Meteorological Optics', in Glazebrook (1923), 3, pp. 518–33, esp. 527–33.

designate the attempt to *reproduce* natural physical phenomena, with all their complexity, in the laboratory.

It is in this mimetic tradition that Wilson's work begins, and the evolution of his thought and his machines must be understood as a continuing dialogue between general theories of matter held at the Cavendish Laboratory and particular demonstrations of remarkable natural phenomena. The Wilson cloud chamber is the material embodiment of the confluence of these two traditions: the analytic and the mimetic. Exploring the origin of the instrument will offer an insight into the intersection and subsequent transformation of the material cultures of both meteorology and matter theory. By focusing on the history of the cloud chamber we will be able to study the formation and disintegration of 'condensation physics', a transitory sub-field of physics. At the same time, an exploration of the construction and deployment of the cloud chamber will deepen our historical understanding of both particle physics and physical meteorology.

THE VICTORIAN CONTEXT OF MIMETIC EXPERIMENTATION

The extremities and rarities of Nature held an endless fascination for the Victorian imagination. Explorers ventured to the ends of the Empire, to the deserts, jungles and icecaps. Painters and poets tried to capture the power of storms and the grand scale of forests, cliffs and waterfalls. Both artists and scientists recognised a tension between the rationalising, law-like image of Nature preferred by the natural philosopher, and the irreducible, often spiritual aspect presented by artists.[6]

There was a similar split in science itself between an abstract law-seeking, often mechanical reductionist approach to the physical world and a natural–historical approach that authors from Goethe to Maxwell had dubbed the 'morphological' sciences.[7] Of these sciences, Goethe took particular joy in meteorology, for 'atmospheric phenomena can never become strange and remote to the poet's or to the painter's eye'.[8] Yet until Goethe was nearly seventy, there had been no systematic

6. Much of the debate was fueled by reforms in scientific instruction. See R.H. Super, 'The Humanist at Bay: The Arnold-Huxley Debate', in Knoepflmacher & Tennyson (1977), pp. 231–45.

7. Maxwell (1885), p. 2: 'What is commonly called "physical science" occupies a position intermediate between the abstract sciences of arithmetic, algebra, and geometry and the morphological and biological sciences'. Goethe, cited in Merz (1965), 2, pp. 212–13. 'Morphology' soon acquires other meanings (on Haeckel see Mayr (1982), but the split between those interested in phenomena for their own sake and those who want to use phenomena to get at abstract explanatory laws persisted.

8. Goethe, cited in Badt (1950), p. 17.

classification of clouds. Then, in 1802–03 a British chemist, Luke Howard, presented a classification system that he modelled on Linnaean taxonomy.[9] Through Goethe, Howard's system entered the cultural mainstream.

Addressing a small philosophical society, Howard sorted clouds with a 'methodical' nomenclature: 'cirrus', 'cumulus', and 'stratus'. Howard chose Latin for his system, in contrast to chemists, who used Greek to label invisible chemical entities. Howard wanted to classify clouds 'by [their] visible characters, as in natural history'.[10] No doubt this affinity with natural history appealed to Goethe. When he discovered Howard in 1815, the poet hailed the new way of seeing clouds: 'I seized on Howard's terminology with joy,' Goethe announced, for it provided 'a missing thread'.[11] From Goethe, the Dresden school of painters learned to view clouds differently. The art historian Kurt Badt surmises that it was Luke Howard's expanded opus of 1818–20 that triggered John Constable's astonishing cloud studies of 1821–22.[12] Meteorology thus fostered cloud studies in painting, in poetry, and later in photography. Clouds became a central figure of romantic thought.

Morphological and abstract science

For Luke Howard the study of clouds was much more what Goethe would call 'morphological' than 'abstract'. Howard never felt at ease with mathematics or the newer, more mathematised forms of chemistry.[13] At the end of the century the historian Theodore Merz commented on this dual aspect of systematic thought; he stressed that the 'abstract sciences' (e.g., optics, mechanics, electricity and magnetism) either involved 'literally a process of removal from one place to another, from the great work- and store-house of nature herself, to the small workroom, the laboratory of the experimenter'; or a process of removal 'carried on merely in the realm of contemplation'.[14] The morphological sciences, by contrast, sought phenomena as they occurred in Nature itself: clouds, currents, flora, and cliffs.

According to many matter physicists, abstraction leading to general laws – not characterisation of the 'countenance of the sky' or the Earth – was the true goal of physics. William Thomson, for one, left natural history a merely preliminary role:

9. Howard (1803).   10. Ibid., I, p. 98.   11. Badt (1950), p. 18.   12. Ibid., esp. Chpt. 6.
13. Howard's autobiographical account, 'Luke Howard on Goethe [1822]' in Goethe
    (1960), pp. 823–32.
14. Merz (1965), p. 200.

in the study of external nature, the first stage is the description and classification of facts observed with reference to the various kinds of matter . . . this is the legitimate work of *Natural History*. The establishment of general laws in any province of the material world, by induction from facts collected in natural history, may with like propriety be called *Natural Philosophy*.

Moreover, because Thomson set natural history on a lower rung, if natural history and natural philosophy ever clashed he thought it the worse for natural history.[15]

For the 'abstract' or 'natural philosophical' investigator, the goal of experimentation was to *extract* the universal law from the particular description – and therefore to achieve Thomson's 'transcendental' truth. The success of the natural philosophical approach was present for all to see in Maxwell's dynamical theory of electrodynamics and Kelvin's dynamical theory of heat. Both took systems of known phenomena and gave them a mechanical, dynamical basis not linked to any single phenomenon.

But scientists at the time realised that despite the manifold benefits of the 'chemical and electrical laboratories with the calculating room of the mathematician on the one side, and the workshop and factory on the other',[16] there were times when the 'abstract' scientific method was inadequate. In 1887 Franz Boas, for instance, vigorously defended the 'study of phenomena for their own sake', as in no way inferior to the physicists' 'deduction of laws'. Citing Goethe, Boas contended that the two kinds of thought satisfied different and complementary 'desires of the human mind': a desire to attend to regularity and a desire to study Nature bit by bit, 'lovingly try[ing] to penetrate into its secrets until every feature is plain and clear'. Thus the physicist risked losing 'sight of the single facts [seeing] only the beautiful order of the world'.[17]

Soon afterwards, Merz captured a like sentiment when he wrote that natural philosophers who exclusively exploited analysis and abstraction were often 'forcibly reminded that he [was] in danger of dealing not with natural, but with artificial things. Instances are plentiful where, through the elaboration of fanciful theories, the connection with the real world has been lost . . . '[18] When natural philosophers invented a dynamical basis for phenomena, they risked inferring from the success of a model to a spurious postulation of hypothetical entities.

15. Thomson cited in Wise & Smith (1988). Maxwell (1885), pp. 2–3 makes a virtually identical division. On the placing of natural history on a lower rung of the sciences see Brush (1978).
16. Merz (1965), p. 202.　　17. Boas (1887).　　18. Merz (1965), p. 201.

Opposing the 'one-sided' working of abstract science lay another ideal of investigation, embodied in the morphological sciences. These were motivated, as Merz rhapsodically put it, by

the genuine love of nature, the consciousness that we lose all power if, to any great extent, we sever or weaken that connection which ties us to the world as it is – to things real and natural: it finds its expression in the ancient legend of the mighty giant [Antaeus] who derived all his strength from his mother earth and collapsed if severed from her.[19]

The morphological scientist, rather than searching for general laws 'look[s] upon real things not as examples of the general and universal, but as alone possessed of that mysterious something which distinguishes the real and actual from the possible and artificial'.[20] Explaining that he is borrowing and extending Goethe's term, Merz included the large-scale study of landscape – mountains and valleys, glaciers, land and water, stratification of rocks and formation of clouds – under the rubric of morphological sciences.[21] Alexander von Humboldt, the great explorer and measurer of natural phenomena like wind and air pressure, was in Merz's eyes 'the morphologist of nature on the largest scale'.[22]

Susan Faye Cannon thus echoes Merz when she identifies the 'great new thing in professional science in the first half of the 19th century' as 'Humboldtian science, the accurate, measured study of widespread but interconnected real phenomena' such as 'geographical distribution' terrestrial magnetism, meteorology, hydrology, ocean currents, the structures of mountain-chains and the orientation of strata, [and] solar radiation'. By their commitment to work on location, the Humboldtians tians opposed 'the study of nature in the laboratory or the perfection of differential equations'.[23] The laboratory dealt with artificially isolated phenomena – not Nature – and differential equations disembodied the variegated reality of actual things. Though there was a long tradition of geological experimentation, many geologists were suspicious of the laboratory's artificiality. Mott Greene reports that two of the most prominent early nineteenth-century geologists were 'resistant to the idea that laboratory testing could recreate or duplicate natural conditions'.[24] According to Martin Rudwick, fieldwork (not laboratory work) was 'the mark of the true [nineteenth-century] geologist; its sometimes

19. Ibid., p. 202.   20. Ibid., p. 203.   21. Ibid., p. 219.   22. Ibid., p. 226.
23. Cannon (1978), chapter 3, p. 105.
24. Greene (1982), p. 53. This refers to Hutton & Werner on the problem of duplicating temperature and pressure effects on material of the subcrust.

arduous nature was the test of his apprenticeship and the badge of his continuing membership in the 'brethren of the hammer'.[25]

Later in the nineteenth century a branch of this tradition of attention to phenomena as they 'really occurred' in the world, cautiously endorsed experiment. In particular some (not all) late nineteenth-century morphological scientists began to use the laboratory to *reproduce* natural occurrences. These 'mimeticists' strove to *make* laboratory versions of real phenomena such as cyclones or glaciers, with all their intrinsic richness. E. Reyer, a geologist writing in 1892 from Vienna, heralded the beginning of a new experimental physical geology. In the past, Reyer wrote, workers had given up either because quantitative experiments seemed impossible or because experiments had been unable to imitate (*nachbilden*) natural conditions. Now, by reproducing these phenomena, at least partially, much could be learned: masses of lava could be replicated with the flow of limited quantities of pulpy substances; chalk banks and soil sediments could be adequately represented by clay and other malleable materials; similarly Reyer found it possible to model successive stages of deformation leading to fold-mountains. Should the programme succeed, Reyer wrote, 'we will have to assign the long-ignored geological experiment a deep significance'.[26]

In this same period meteorology, like geology, welcomed experiment only if it could imitate Nature without gross distortion. An advocate of the mimetic approach was John Aitken; as we will see, it was on Aitken's miniature cloud-building that Wilson most liberally drew. Aitken was born at Falkirk, Scotland on 18 September 1839, son of the head of an established legal firm. After studying engineering at the University of Glasgow, he served an apprenticeship in Dundee, followed by three further years with Messrs Napier and Sons, shipbuilders, Glasgow. Abruptly, his years as a marine engineer ended with a breakdown in his health. Instead, and in part inspired by William Thomson's lectures on Natural Philosophy, the young engineer transferred his manual skills to building a laboratory and workshop, including lathes, blow-pipes, and glass-work. Throughout his life Aitken pursued mimetic experiments. For example, he designed large-scale models of vortices in order to present the dynamics of cyclones and anticyclones to the Royal Society of Edinburgh.[27]

To explain the circulation of ocean waters Aitken built a glass trough, filled it with water, and used the motion of tracer dyes to track the

25. Rudwick (1985), p. 41.   26. Reyer (1892), pp. 3–4.
27. Aitken (1900–01, 1915). Details on Aitken's life: Knott, 'Sketch of John Aitken's Life and Scientific Work', in Aitken (1923).

effects of air jets (miniature windstorms).[28] When seeking to account for glacier motion, Aitken noted that freezing water in glass tubes offered a 'mimic representation of glacier motion'.[29] Throughout his work Aitken imitated Nature without offering abstract laws. He proceeded by *making* a whole cyclone – not by deducing the cyclone's dynamics from the 'application' of interaction laws for fundamental particles. Aitken's collected papers contain not a single equation.

C.T.R. WILSON

C.T.R. Wilson's fascination with science was grounded, like Aitken's in the remarkable, real phenomena of the natural world. Charles was born the eighth and youngest son of John Wilson, a Scottish sheep farmer. John died when Charles was four, leaving Charles' mother, Ann Wilson, to care both for her own three children and four step-children. Taking all seven with her, she moved to Manchester where the Wilsons led a precarious existence. Charles' step-brother, William, helped support the family from Calcutta where he worked as a businessman. Although William never received what he considered an adequate education, he was determined that Charles should have one.[30]

At fifteen Charles entered Owens College, then part of Victoria University in Manchester, to prepare for a medical career. The young institution had equipped itself for science by drawing on middle-class manufacturers to sponsor its scientific and technological facilities.[31] Before he could begin medicine, Wilson had to complete studies that included botany, zoology, geology and chemistry. Upon graduating with his B.Sc. in 1887, and then spending an additional year studying philosophy, Latin and Greek, Wilson won a scholarship at Cambridge. William Wilson, obviously proud of his younger sibling, wrote from India in January 1888 that

one of the pleasures of my life was imparted to me on Sunday morning when I heard that you had been successful at Cambridge. I was *very* pleased to hear it, and hope the acquisition of this scholarship may eventuate in much real advantage to you. Now that the impetus of success has set in I expect it won't expend its energy until it finds you seated in the presidential chair of the British Association!'[32]

28. Aitken (1876–77).
29. Aitken (1873), p. 4. All page references to works by Aitken given in the footnotes and figure captions will be to the pages as reprinted in Aitken (1923).
30. See 'Biography', note 2, and C.T.R. Wilson (1960) for biographical details.
31. See Sviedrys (1970). This reference, 'Commentary', p. 148.
32. William Wilson to C.T.R. Wilson, 24 January 1888. Wilson papers (misc.) held in archives of Churchill College, Cambridge.

Charles completed Cambridge's Natural Science Tripos in 1892, keen to pursue science but fearful that he would be unable to support the other members of his family. One route that appeared open was the vocation of a mapper: 'I felt I might be of use as an explorer as I had some knowledge of a wide range of sciences and powers of endurance tested on the Scottish hills'.[33] It was a career entirely in keeping with the Humboldtian tradition – exploration in the service of science, precise knowledge of the variety of Nature by examining it *in situ*, and reproducing it to scale. At Owens, Wilson had spent his school vacations exploring the Scottish countryside, his eye opened by a trip to the North High Corrie in Arran (off the West Coast). There he was 'strongly impressed with the beauty of the world', and 'spent all my spare time looking for and studying beetles and pond-life which I also learned to love'.[34]

Like many Victorians, the Wilson brothers took up photography. The depiction of Nature must have appealed to C.T.R., as he specialised in pictures of landscapes and clouds, from which his mother would paint.[35] (Fig. 8.2) 'Even in Calcutta, William was devoting his spare evenings . . . to photography. I have got my enlarging lantern and have been experimenting'.[36] About the same time, C.T.R. and his brother George reported to their elder sibling on their own first steps in the new art. William guided his novices by mail, counselling them even in the details of their selection of a lens:

But I must hasten to give you my opinion of your first pictures. Your exposure has perhaps been full, but your development has been first rate, I think, judging from the pictures. You have printed them very well. I would strongly recommend you to soak your negatives in every case in a saturated solution of alum after they have been fixed . . . . I would advise you to keep a supply of Manchester or Wratten and Wainwrights [instead of Ilfords] beside you for first class pictures . . . .[37]

Thus, long before C.T.R. Wilson turned his camera on the microphysical world, he had used it to recreate the natural world, especially the crags, cliffs and clouds of Scotland. Later, Wilson took his camera with him on his hikes to Ben Nevis and elsewhere, and alongside scientific notes, inscribed his notebooks with the circumstance of each individual

33. Wilson (1960), p. 165.   34. Ibid., P. 165.
35. Interview, J. Wilson (Wilson's daughter).
36. William Wilson to C.T.R. Wilson, 28 November, 1887. Wilson family private papers. We would like to thank Miss J. Wilson for making this and other documents available to us.
37. William Wilson to C.T.R. Wilson (1887?). Wilson family private papers.

Figure 8.2. Early photograph (c.1890) by C.T.R. and George Wilson (Wilson family private papers)

exposure.[38] The immense popularity of such amateur nature photography was a notable feature of Victorian Britain, and in general British photography occupied 'a stylistic and conceptual midpoint between French and American photography of the nineteenth century'. Where on the whole American photographers were scientists or entrepreneurs and the French tended to come from the ranks of painters, the Victorian amateurs 'compromise[d] between these two extremes . . . successfully manag[ing] to blend emotional evocation with an objective assertion of sheer physical fact'.[39]

REPRODUCING THE DRAMATIC IN NATURE

The British desire to reproduce 'sheer physical fact' had no better object in the 1880s than the effects caused by the violent eruption of Krakatoa

38. Wilson notebook A21, e.g., 9–16 April, 1907. Wilson's labooks are housed at the Royal Society, London. They have been indexed by P.I. Dee & T.W. Wormell (1963). Dee & Wormell divided Wilson's notebooks into two groups: A, concerning condensation phenomena and including the notes on the development of the cloud chamber and B, treating the Earth's electric field and thunderstorms. Hereafter references to the Wilson notebooks are by letter, number and date of entry.

39. Millard, 'Images of Nature: A Photo-Essay', in Knoepflmacher & Tennyson (1977), pp. 23–4.

on 26 and 27 August, 1883. Sounds of the explosion echoed through Rodriguez and Diego Garcia, respectively 3080 and 2375 miles from the volcano. Windows burst and walls cracked a hundred miles distant.[40] Amidst the staggering mass of particles shot into the upper atmosphere, strange optical phenomena appeared around the world. From Honolulu an observer saw a 'peculiar lurid glow, as of a distant conflagration, totally unlike our common sunsets';[41] Herr Dr. A. Gerber from Glückstadt recalled in the first volume of the *Meteorologische Zeitschrift* how 'the sailors declared, "Sir, that is the Northern Lights'" and I thought I had never seen Northern Lights in greater splendour. After 5 minutes more the light had faded . . . and the finest purple-red rose up in the S.W.; one could imagine oneself in Fairyland'.[42] Reports from Italy, Ohio, Switzerland, Lisbon, India, Japan, and Australia cascaded into newspapers; unknown observers and renowned scientists posted notices to all the major scientific journals. Even Hermann von Helmholtz in Berlin took the time to report on 'cloud-glow' and the remarkable illusion of green clouds.[43]

Photographers, artists, and scientists all tried to capture the extraordinary visual events: mere black and white photographs could not do them justice. At ten-minute intervals William Ascroft drew quick pastels of the yellow-green sunset over the Thames (Fig. 8.3) and the extraordinary Bishop's Ring (corona) he saw on 2 September, 1884.[44] When in 1886 the Royal Society published its comprehensive report on the events of Krakatoa, they included Ascroft's sunset sketches as scientifically useful. Throughout the years following the eruption, debate raged over whether the Sun's strength was changing, whether the effects really were correlated with the eruption, and if so what mechanisms could account for them. Even if no explanation was absolutely persuasive, at least scientists could *mimic* this extraordinary event. Karl Kiessling attempted to reproduce the dramatic coronae of late 1883 with a 'diffraction chamber' filled with dust, ordinary air, and filtered vapor.[45]

Aitken was well prepared for the Krakatoa debate. In 1883 he was continuing the important work he had begun with his article, 'On Dust, Fogs, and Clouds',[46] demonstrating the role of dust in nucleating fog and cloud droplets; his observations on the extraordinary sunsets fol-

40. Symons (1888), p. 27.   41. Ibid., p. 153.   42. Ibid., p. 157.   43. Ibid., p. 171
45. Kiessling's experiments are described in Symons (1888), pp. 196–8, 253–4, 258–61.
44. T.A. Zaniello locates the Krakatoa sunsets as a meeting point for scientific and artistic concerns in Britain during this decade. See 'The Spectacular English Sunsets of the 1880's', in Paradis and Postlewait (1981), pp. 247–67, esp. 248–51.
46. Aitken (1880–81).

Figure 8.3. Ascroft's Pastels: *(a)* Twilight and afterglow effects at Chelsea, London, 26 November, 1883. (Symons *Krakatoa* (1888), frontispiece); *(b)* Bishop's Ring, 2 September 1884. (Science Museum Library, South Kensington)

lowed as a natural sequel. To avoid Britain's cloud cover, Aitken voyaged to the south of France where he repeatedly witnessed the white glare of the daytime sun, the yellow–orange–red sequence of colours on the western horizon at sunset, and then the brilliant afterglows some 15 and then 30 minutes after sunset.[47] A decade later Aitken was still struggling to understand those cataclysmic events by *reproducing* the Krakatoa green sunsets in his laboratory using electrified steam: 'The colours produced by such simple materials as a little dust and a little vapour are as beautiful as anything seen in nature, and well repay the trouble of reproducing them'.[48]

Victorian England was fascinated with all kinds of reproductions of the dramatic in nature: through painting, poetry, photography, and even laboratory recreation. In addition, there were immediate, practical issues at stake. Weather affected transportation, fishing, public health, military affairs, agriculture and communication.[49] Aitken, among others, frequently stressed this practical side of meteorology, arguing that Britain's dusty industries produced fogs that endangered its citizenry: 'All our present forms of combustion not only increase the number and density of our town fogs, but add to them evils unknown in the fogs which veil our hills and overhang our rivers'.[50] Such evils, Aitken asserted, force our attention to the importance of dust in the origin of clouds:

As our knowledge of these unseen particles increases, our interest deepens, and I might almost say gives place to anxiety, when we realize the vast importance these dust particles have on life, whether it be those inorganic ones so small as to be beyond the powers of the microscope, or those larger organic ones which . . . though invisible, are yet the messengers of sickness and of death to many – messengers far more real and certain than poet or painter has ever conceived.[51]

Due to these varying interests, the world-wide establishment of weather networks, observatories and societies had many roots.[52]

47. Aitken (1883–84b).   48. Aitken (1892b), p. 283.
49. For example an international conference on the study of weather at sea was suggested by Lieut. M.F. Maury of the US Navy in 1853 and the (British) Board of Trade established a Meteorological Department in 1854. See Shaw (1931). The Scottish Meteorological Society had close relations with fisheries organisations which substantially subsidised the Council's research – see, for example, Scottish Meteorological Society (1884), 57p. Also, as Sir Ernest Wedderburn mentions (Wedderburn (1948), p. 235): 'Amongst [the Society's] subsidiary objects were . . . the bearing of meteorology on public health, agricultural [sic] and horticulture, the alleged periodical recurrence of wind storms, and the general laws regulating atmospheric changes, the discovery of which might lead to a knowledge of the coming weather'.
50 Aitken (1880–81), pp. 49–50.   51. Aitken (1883–4a), p. 84.
52. For an excellent meteorology bibliography (especially on national weather systems), see Brush & Landsberg (1985).

Wilson's career owes much to an observatory perched on the summit of Ben Nevis, the highest mountain of Northern Scotland, indeed of the British Isles. Impressed by the new style of meteorological research pioneered in Germany and the United States, many Britons had enthusiastically supported the establishment of an observatory. With the help of devoted amateurs and private subscriptions, the Scottish Meteorological Society founded their station to track 'vertical meteorological sections of the atmosphere'.[53] One Clement L. Wragge volunteered to take weather readings from Ben Nevis by himself to show the value of careful observation. A 'king among eccentrics', Wragge assigned Christian names to cyclones, and even edited his own journal *Wragge – For God, King, Empire and People*[54] On the heels of the wide publicity given to Mr. Wragge, contributions arrived from such diverse benefactors as Her Majesty, and the Worshipful Company of Fishmongers, London. These funds allowed the Ben Nevis Observatory to open in 1883. Until it was closed down in 1904, resident observers telegraphed daily reports to Scottish newspapers. For years Wilson's research was shaped by his experiences at this distant meteorological outpost.

DUST, RAIN AND IONS

Wilson's career as a professional physicist began haltingly.[55] After graduating from Cambridge in 1892, he stayed on, demonstrating at both the Cavendish and Caius Chemical Laboratories. Hoping to obtain a Clerk Maxwell fellowship (he did not succeed) Wilson wrote to J.J. Thomson in November 1893 about his work on the distribution of a substance in solution that was kept hot on the top and cold on the bottom: 'Very few experiments appear to have been made on the subject, and it seems of considerable importance in connection with theories of solution and osmotic pressure'. The aspiring physicist 'would determine the concentration of different parts of the solution optically'.[56] In his later work Wilson used theoretical calculations of vapor pressures to try to understand the condensation of water vapor, and he pursued the problem experimentally with an optical method, the cloud chamber. His earlier work with non-equilibrium systems contributed to his interest in thermodynamic instability – the fundamental

---

53. Report of the Council of the Scottish Meteorological Society for 1875, cited in Paton (1954), p. 292.
55. Biographical details from C.T.R. Wilson (1960).   54. Ibid., pp. 293–4.
56. Letter from C.T.R. Wilson to J.J. Thomson, 8 November 1893. Churchill College, Cambridge, (note 32).

feature of cloud condensation. But the immediate catalyst for Wilson's work came from his stint in the Fall of 1894 at Ben Nevis.

By the early 1890s, the mountain-top observatory was flourishing and its staff welcomed volunteers to work during the easy observing periods of summer and autumn. After graduating from Cambridge, Wilson's affection for the mountains led him to the small station many times. He made his first trip on 8 September 1894, which began as a cloudless day. Soon a thick haze embraced patches of fog which gradually became continuous. The next evening, at 9 p.m., the observers sighted first one Lunar Corona, then another at 10 p.m., and at least seven more during the next two weeks. On the 15th the logbook records: 'Solar fogbow & glories at 16^H & Lunar Corona at 23^H'. Just hours before Wilson descended from the heights on 22 September, light and clouds performed spectacularly − even the dry tone of the logbook rises, in the remark that 'some beautiful Triple Lunar Coronae were seen this morning through thin passing fog'.[57]

On descending from the mountaintop station Wilson, in tune with Aitken and so many contemporaries, wanted to *mimic* the wonders of Nature:

In September 1894 I spent a few weeks in the Observatory [of Ben Nevis.] The wonderful optical phenomena shown when the sun shone on the clouds surrounding the hill-top, and especially the coloured rings surrounding the sun (coronas) or surrounding the shadow cast by the hill-top or observer on mist or clouds (glories), greatly excited my interest and made me wish to imitate them in the laboratory.[58]

Returning to the mountain peak just nine months later, in June 1895, Wilson recorded that he 'walked along Spean Bridge road. Saw lightning lighting up mist in big corrie & heard thunder rolling in that direction'. The power of the electrical storm was evident: 'Saw the damage done by lightning yesterday. Telegraph instrument fused at various places, including part of one of the steel keys . . . . Boxes on shelf above [lightning] arrester thrown on to floor'. After recording the events of a few days mostly spent hiking in the hot, blazing sun, Wilson noted on 26 June an abrupt alteration of weather.

Mist suddenly began to pour down a gully between B[en] Nevis and Carn Dearg, & afterwards hid upper part of cliff . . . . Heard continual muttering of thunder in distance. Walked along ridge to top of Carn Mòr Dearg. After a

57. Logbook from Ben Nevis Observatory, from Meteorological Office, Edinburgh. We would like to thank Marjory Roy for making this material available to us.
58. Wilson (1965), p. 194.

240   *Peter Galison & Alexi Assmus*

minute or two there, suddenly felt St. Elmo in my hair, & in my hand on holding it up. Ran down into corrie. Bright lightning & loud thunder . . . .[59]

The optical and electrical phenomena that Wilson witnessed during those two trips to Ben Nevis set the outlines of his life-long scientific goals. Meteorological optics and atmospheric electricity remained central to him until his death in 1959.

Between demonstrating physics and tutoring students, young Wilson earned a living in Cambridge, but had no time to do any research. To improve the situation, he had tried teaching for a few months at the Bradford Grammar School, but this left him no freer than before. Returning to Cambridge, he felt lucky to land a job demonstrating physics to medical students at the Cavendish. 'With this I had just enough to live on, a connexion with the Cavendish, and at last time to do some work of my own just when I had ideas which I was impatiently waiting to test'.[60] His dramatic encounter with meteorological effects at Ben Nevis left him with a burning desire 'to reproduce the beautiful optical phenomena of the coronas and glories I had seen on the mountaintop'.[61] With this in mind, Wilson began a notebook in which he recorded speculations as well as research notes. Immediately preceding the notes on his first 'cloud' experiment, we find a page of questions addressed to himself. Under 'Cloud Formations & c.' he pondered:

1. Are the rings of corona and glory formed simultaneously in same cloud, equal in radius? Try monochromatic light.
2. Condition of formation. When best formed. (In dusty or tolerably pure air or in presence of soluble substances. [)]
3. Are ice particles ever formed instead of water drops. (Halos & c.)
4. Are coronae & c. formed when one liquid separates out in milky form from another. (Mixture of ether and water allowed to cool & c.) Also are halos ever formed when crystalline precipitates are formed. Applications.[62]

On what tradition did Wilson draw when he decided to perform these experiments to reproduce clouds, glories and coronas in his laboratory? Other scientists were working on similar problems; several had even tried to reproduce clouds or fogs, and early in his notebook Wilson reviewed the extensive literature on cloud formation. Aitken, Jean Paul Coulier and Robert von Helmholtz had all tried to condense water vapor – either by exploiting an air pump or an india rubber ball to expand saturated air in a glass vessel, or by watching a steam jet

59. C.T.R. Wilson, A21, 19, 20, 26 June, 1895.   60. C.T.R. Wilson (1960), p. 166.
61. Ibid., p. 166.   62. Wilson, A1, before dated entry of March 1895.

abruptly expand as it escaped from a nozzle.[63] The theoretical justification for this method is this: an adiabatic expansion of saturated gas lowers the temperature[64] of the gas, causing supersaturation which can lead to condensation. In the term 'supersaturated air', supersaturation is usually defined to be the ratio of actual vapor pressure to the equilibrium vapor pressure above a *flat body of water*. The obvious way to study condensation is with supersaturated vapors, but it is not always easy to precipitate the liquid. Under certain circumstances it is possible to have supersaturation without condensation; for example, if the vapor is *not* over a flat body of water.

Aitken's cloud experiments

Among those predecessors who expanded air to investigate condensation, Aitken was the most important for Wilson. Also a Scotsman, Aitken initiated research work at Ben Nevis with apparatus on which Wilson would model his own. The remarkable similarity between Aitken's and Wilson's instruments, and the fact that the only reference in Wilson's first published paper is to Aitken, suggests that Aitken's meteorology merits a closer look.[65]

Aitken's first conclusion from his cloud experiments was that dust particles acted as nuclei for water droplets in supersaturated air. Without dust there was no condensation under what Aitken considered 'normal' conditions. In 1880 he pointed out that '[d]usty air – that is, ordinary air, gives a dense white cloud of condensed vapour'.[66] He decided to use his dust by designing an instrument to measure the number of particles in the air. 'Powerful as the sun's rays are as a dust revealer, I feel confident we have in the fog-producing power of the air a test far simpler, more powerful and delicate, than the most brilliant beam at our disposal'.[67] Aitken's motivations for studying dust were thus meteorological; he believed in the 'possibility of there being some relation between dust and certain questions of climate, rainfall, etc.'[68] and hoped to settle the 'great fog question' the problem of town fogs, 'whose increased frequency and density [calls] for immediate action'.[69]

63. Coulier (1875); Aitken (1880–81); Helmholtz (1886).
64. In practice the expansion was quick enough to give a good approximation to an adiabatic system and slow enough so entropy remains constant. Therefore the gas obeys the equation $P(V)^k$ = constant, where k is the ratio of specific heats. Using the ideal gas equation we obtain a fall in temperature as the air expands,
$$T_1/T_2 = [V_2/V_1]^k - 1.$$
65. C.T.R. Wilson (1895). 66. Aitken (1880–81), p. 35. 67. Ibid., p. 41. 68. Ibid., p. 41. 69. Ibid., p. 48.

By 1888 Aitken had a method of counting dust particles in the air based on his observation that they were a source of condensation in supersaturated air. His instrument is reproduced in Figure 8.4. The condensation occurs in the ordinary glass flask (*A*), the receiver, and the water droplets (each surrounding a single dust particle), are counted by means of the compound magnifying glass (*S*). The glass flask (*G*) contains the air to be tested, which is kept saturated by water in the flask. *D* is a cotton-wool filter which clears ordinary air of dust particles. Air from *G* and *D* is mixed so that 'too much dusty air [is not] sent into the test receiver at one time, or the drops will be too close for counting'.[70] Aitken wanted to ensure that the dust particles would be so far apart that each particle would serve as a centre of condensation and all would be counted. After this mixture of air is in receiver *A*, and stopcock *F* has been closed, the experimenter makes one stroke of the pump, while watching stage *O* very carefully. As the air in *A* expands, it supersaturates; condensation occurs on the dust particles which then fall onto *O*. Counting is facilitated by a square grid where the rulings are one millimetre apart (on *O*).

Aitken spent 1889–94 counting dust particles in the British and continental air with a pocket version of his counter. In a three-part paper published during this period, 'On the Number of Dust Particles in the Atmosphere of Certain Places in Great Britain and on the Continent, with Remarks on the Relation between the Amount of Dust and Meteorological Phenomena',[71] Aitken compared measurements taken on Ben Nevis with others from a low-altitude station at Kingairloch. Mr. Rankin, the observer who took the readings on Ben Nevis, recorded that two dust counters were bought for the observatory in 1890, 'one, a portable form of the instrument mounted on a tripod stand, for use in open air; the other, a much larger form, for use in the laboratory . . . . Both instruments were made from plans and specifications prepared by Mr. Aitken'.[72] The last measurement recorded on Ben Nevis was in 1893, just a year before Wilson arrived for his first visit of September 1894. Wilson must have seen the Aitken apparatus on his visit; the observatory was small and crowded and there was little to do once the sun went down.

---

70. Aitken (1888), p. 193.
71. Aitken (1889–90, 1892a, 1894).
72. Rankin (1891), p. 125. The dust counter remained at Ben Nevis until at least 1901. In a deposition of property the counter is specifically mentioned: Ben Nevis (1903), p. 162.

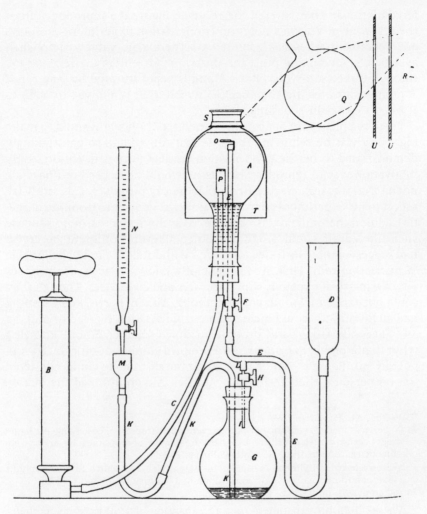

Figure 8.4. Aitken's dust chamber of 1888. (Aitken, 'Dust Particles', p. 190)

Wilson's early experiments

After his short stay in Scotland, Wilson began to study the optical effects of clouds. By condensing water vapor in an expansion apparatus remarkably similar to Aitken's (see Fig. 8.3), Wilson was able to produce the rings of colour that had delighted him on Ben Nevis. Though Wilson obviously derived his method from Aitken's work, their procedures differed crucially. Instead of mixing purified and dusty air, Wilson filtered *all* the air that entered the receiver. Figure 8.5 reproduced from the first page of Wilson's notes on condensation experiments, displays an apparatus that tests *only* filtered air. There is no valve system, such as Aitken's, for mixing pure and dusty air. Air enters the receiver (*V*) through the cotton-wool filter (*F*) and is kept saturated by water in *V.* A pump evacuates the chamber (*R*) which then is allowed to open to *V* when expansion is desired.[73]

There is a puzzle in that cotton-wool filter. If Wilson wanted to make fogs, why was he removing the dust that Aitken had so painstakingly demonstrated to be the condensation nuclei? Surely, if Wilson's only motivation was to reproduce clouds he would have used ordinary air, not air that was *specially prepared* for laboratory purposes. This 'artificial' aspect to his experiments signals a significant departure from the mimetic tradition that initially inspired him. Was there any reason to suppose such filtered air could produce droplets? Several different observers had suggested that electrified air – even without dust – might be capable of nucleating rain. First, it was generally known that as steam hisses out of a nozzle it expands, supersaturates and condenses. Though they could not explain *why*,[74] R. von Helmholtz, Aitken, Barus, and Kiessling had all found they could increase that condensation by electrifying the jet.[75] Second, Aitken and others had noted (again without knowing why) that a high expansion would produce condensation even in supposedly purified air.[76] After satisfying himself that he could eliminate this bothersome effect in his dust counter, Aitken dropped the matter.

73. Wilson, A1, 26 March, 1895.
74. Aitken reasoned that electricity would prevent small drops from coalescing into larger drops because of electrostatic repulsion, so the number of drops must be large, promoting dense condensation. Aitken:(1892b) pp. 258–9.
75. Wilson referred to Helmholtz's paper in his notes (A1, before March 1895), Helmholtz (1886), and to Barus, Kiessling, and Aitken in a later paper: Wilson (1897a).
76. Aitken (1888), on p. 201:

We see from his experiment that an expansion of 1/50 is nearly, perhaps, quite sufficient to cause condensation to take place on even the smallest particles in the air tested; from which we may conclude that the showers which unexpectedly took place from time to time in the experiments described, where high expansions were used, were not due to the presence of extremely small particles which had become active with the high degree of supersaturation.

Aitken got rid of this problem by pumping slowly, thinking the 'shock' of expansion was causing unwanted condensation. Actually because his expansion was slow, and therefore not as adiabatic as a fast expansion, the supersaturation was smaller and therefore the unexpected condensation did not take place.

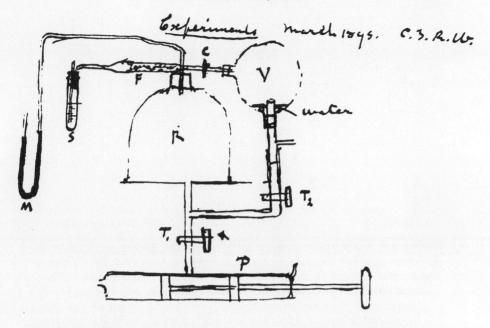

Figure 8.5. Wilson's cloud chamber apparatus of 1895. (Wilson, notebook A1, March 1895, see note 38)

Wilson was aware of both phenomena,[77] though it would be years before he would be certain of the connection between the two.

Thus, suspecting that there might be several ways for droplets to form, Wilson filtered the air to identify the best conditions in which to make the 'wonderful' optical effects: coronas and glories. 'When [are the optical effects] best formed [?] (In dusty or tolerably pure air . . . [?])[78] Soon, the notebooks reveal, he began experimenting *exclusively* with filtered air. For Wilson dust became just an annoyance to be removed; in constrast, dusty air was Aitken's prize specimen, and condensation in purified air was a background problem to be eliminated, then forgotten. Wilson created *artificial* laboratory conditions to 'dissect' Nature; Aitken wanted to remove the artificiality from his experiment. It is at first puzzling that Wilson used an apparatus almost identical to Aitken's, yet the two fastened on mutually exclusive phenomena – one on dust, the other on purified air. Wilson's apparently simple move of filter *F* in Figure 8.6 marks a profound shift in material

77. Wilson, A1, before dated entry of March 1895.   78. Ibid.

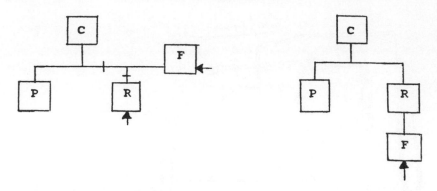

Aitken 1888   Wilson 1895

Figure 8.6. Schematic representation of the Aitken dust chamber and Wilson cloud chamber apparatuses depicted in Figs. 8.4. and 8.5. In the Aitken apparatus (left) air from the outside enters through the reservoir (R) and is mixed with a certain amount of filtered air coming through the filter (F). The filtered air merely serves to keep the air in the chamber from being too dusty. By contrast, in Wilson's apparatus *all* the air must pass through *F before* it can enter the reservoir (R), the main chamber (C), or the pump system (P). Thus Wilson had to find it plausible that water vapour might condense on dust-free air.

culture and conceptual structure. The source of this change in traditions lay squarely in the scientific programme of the Cavendish.

ANALYSIS AND MIMESIS

Through J.J. Thomson and his collaborators, Wilson learned the Cavendish style of analytic physics. Thomson, appointed to the Cavendish chair in 1884, was a firm believer in analytic solutions to problems: '[the] principal advances made in the Physical Sciences during the last fifty years . . . [have] itensif[ied] the belief that all physical phenomena can be explained by dynamical principles and [stimulated] the search for such explanations'.[79] As a student at Cambridge, from 1888 to 1892, and as a researcher at the laboratory starting in 1895, Wilson absorbed much of the matter-physicist's style of physics. In addition, the Cavendish was unique in its firm programmatic commitment to the assumption that electric charge came in discrete bits (ions). Wilson borrowed heavily from his Cambridge colleagues; ion physics had been essential

79. Thomson (1888), p. 1.

to his invention of the cloud chamber. In return, Wilson's work became an integral part of the Cavendish investigations into the electrification of gases. Eventually, under Rutherford, his instrument became the Cavendish Laboratory's primary research tool. The cloud chamber, developed under one research programme, would itself (in its role as a piece of experimental equipment) establish the boundaries of another, completely different research project.

Evidence of the effect of the Cavendish research programme on its students is easy to find. As a Natural Sciences Tripos student (Wilson took the new laboratory course rather than the more formal Mathematical Tripos) Wilson faced questions such as: 'Give some account of the phenomena observed in the neighbourhood of the negative electrode when an electric discharge passes through a tube containing gas at low pressure'.[80] Interest in this type of discharge was central to Cavendish research, which sought to understand the structure of matter through the investigation of cathode rays. One of the programme's great triumphs followed not many years later when Thomson, in 1897, discovered that the electron was a subatomic particle.[81]

Since the mid-nineteenth century, when it became possible to produce a good vacuum, physicists across Europe had been exploring the effects of electrical discharges. Germans led the effort to probe the invisible rays emitted from a hot cathode contained in a tube with rarified gas. At first, attention focused on light and dark bands in the tube; as higher and higher vacuums were achieved, scientists saw the glowing in the tube dwindle until, at very low gas pressure, light only appeared on the walls of the tube. Since the rays seemed to originate from the cathode (rather than the anode) they were named cathode rays.[82] In 1879, William Crookes demonstrated that magnets deflected the rays, which led him to identify the rays as streams of charged particles.[83] Crookes thus added support to a view, held intermittently since Faraday's electro-chemical work in the 1830s, that electricity comes in discrete 'atoms'. Opposing the particulate view were scientists who viewed all electromagnetic effects, including charge itself, as deformations of the ether.[84] By the 1880s English though not continental scientists had decided in favour of the corpuscular nature of electricity. One of the chief proponents of the ion picture was J.J. Thomson.

80. D. Wilson, 'Among the Mathematicians', (1982), pp. 346–7.
81. Much of the following on early atomic research is based on J. Heilbron (1964), first two chapters. Heilbron: 'the "great discoveries" [of X-rays, radioactivity, and the corpuscular electron] developed mainly out of experimental investigations of the phenomena accompanying the discharge of electricity through rarefied gases', p. 59.
82. Wheaton (1983), p. 5.   83. Ibid., p. 6.   84. See e.g., Harman (1982).

Thomson began his study of the discharge of electricity through gases in 1886[85] and soon most of the lab was working on this problem. From 1893 to 1895, the year Wilson arrived at the Cavendish, fully half of the papers published by Cavendish scientists were concerned with the discharge of electricity.[86] As a discharge effect, possibly explicable in terms of ions, Aitken's electrified steam experiments would have seemed relevant to Thomson. And in 1893, two years before Wilson published his paper on condensation in the absence of dust,[87] Thomson provided a theoretical justification for the growth of a drop in the presence of a non-uniform electric field.[88]

Thomson first argued that because surface tension puts a drop under pressure, the equilibrium pressure surrounding the drop is very high for small drops. Quantitatively the presence of a drop of radius r increases the equilibrium vapor pressure by a factor of $1/r$ promoting evaporation. The smaller the drop, the greater this effect. As Thomson put it,

It is evident that this property makes the growth of drops from smaller ones of microscopic dimensions impossible, for these small drops would evaporate and get smaller, and the smaller they get the faster will they evaporate. They are in the position of a man whose expenditure increases as his capital decreases, a state of things which will not last long.[89]

Thomson next argued on thermodynamic grounds that a non-uniform electric field *decreases* the vapor pressure as $1/r^2$. Energy considerations provide a simple way to understand this: because the dielectric constant of water is about 80 and that of air about one, the energy in the electrical field decreases when the charge is surrounded by water. Assuming the air acts as a thermal reservoir, condensation of water is favoured, since the Gibbs free energy of the systems tends to a minimum. Because surface tension increases equilibrium vapor pressure by $1/r$ and the presence of charge decreases it by $1/r^2$, for small drops the electric field wins and the drop can grow.

Thomson's programme provided a quantitative model of the physics of ionic condensation in dust-free air which allowed Wilson to consider

85. Thomson's first paper on electrical discharge in gases is Thomson (1886).
86. *History* (1910), pp. 297–8. We have taken the number of papers published on electrical discharge phenomena from the list (given by year) of all Cavendish papers provided at the end of this volume.
87. C.T.R. Wilson (1895).
88. Thomson (1893). Cf. Thomson's earlier book, Thomson (1888), esp., chapter XI, 'Evaporation', pp. 158–78.
89. Thomson (1937), p. 416.

experiments that would have seemed pointless without Thomson's detailed scheme of ion drop formation. Just a week after Wilson began his cloud experiments, he explicitly used Thomson's formula to calculate the magnitude of the electric charge he suspected he might be seeing in his chamber. 'If nuclei be present in [the] shape of small electrified drops of radius 2 time 10^{-7} [centimetres] each charged with atomic charge, we can calculate magnitude of this charge necessary to neutralize effect of S. T. [surface tension]'.[90] As Wilson turns his thoughts to the world of sub-visible ions, one sees the impact of Cavendish research on his programmatic goals.

During the remainder of 1895 Wilson used his cloud chamber daily. Each morning he would fill his chamber with air and make several expansions to remove dust, a method he found more satisfactory than using a cotton-wool filter. Then for the rest of the day he used this purified air to determine expansion ratios for condensation. (The expansion ratio was defined to be V_2/V_1 where V_2 was the volume after expansion and V_1 the volume before.) On April 3rd alone he made 115 runs.[91] In a one-page paper, his only article of 1895, Wilson stated the critical expansion ratio given an initial temperature of 16.7°C: $V_2/V_1 = 1.258$.[92] Such precision would become important to Wilson later when he turned to the new 'rays' that were soon to be discovered on the Continent.

As he developed this cloud chamber, Wilson never left meteorology, constantly interjecting weather questions into his notebook. During the spring of 1895 he speculated on the significance of his experiments in a section labelled 'Meteorological'. Beginning with optics, Wilson claimed that the 'existence of coronae shows uniformity in size of drops in the clouds showing them'.[93] The drops, Wilson speculated, might oscillate in a cloud: as drafts carry them up into regions of high supersaturation they grow bigger until their weight causes them to descend, thus reaching an area of lower saturation where they begin to evaporate until they are light enough to begin the cycle again. In addition, Wilson began to think that weather might be affected by ions, and wondered, *'when drops are suddenly formed do they throw off small electrified drops or free ions?* (Thunderstorms & c.) Either way, air should be electrified'.[94]

During the New Year celebration of 1896 the world was delighted by Röntgen's photographic demonstration of his discovery of X-rays; Röntgen was even called to Potsdam to give Kaiser Wilhelm a demon-

91. C.T.R. Wilson, A1, 3 April 1895. 92. C.T.R. Wilson (1895).
93. C.T.R. Wilson, A1, 27 March, 1895.
94. C.T.R. Wilson, A1, 30 March, 1895. (Wilson's emphasis.)

stration.[95] At the Cavendish, experiments with X-rays began at once. Ernest Rutherford, a new research student at the lab, wrote on 25 January 1896, 'The Professor [J.J. Thomson] of course is trying to find out the real cause and nature of the waves, and the great object is to find the theory of the matter before anyone else, for nearly every Professor in Europe is now on the warpath'.[96] Thomson used the new rays in conjunction with a classic Cavendish experiment. He and Rutherford watched the effect that Röntgen rays had on the passage of electricity through gas.[97] They soon found that the rays enhanced conduction and explained this by suggesting that the Röntgen rays produced ions in the gas as they passed through it.

When Wilson heard this he was anxious to shine the new rays into his cloud chamber. He borrowed an X-ray tube from Ebenezer Everett,[98] Thomson's assistant and, turning it on his chamber, was thrilled to find that 'no effect is produced by the X-rays unless the expansion is great enough to produce condensation in any case. When it is sufficient to cause condensation without the rays, they produce a very great increase in the number of drops'.[99] Wilson had carefully determined the expansion ratio that caused condensation around nuclei produced by Röntgen rays and found that exactly the same expansion ratio condensed vapor about nuclei naturally present in dust-free air. Therefore, Wilson reasoned, 'It seems legitimate to conclude that when the Röntgen rays pass through moist air they produce a supply of nuclei of the same kind as those which are always present in small numbers, or at any rate of exactly equal efficiency in promoting condensation'.[100] Quantitative experimentation thus let Wilson identify the two nuclei as one and the same.

Becquerel's discovery of uranium rays in March 1896 added yet another dimension to Wilson's experiments. Wilson found that, like X-rays, uranium rays increased condensation at the established expansion ratio of $V_2/V_1 = 1.25$. By late 1897 Wilson was willing to go to press with the assertion that, 'the electrical properties of gases under the action of Röntgen rays and Uranium rays point to the presence of free ions'.[101] Wilson's patience and precision in determining expansion ratios led to a modification in the ion theory of condensation. By 1898 he could show that precipitation in the cloud chamber fell into four

95. Keller (1893), p. 57.   96. E. Rutherford quoted in letter, 25 January to fiancée, ibid.
97. Thomson & Rutherford (1896).
98. C.T.R. Wilson, interview on Scottish Home Service, 'Wilson of the Cloud Chamber', February 16, 1959. Voicewriter transcription, Edinburgh University Library, Appleton papers, D57.
99. C.T.R. Wilson, A2, 17 February, 1896.   100. C.T.R. Wilson (1896), p. 339.
101. C.T.R. Wilson (1897b), p. 337.

regimes bounded by three expansion ratios. Below $V_2/V_1 = 1.25$ there
is no condensation in dust-free air, between $V_2/V_1 = 1.25$ and 1.31 there
are distinct 'rain' drops, at 1.31 there is a sudden increase in the number
of drops, and finally above 1.37 a dense fog is seen. Wilson supposed
that the fog was formed by the statistical aggregation of water molecules
because of extremely high supersaturation; no nuclei were needed. He
did not, however, understand the reason for the existence of the *two*
lower expansion ratios. In March 1898 he speculated, in his notebook,
that the difference in condensation in the two domains could be due
to the *quantity* of charge acting as precipitant, for a larger charge would
promote condensation at a lower supersaturation.[102] He thought that
two sets of ions might exist, one with a charge twice that of the other:
the double charge was strong enough to pull in water at 1.25 while
the single charge would only precipitate water at the ratio of 1.31. At
this stage Wilson drew his ion models from chemistry. His doubly-
charged carrier (ion) was an oxygen atom. He did not suggest that
particles smaller than atoms carried charge.

By July 7, 1898, Wilson was considering another reason for the two
distinct expansion ratios. His notebook entry of that day began with
Thomson's idea: 'J.J.T. suggests that if the expansion required to catch
positive and negative ions is different (say less for the negative than
for the positive) gas would be left charged if expansions were only
sufficient to catch the negative but not the positive'.[103] Wilson saw
immediately that 'this would have obvious meteorological application
if atmospheric air were ionised to even very small extent'.[104] What did
Wilson mean? It was generally known that the Earth is negatively
charged and that a potential gradient exists between the Earth and the
ionosphere. (On an ordinary day this is 100 volts per metre, amounting
in total to a change in voltage of 400 000 volts).[105] If there were no
external forces acting on this system such a huge potential difference
would discharge the Earth in approximately half an hour. The problem
was that no mechanism was known that could maintain this voltage.

Wilson quickly realised that if negative ions were more likely to cause
condensation, rain would bring down negative charge to the Earth,
thus maintaining the fair-weather gradient. Immediately he began to
speculate on meteorological consequences of Thomson's suggestion.
Under a section heading, 'Condensation problems suggested during
previous work with *meteorological* bearings', Wilson asked himself, '[1]
are there any free ions in ordinary moist air in absence of external

102. C.T.R. Wilson A8, 4 March, 1898.   103. C.T.R. Wilson, A8, 7 July, 1898.
104. Ibid.   105. Feynman (1964), 2, chapter 9.

influences? , . . . [2] are there any agents capable of ionising air? . . . [3] are negative ions more easily caught than positive ones?'. He concluded, 'the three questions are all connected with the theory of atmospheric electricity suggested on the previous page. That is that the ground is kept negative by the fall of rain, each raindrop containing one negative ion'.[106] Another of Wilson's notebooks for this period offers a detailed discussion of balloon experiments that measure the potential gradient accurately and records ideas for laboratory experiments 'to find probable nature of the carriers of atmospheric electricity'.[107]

The Cavendish adopts Wilson's cloud chamber

Between January 7th and 23rd of 1899, Wilson carried out the experiments that confirmed Thomson's hunch about the different expansion ratios for negative and positive ions. He used a modified cloud chamber with a brass wire down the centre that could be held at a constant potential. Wilson's conclusions were inescapable:

This showed differences plainly, and to some extent the distribution of the ions in the tube appeared to be made visible; such fogs as were obtained with EMF arranged to drive positive ions outwards, being concentrated around the wire, while the much more marked fogs obtained with the EMF in the other direction extended throughout the tube. The expansions available were scarcely sufficient to catch positive ions.[108]

By this time Wilson's Cavendish colleagues were using the cloud chamber. Most notably, J.J. Thomson resolved to establish a value for the charge of the electron, (*e*), and thus add support to his growing conviction that the electron was a fundamental particle. Just a year before, in 1897, he had claimed that cathode rays were streams of elemental particles[109] and had shown that the charge-to-mass ratio was a constant for these particles. This was no guarantee, however, that *e*, and the mass, *m*, were separately constant and some scientists judged it insufficient proof for the existence of an electron.[110] At the Cavendish, however, there were few doubters, and research to accurately determine the value of *e* started immediately. Thomson's first attempt was crude, but it established a method that Robert Millikan later exploited in his famous oil-drop experiments.[111] Thomson determined the number of

106. C.T.R. Wilson, A8, 7 July, 1898. 107. C.T.R. Wilson, B1, 4 October, 1898.
108. C.T.R. Wilson, A3, 7–23 January, 1899. 109. Thomson (1897).
110. Gerald Holton, 'Subelectrons, Presuppositions, and the Millikan-Ehrenhaft Dispute', in Holton (1978), pp. 25–83. Cf. Franklin (1981).
111. Ibid., pp. 25–83.

ions times the electric charge (ne), from the measurement of current through gas (a procedure he had been using for a long time); he only had to determine n to get a value for the electric charge. Quoting Thomson: 'the method I have employed to determine n is founded on the discovery made by Mr. C.T.R. Wilson'.[112] His method was to take the gas for which ne had been determined and subject it to expansion. Assuming that each water droplet caused by the expansion contained a single ion, n would be equal to the number of droplets. By observing how quickly the cloud, caused by the expansion of the gas in a cloud chamber fell under the influence of gravity, he was able to estimate the number of drops. Take the mass of water in the air to be fixed. (The greater the number of drops, the smaller each drop must be.) Stokes' law then predicts that the cloud falls more slowly for smaller drops. These early experiments gave the value of 7.3×10^{-10} electrostatic units for the electron charge.[113]

Wilson, borrowing from Thomson, adopted the falling cloud method. In a more complicated version of the experiment, to test whether negative charge induces condensation before positive, Wilson divided his chamber vertically into two regions with three brass plates.[114] The middle plate was grounded, the left plate was kept at a positive potential and the right at a negative potential. Oppositely charged ions were drawn into different parts of the chamber so that Wilson could determine whether negative charges differed from positive in their ability to condense water vapour. After making an expansion of 1.25, he 'counted' the number of charges in the two sides of the chamber using Thomson's method and found that negative charge precipitated water but positive charge did not.

Soon H.A. Wilson, also at the Cavendish, joined Thomson in his determination of e. Together, they added electrically charged plates (set horizontally, not vertically as in C.T.R.'s experiments) to Thomson's original experiment and compared the fall of clouds under the action of an electric field with that of no field. This allowed them to obtain an estimate of e without resorting to counting the number of droplets n.[115] Eventually Millikan embraced the Thomson–H.A. Wilson method, modifying it until he could balance an oil drop in an electric field and watch its progress as it picked up charges.[116]

112. Thomson (1898), p. 528.
113. Ibid., p. 542.
114. C.T.R. Wilson (1899b).
115. Holton, 'Subelectrons', (note 110), esp. pp. 42–3.
116. Ibid., (note 110), esp. pp. 42–3.

Was Nature in the laboratory?

It is clear that C.T.R. Wilson's work was now assimilated into the heart of Cavendish research. On the one hand, many of his experiments were interpretable only in terms of ion physics and Thomson was quick to suggest ionic explanations of cloud-chamber phenomena. On the other hand, Thomson and other Cavendish physicists used Wilson's chamber to bolster a major premise of ion physics, the existence of an elemental charge. Yet Wilson hardly contributed to these experiments on matter. Why not? Wilson's attention fell elsewhere. To understand why, we need to specify more fully the goals of Wilson's self-defined sub-field and the relation in it of laboratory to atmospheric phenomena.

It was precisely in the link between laboratory and sky that Wilson's identification of condensation nuclei with electrical ions ran into trouble. In 1899, before his experiment on positive and negative ions, Wilson had been able to demonstrate conclusively what he could only suggest in 1897: condensation nuclei were ions. To prove his point, Wilson applied an electric field to his chamber after it had been exposed to X-rays or other sources of ions but before the expansion was made. He expected the field to sweep away the electrical ions, and to his satisfaction the plates did, in fact, diminish the condensation. At the time, Wilson had felt confident enough to state categorically: 'this behaviour of the nuclei proves them to be charged particles or "ions"'.[117] Unfortunately Wilson's identification procedure soon encountered two striking anomalies. Ultra-violet light produced condensation, but an electric field had no effect on droplet formation. To make matters worse, Wilson found that his electric fields seemed powerless to sweep away the nuclei continuously present in dust-free air. Both effects threatened the ion explanation of dust-free condensation.

Thus when Wilson turned to the experiments with positive and negative ions, he again tried to prove that ions caused condensation both in filtered air and in air exposed to ultra-violet light. But even with a stronger electric field and a more sophisticated method of determining if there was any effect, Wilson failed completely. An electric field produced no noticeable diminution of condensation. He was now faced with an almost impossible choice: the enormously productive ion hypothesis appeared pitted against two clear experimental counterexamples:

The slight rain-like condensation which takes place, when V_2/V_1 lies between 1.25 and 1.38, in the absence of all radiation, as well as the much denser

117. C.T.R. Wilson (1899a), p. 129.

condensation produced by the same expansions when the air is exposed to weak ultra-violet light, are thus essentially different phenomena from the apparently similar condensation produced in air ionised by Röntgen rays.

From such an observation, Wilson rhetorically asked if perhaps ions were not involved at all. But that hypothesis carried with it

the unlikelihood of two entirely different classes of nuclei being so exactly identical in the degree of supersaturation necessary to cause water to condense on them. The apparent existence of a second coincidence (an increase of the number of drops when V_2/V_1 exceeds 1.31) is still harder to explain on this view.[118]

Wilson could account for the electric field's failure to remove the ions by supposing that ions were artifacts of the expansion process. Since expansion occurred *after* the electric field was applied, Wilson's conjecture would explain why the field had no effect. He seemed happy with this *ad hoc* story. To resolve the conflict between experiment and theory Wilson modified the hardware – he tried to create a continuously operating chamber that would not require sudden expansions.[119] It was a short-lived attempt, however, and Wilson soon realised that he would have to give up the cloud chamber altogether if he wanted to find out if there were ions present in undisturbed air.

Wilson's primary purpose in designing the cloud chamber had been to reproduce atmospheric phenomena faithfully, yet it now threatened to be a repository of mere artifacts. Two problems threatened the link between the chamber and the Earth's atmosphere. First, when used on unmolested air, the device caused condensation – but the nuclei did not respond as to an electric field. Were the ions produced by the machine itself? Or were they another kind of nuclei that would not respond to electric fields? Second, while Wilson could show that Thomson's ions (generated, for example, by X-rays) *could* cause condensation, he could not show that the real atmosphere contained any such ions. Thus Wilson feared that he had entirely lost contact with the authentic conditions of Nature – dust-free condensation might have *nothing* to do with fog, rain, or atmospheric optics. It was the mimetic physicist's nightmare.

In a paper written in 1901, Wilson explained his quandary and described the electroscope he would now adopt as his preferred instrument. After months trying to devise a scheme to provide a continuous stream of drops, Wilson

118. C.T.R. Wilson (1899b), p. 305.
119. C.T.R. Wilson, B2, June 1900.

abandoned the condensation method, and resolved to try the purely electrical method of detecting ionisation. Attacked from this side the problem resolves itself into the question, Does an insulated-charged conductor suspended within a closed vessel containing dust-free air lose its charge otherwise than through its supports, when its potential is well below that required to cause luminous discharges?[120]

It seemed the cloud chamber had carried him to a dead end.

TRACKS

To save his condensation programme Wilson had to discover if his cloud chamber ions really existed in the atmosphere. In desperation he turned to an entirely different class of devices – electroscopes – which, although they could not visually reproduce atmospheric phenomena, could at least monitor atmospheric electricity continuously without creating artifacts. Having built an electroscope, Wilson turned to the well-known problem of leakage. Two German physicists, J. Elster and H. Geitel had shown that an electrified body lost its charge even in dust-free air; moreover it did so in night as in day, as much for positive as for negative charges on the electrodes of the apparatus, and the leak did not seem to depend on the voltage.[121] Wilson confirmed these findings but added two original contributions: the rate of leakage was roughly proportional to the pressure, and at atmospheric pressure, in the language of the Cavendish, the leakage corresponded to the production of 20 ions/cc/sec.[122] Wilson briefly toyed with the suggestion that the vessel walls might be radioactive. When that would not stand up to experiment, he wondered if the production of ions in dust-free air was 'due to radiation from sources outside our atmosphere, possibly radiation like Röntgen rays or like cathode rays, but of enormously greater penetrating power'.[123] Lugging his electroscope into a deep rock tunnel owned by the Caledonian Railway, Wilson found no evidence that his instrument leaked more slowly underground than it did on the surface. While later observers attributed his failure to detect cosmic rays to the Earth's radioactivity, at the time he saw no choice but to agree with Geitel that the ineradicable ionisation was due to 'a property of the air itself'.[124]

120. C.T.R. Wilson (1901), p. 152.
121. Wilson first read Geitel on spontaneous leakage in Geitel (1900–01), which Wilson received from Geitel; Wilson to Geitel, 28 November 1900, courtesy of Staatsbibliothek Preussischer Kulturbesitz, Berlin. Wilson rejected extra-terrestrial particles: 'the ionisation does not seem to be due to very penetrating rays which have traversed our atmosphere'. Wilson to Rutherford 20 April, 1901, Rutherford papers, Add 7653. (Microfilm, American Institute of Physics).
122. C.T.R. Wilson (1901), p. 153.   123. Ibid., p. 159.   124. Ibid., p. 161.

For three years Wilson pursued his radioactive air hypothesis by studying the radioactivity brought to Earth by rain and snow.[125] Meanwhile his Cavendish colleagues exploited the cloud chamber to explore more central questions of ion physics. Thomson and H.A. Wilson, for example, used it to determine the unit of elemental charge. In 1903 in a short review article on atmospheric electricity for *Nature*, C.T.R. Wilson wrote, 'It is quite conceivable that we may be driven to seek an extra-terrestrial source for the negative charge of the earth's surface'.[126] But Wilson never offered concrete arguments for the existence of cosmic rays. Instead, he concentrated his efforts on atmospheric electricity, speculating that the ionisation might come from meteorological processes, such as charges asymmetrically scattered during the splashing of raindrops.[127] A decade later, Victor Hess took three electroscopes on a balloon ascent and proved that the ionisation effects did not decrease with height as they should have were the radiation emanating from the Earth's crust.[128] It may be that Wilson's non-discovery of cosmic radiation reflected his preoccupation with meteorology, a world view that kept his mind's eye fastened on the sub-lunary sphere.

While J.J. Thomson pushed the cloud chamber unambiguously towards reductionist natural philosophy, Wilson's concerns circled around his self-constructed complex of condensation problems: atmospheric electricity, rain, hail, fog, ion properties, and atmospheric optics. His notebooks record a constant movement back and forth between questions of ionic charge and the nature of atmospheric phenomena. Again and again he returned to these core problems: his notes from 1908, for example, include a carefully lettered section entitled, 'Theory of Corona'.[129] Wilson's theory of rain formation involved several stages, First, water vapour condensed around negative ions – just as it did in the cloud chamber. If these small drops rose more slowly than the updraft of supersaturated air, the drops would rapidly grow, perhaps providing the mechanism by which a cumulus cloud metamorphoses into a cumulo-nimbus. Wilson speculated that a thin cloud cap might form suddenly 'over the head of the cumulus and sink rapidly into

125. C.T.R. Wilson (1902a, b, 1903a).   126. C.T.R. Wilson (1903b), p. 104.
127. In Wilson's (1903b), he favoured two purely meteorological explanations for atmospheric electricity: i) differential condensation on positive and negative charges could bring down the negative charges while leaving the positives, ii) raindrops, with their charges polarised in an ambient electric field, could hit other drops on their way down, driving one kind of charge up and the other down.
128. On Hess see for example, R. Steinmaurer, 'Erinnerungen an V.F. Hess, den Entdecker der Kosmischen Strahlung, und an die ersten Jahre des Betriebes des Hafelekar-Labors', in Sekido & Elliot (1985), pp. 17–31.
129. Wilson, A9, before dated entry of April 1908.

it . . . . The drops formed on the ions may themselves fall through the lower cloud and reach the ground as rain'.[130] In addition, the separation of positive ions (left in the air) from the negative ones (borne downwards by rain) would account for the intense electric fields found in thunderstorms.

Wilson's concern with the real condition of Nature drove him out of the laboratory. To test if beta rays from the atmosphere might be charging the Earth, he set his electrometer under the 'natural screening' of trees, or under sections of sod; he also used wire screens in the garden: 'Jan 16 [1909]. Experiments in garden. Cambridge. Electrometer placed on table under bare apple trees. Electrical field too small to detect . . . . Cloudless sky. [F]air breeze'.[131] Apparently to continue the study of very small drops that would initiate thunderstorms, Wilson returned to condensation experiments to count drops when they were numerous. He set up a Nernst lamp and inserted an eyepiece, but was distracted by the coronae in the new apparatus.[132] When Wilson resumed his rain experiments in March 1909, he wanted to know if his ion-condensed droplets grew into raindrops. At least since reading Osborne Reynold's work, he knew that coalescence was a strong candidate for the growth mechanism of both rain and hail. Simple inspection of the ice crystals' cross-sections demonstrated that hail could not be frozen rain. Furthermore, in both rain and hail the heat of condensation is too great for the growth to have been entirely due to condensation. 'Do ordinary cloud particles coalesce spontaneously? If not do they when rather larger drops fall through the cloud[?] Effect of electrical field[?]'[133] Using a vertical tube Wilson hoped to make visible the process of drop formation and growth.[134]

Obviously persuaded that condensation alone would not suffice to produce real rain, Wilson intensified his efforts to *see* the drop-formation process itself. As he speculated on droplet dynamics in his notebook, Wilson recalled the astonishing high-speed pictures of drops and splashes recently publicised by Worthington and Cole.[135] Worthington's

130. Ibid., April 1908, subsection entitled, 'Atmospheric Electricity [and] Condensation on Negative Ions'.
131. C.T.R. Wilson, A9, 16 January, 1909.
132. Ibid. Notes describing these experiments are between dated entries of 16 January, 1909, and 6 February, 1909.
133. C.T.R. Wilson, A9, between the dated entry of 6 February, 1909, and the dated entry of 10 April, 1909. Section titled, 'Meteorological Experiments'.
134. Ibid. Wilson then took careful reading notes on Reynolds (1879).
135. Worthington (1908); in Wilson's unpublished work see A9, section marked 'Atmospheric Electricity [and] Condensation on Negative Ions', immediately following dated entry of April 1908.

book, *A Study of Splashes* had, in 1908, just appeared and held a two-fold fascination for Wilson. First, the photographer had brilliantly exploited the light of sparks that endured mere *millionths* of a second to capture on film events ordinarily too fleeting for a human observer. (see Figure 8.7) The high-speed method offered Wilson the technical means to reveal the elementary processes of condensation and coalescence. Second, Worthington's splashes might indicate the mechanism that separated positive and negative particles, causing the electrical gradient in rain clouds. Again Wilson referred to photography as a means of reproducing cloud phenomena, though now at a more fundamental level. Wilson soon adapted Worthington's spark illumination system. In a section of his notebook labeled, *'On drop counting methods,'* he reasoned that a high-speed camera would enable him to count droplets more accurately; 'Methods depending on instantaneous photography of drops immediately after their production are superior to those in which drops falling through an illuminated layer are counted – the possibility of photographing the drops being presupposed,.[136] This was a crucial and original step away from the falling cloud methods of H.A. Wilson and Robert Millikan; by May 1909 Wilson had his first good negatives.

As was so often the case, within his own special science of condensation physics Wilson oscillated between meteorological and ion questions. Now his technical success with the photography of rain formation led him from meteorology into ion physics. His notebooks display an attempt to use a permanent film record to improve Thomson and H.A. Wilson's ion-charge experiments.[137] Despite some inventive schemes recorded in the notebook, Wilson never advanced the cause of charge determination, and soon returned to the photography of his laboratory clouds.

The day before Christmas, 1910 Wilson summarised the 'present state of work on expansion apparatus'. There he recorded his satisfaction with newly discovered tricks of the trade, especially his procedure for coating the glass bulb with gelatine. This prevented the glass from fogging (as water vapour deposited in a dew-like form) and offered a conducting surface for use in electrical experiments. Experimental skills like these could be used in conjunction with a variety of chambers. As he noted, the 'simple flash form' chamber would provide a stage for the photographing of individual drops condensing on 'spontaneously' produced ions. A second chamber would sport an electrode down the centre and would measure the effect of fields on spontaneously pro-

136. C.T.R. Wilson, A9, between 10 April, 1909, and 17 July, 1909, probably May 1909.
137. C.T.R. Wilson, A9, 13 September, 1909, and 16 January, 1910.

Peter Galison & Alexi Assmus

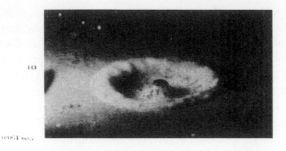

Figure 8.7. High speed splash pictures of milk on water. (Worthington, 1908, p. 19)

duced ions – and, with luck, help find the electric charge, *e*. Yet a third chamber is listed simply with 'applications' to search for particle tracks: 'Assuming that the difficulties of photographing the drops were overcome, one might see tracks of [alpha] rays marked out by narrow core of drops formed on positive ions with more scattered cloud of those formed on negative ions round [the positive ions.]'[138]

Learning to see with the cloud chamber

With all the means at his disposal, Wilson approached the elementary condensation process. Pursuing his quarry, he designed an electrometer that might reveal the presence of a charge of no more than 'about 30 ions'. On the very next page, in the entry of 18 March, 1911, Wilson returned to the cloud chamber. He was hoping to drive his photographic vision ever closer to the single ion:

Cloud chamber exposed to X rays (from above). Nernst lamp illumination. Cloud was discontinuous, showing numerous knots. Are these cross sections of tracks of rays? They showed up without field (exposure for short time only) Next expansion (without rays) showed uniform cloud (on residual nuclei) without any such knots. Expansion apparatus working perfectly. Field was due to single secondary cell.[139]

Was his striking first result an artifact? Wilson increased the electric field to encourage drop formation; within days there was no doubt: '[R]ays showed up much better. They were extremely sharply defined . . . . Look[ed] at from above. The individual rays were seen, in many cases as extremely fine lines, chiefly radiating from aperture – but in some cases running in other directions (secondary rays)'.[140] Photography, by then second nature to Wilson, lent credence to the phenomena. From the splash and drop-formation studies he knew the photographer's art could probe where the eye could not. Soon Wilson had the knack of photographing tracks. He recorded that he could produce images of beta rays taken 'as before'. Almost simultaneously he successfully found the footprints of X-rays and used gamma rays from radium to produce tracks even through a lead plate. By 29 March, 1911, Wilson had enough of a feel for the 'ordinary' images of radioactive processes to record an extraordinary one, shown in Figure 8.8: 'On one occasion in addition to ordinary thread-like rays, one large finger-like ray was seen, evidently a different form of secondary ray – giving rise to enormously more ionisation than even ordinary [alpha] ray'.[141]

138. C.T.R. Wilson, A9, 24 December. 1910.   139. C.T.R. Wilson, A9, 18 March, 1911.
140. C.T.R. Wilson, A9, 20 March, 1911.   141. C.T.R. Wilson, A9, 29 March, 1911.

Figure 8.8. The first extraordinary cloud chamber event. (Wilson notebook A9, 29 March, 1911, see note 38)

In mid-April Wilson submitted his first track paper, 'On a Method of making Visible the Paths of Ionising Particles through a Gas'.[142] There he exhibited his photographs of alpha rays and the 'clouds' that precipitated on X-ray-produced ions. 'The photograph', Wilson wrote, conveys 'but a poor idea of the really beautiful appearance of these clouds'.[143] It was clear that X-rays ionised by producing secondary, charged corpuscles. Still 'undecided' was the question whether an X-ray was a 'continuous wave front, or is itself corpuscular as Bragg supposes, or has in some other way its energy localised . . . in the manner suggested by Sir J.J. Thomson'.[144]

In this and in many other questions regarding the neutron, the wave-particle duality, nuclear scattering, nuclear transformations, the cloud chamber played a central part in Cavendish physics. On all fronts physicists now wanted to 'see' the processes they earlier had to study by less tangible inference. Rutherford, Thomson's successor as director of the Cavendish, steered the lab into a new channel of physics research. The goal was to understand the nature of sub-atomic matter. Although it followed from Thomson's goal of explaining the ionic structure of Nature it was both broader in scope and finer in detail. Neutral particles were admitted into the game and the point was to try to smash and bombard the more complex manifestations of matter until the simplest sprayed forth, bringing with them the secret of the atom. Wilson's

142. C.T.R. Wilson (1911).   143. Ibid., p. 286.   144. Ibid., pp. 287–8.

cloud chamber was indispensable for this task; having been developed under the aegis of ionic physics it now led the way into the sub-atomic realm. Yet Wilson neither led nor followed his Cavendish colleagues, withdrawing ever more toward weather phenomena. During the heyday of Cavendish cloud-chamber atomic physics Wilson's separation was as physical as it was scientific: he moved his apparatus to the Solar Physics Laboratory, geographically distinct from the Cavendish. There he found the solitude he needed to begin innovative meteorological studies, culminating in his final paper at the age of ninety – the first paper to give a dynamical explanation of thunderstorms.

From high on mountain-top observatories to the protection of underground laboratories, cloud chambers became the means to the discovery of new particles. So successful was the instrument, and its close cousin the nuclear emulsion, that in 1951 Blackett reported that 'All but one of the now known unstable elementary particles have been discovered by one or other of these techniques'.[145] Both methods produced *visual* evidence, giving 'us pictures of what single particles do'. The skills involved in identifying characteristic tracks soon became a quasi-autonomous branch of experimental science. Leaders in the field created a new species of literary object, the cloud-chamber atlas. This had as its stated function the teaching of pattern-recognition skills. In the foreword to one, Blackett puts its purpose succinctly:

An important step in any investigation using (the visual techniques) is the interpretation of a photograph, often of a complex photograph, and this involves the ability to recognise quickly many different types of sub-atomic events. To acquire skill in interpretation, a preliminary study must be made of many examples of photographs of the different kinds of known events. Only when all known types of event can be recognised will the hitherto unknown be detected.[146]

The creation of this domain of skills would soon constitute a bridge to other instrument systems because physicists trained in cloud-chamber techniques could move easily to emulsion groups and emulsion physicists could transfer their abilities to the bubble chamber. By creating an example of a visual, single-particle detector, Wilson had laid the groundwork for a profoundly new class of instruments; at the same time he introduced a new *class* of physical evidence: pictures of interactions, and occasionally the single, decisive 'golden event'.

145. P.M.S. Blackett, foreword to Rochester & Wilson (1952), p. vii. Cf. Gentner, Maier-Leibnitz & Bothe (1940, 1954).
146. Blackett, (*ibid.*)

THE ORIGIN OF THE CLOUD CHAMBER AND THE IDEALS OF EXPERIMENTATION

Historically and philosophically the cloud chamber lies at a crossroads. The device issued from two quite distinct branches of natural science: the tradition we have labelled analytic, located at the Cavendish, and the very different morphological tradition, located in the field. C.T.R. Wilson straddled the two, and for almost two decades succeeded in creating a hybrid subject at which he excelled. We call this 'bridge' field 'condensation physics', a domain that drew for its practitioners, devices, techniques, theoretical entities, and goals from both branches of knowledge. (see Figure 8.9)

During the period 1895–1910 both morphological and analytical physics were transformed by Wilson's work. In the early nineteenth century, meteorology along with geology, was predicated on accurate classification. Successful research drew its strength from scrupulous analysis in the field. Humboldt set the tone with his wide-ranging explorations and systematic use of careful data-taking. Through his extensive use of isobaric and isothermal charts, he lifted meteorology from the chaos of conflicting local reports to reveal an order only visible on scales of tens, even hundreds of miles. A similar, field-based effort around the globe led to a widely accepted stratigraphy and its accompanying geological history of the Earth. Cloud physics could only begin after Luke Howard's remarkable classification.

At the time Wilson came of scientific age a second transformation of the morphological sciences was under way, one in which *in situ* explorations could be complemented by *in vitro* investigations. For decades the morphologists had held the laboratory in some suspicion – after all the great gains of recent decades had been made by looking at Nature as it really was, not as artfully constituted in the glass, chemicals, and springs of the workshop. To achieve legitimacy in the eyes of the morphologists the new experimentation of the 1880s and 1890s had to promise to connect to the real world if it was to have any persuasive power. It was here that experiments on synthetic lava flows, miniature glaciers, and scaled-down mountain-building had a role to play.

John Aitken, whose work stood most immediately in Wilson's vision, was a perfect exemplar of this 'mimetic' transformation of the morphological sciences which attempted to recreate rather than dissect Nature. The causal structure of the miniature cyclone was supposed to be *just like* the vast entity that could destroy towns and flood fields. Most importantly, Aitken's dust chamber could recreate some of the

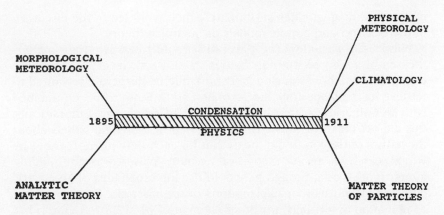

Figure 8.9. The creation and disintegration of condensation physics.

most striking meteorological phenomena of the times. In a controlled, but faithful fashion, it could produce the fogs that had descended threateningly on England's industrial cities; it could make rain and, most strikingly, the amazing green sunsets, the afterglow of Krakatoa. Aitken's and others' related work provided the basic material culture for Wilson's entry into science. From Aitken, Wilson drew the mechanical design of the cloud chamber: the pump, reservoir, filter, valves and expansion mechanisms all issued directly from the dust chamber. Indeed Wilson's desire to reproduce clouds meshed well with Aitken's own fascination with the recreation of dramatic aspects of Nature. At the same time Wilson's youthful enthusiasm for the photographic imitation of Nature, the building of models, the drawing of maps, and the taxonomic collection of insects flowed smoothly into his mature interest in the production and reproduction of clouds.

When Wilson went up to Cambridge, he entered a community with utterly different ambitions. Under J.J. Thomson, Cavendish physics held classification and reproduction as 'mere' Natural History. For Thomson, the new physics would *explain* the descriptive feature of Nature dynamically, in terms of more basic, simpler units. It would do so by taking Nature apart, not by imitating it. Thomson himself had contributed significantly to knowledge of the ion – the charged particle that would, in aggregate, constitute bulk matter. Using electrified gases, vacuum pumps, and electrometers, the ion physicists had by the late 1880s and early 1890s began to articulate some of the entity's properties. Thomson and his colleagues produced and manipulated the ion using electric and magnetic fields. Ultimately their work led to the discovery of the electron and atomic models of quantum theory.

Wilson did *not* follow the Cavendish route towards purely matter-theoretical goals, yet the ion figures in his work from the very first page of his notebook on clouds. Even while he desperately wanted to retrieve in the laboratory the coronae and St Elmo's fire, he imbibed with his Cambridge education the tradition of ion physics. Emphatically, this did *not* mean he was committed to particular, rigid beliefs about the nature of the ion. It did mean that for each new phenomenon, he would search for an ion-theoretical account. Wilson was not building atomic models and, though he would use the broad idea of an electric particle, had no use for explanations of spectroscopic lines, chemical combination or virtually any other element of what John Heilbron has called Thomson's "program".[147]

The cloud chamber was born and raised in the tension between the two programmes. In good Thomsonian fashion Wilson used X-rays to make rain nuclei, just as Thomson exploited X-rays to make conduction ions. Wilson built charged parallel plates to move condensation nuclei, just as Thomson had installed them to displace conduction ions. But in stark contrast to Thomson, Wilson then directed his energies towards the reproduction of phenomena that had drawn him into physics in the first place: thunderstorms, coronae, and atmospheric electricity.

Condensation physics, 1895–1911

Was Wilson's cloud-chamber research of 1895–1910 meteorological, or was it matter physics? Is it reasonable to assess meteorology as a 'practical inspiration' that launched Wilson's 'purer' science? Or should we instead deduce that the meteorology was no more than 'pure' Cavendish ion physics 'applied' to the complicated, great outdoors? In light of what we have seen, both the 'application' and 'inspiration' accounts appear thoroughly inadequate: we need better posed questions.

Wilson wanted to know how condensation worked *in the world*. Thus his work certainly had ties to the more general Cavendish program – there was a shared concern with the nature of the ion, its charge, the relation of positive and negative ions. But Wilson was also tied to physicists altogether outside that tradition. He drew heavily on the work of the American physicist Carl Barus, on the dust-master John Aitken, on the student of high-speed splashes, Worthington, and on the many others who had manipulated droplets, hailstones, steam jets,

147. Heilbron (1977), 52p.

and splashing water. Perhaps the most striking evidence of Wilson's participation in the morphological tradition was his abandonment of the cloud chamber when he suspected that it was generating condensation nuclei not normally present in the atmosphere. For Wilson, the morphological physicist, such a discrepancy with real atmosphere was an artifact of the most subversive kind.

Our notion of a hybrid problematic drawing on two previously distinct areas is explanatory in a positive and a negative sense. *Positively*, Wilson drew on technologies from both predecessor fields, pulling the dust chamber from meteorology and the electric deflection plates from matter physics. Perhaps most importantly – and most subtly – Thomson's ion served as an *enabling notion*: the idea of particulate electricity, not the detailed theory embedded in atomic models, *made possible* the rearrangement of the cotton filter permitting experimentation on dust-free air. In a strong sense Wilson used the notion of an ion to transform the dust chamber into the cloud chamber. Simultaneously, Wilson's morphological commitments exerted a crucial *negative* or constraining influence in the definition of condensation physics. Meteorology imposed boundaries on what Wilson would consider reasonable experimentation. When he obtained results that appeared not to hold good in the bulk of the Earth's atmosphere, he abandoned his cloud-chamber research. When the determination of the charge *e* exceeded in accuracy what Wilson needed to address real atmospheric condensation, he lost interest. And when the spontaneous discharge of electroscopes led to cosmic rays, he observed the research from a respectful distance. Ion theory too imposed constraints, for Wilson demanded that his experiments be reconcilable with Cavendish experiments on conduction ions. Other physicists working at the same time did not feel their work to be under such constraints. For all these reasons there was a brief moment in the history of the physical sciences, from 1895 to 1911, when condensation physics participated in both fields at once: it was an investigation into the fundamental electrical nature of matter *and* an exploration into clouds.

Condensation physics as an overlapping concern of meteorology and ion physics ended abruptly in 1911. In March of that year Wilson turned the cloud chamber towards its new role as a detector. With that move, the physics community seized upon the cloud chamber-as-detector and consequently relegated its earlier functions – its ability to produce all those phenomena that principally intrigued Charles Wilson – into the background. There was no longer any useful sense in which ions, coronae, fogs, and rain raised the same questions as problems in the theory of matter. As the knotty clouds blended into the tracks of alpha particles and the 'thread-like' clouds became beta-particle trajectories, the old sense and meaning of the chamber changed. Previously Wilson had an audience among the matter physicists for his question: how do droplets form? Now, in the water trails, the matter physicists had other, more pressing queries: what are the energies of the gamma rays that produce electrons? How do alpha particles scatter? What does the scattering of a particle imply about it and its target? With the addition of magnets, and then of triggerable expansions, the chambers began to reveal new particles, with their masses and interaction properties deduced from the curvature and density of the curved trajectories. For hunters, snow tracks tell about quarry not snow.

Meteorology and the physics of matter diverged and condensation physics disintegrated. But Wilson's work had permanently changed both. When contributors to the ninth edition of the *Encyclopedia Britannica* reviewed meteorology in 1895, they identified the founding of scientific meteorology with Humboldt's *Isothermal Lines*.[148] Geographic variation and periodic temporal variation of climate and weather were the substance of the field. Essentially the subject amounted to the systematic compilation of daily, weekly, monthly, and yearly variation of temperature, pressure, wind, rain and cloud cover. Sixteen years later the 11th edition began by dividing the subject into the statistical study of the atmosphere known as *climatology* and the theoretical and physical study of meteorology. The latter embraced both the macroscopic, hydrodynamical processes of bulk air and the microscopic explanations of processes that are 'mostly comprised under thermodynamics, optics and electricity'. None of the analytic schemes is given as much prominence as C.T.R. Wilson's theory of dust-free condensation. The encyclopedia presents this as the 'first correct idea as to the molecular processes involved in the formation of rain' and atmospheric electricity.[149] Meteorology thus passed from pure climatology to a split discipline: it embraced a large-scale descriptive function and a dynamic explanatory component that would be known as physical meteorology.

Real particles from artificial clouds

Cavendish-style physics too was transformed. The cloud chamber had rendered the sub-atomic world visualisable – and consequently it took on a reality for physicists that it never could have obtained from the

148. Buchan & Stewart (1895), p. 114.   149. Abbe (1911) pp. 264, 290.

chain of inferences that had previously bolstered the corpuscular viewpoint. Soon the cloud chamber and its close relation, the nuclear emulsion,became the defining instruments of a new field: particle physics. As late as 1951 all but one of the myriad of unstable particles had been discovered in one or the other of these two visual devices. The cloud-chamber experts, scientists including Anderson, Auger, Blackett, Leprince-Ringuet, Rochester, Skobeltzyn and Thomson, all succeeded in large part because they came to have a craft familiarity with tracks themselves. Cloud-chamber studies – or more broadly, track-following experiments – helped usher in particle physics.

In November 1927 George P. Thomson wrote a congratulatory letter to Wilson on his receipt of the Nobel Prize. 'Your work has always seemed to be the *beau idéal* of an experiment, carrying such immediate and complete conviction, and making real and visible what was before only, after all, a theory. It is a perfect example of the peculiar genius of British physics'.[150] Wilson eventually failed in his attempt to persuade others that clouds formed around ions – dust and sea salt capture much more water than electrical particles. His laboratory clouds therefore never transformed meteorology. But with artificial clouds Wilson had made particles real.

ACKNOWLEDGEMENTS

We would like to thank D. Blanchard, A. Davison, P. Forman, G. Holton, C. Jones, M. Jones, Jane Maienschein, S. Della Pietra, M. Roy and J.G. Wilson for helpful comments and the staffs at the library of the Royal the Edinburgh University Library, the Staatsbibliothek Preussischer Kulturbesitz Berlin, the Churchill College Library and the Meteorological Office, Edinburgh. Miss J.I.M. Wilson and Miss R.H. Wilson kindly provided us with unpublished letters and photographs.

References

Abbe, C. (1911). Meteorology. In *Encyclopedia Britannica* 11th edn. vol. 18, pp. 264–91. Cambridge: Cambridge University Press.

Aitken, J. (1873). Glacier motion, *Nature*, February 13; reprinted in Aitken (1923), paper 2, pp. 4–6.

Aitken, J. (1876–77). On ocean circulation. *Proceedings of the Royal Society of Edinburgh*, 9; reprinted in Aitken (1923), paper 5, pp. 25–9.

150. G.P. Thomson to C.T.R. Wilson, November 1927. Private papers, Wilson family. Courtesy, Miss J. Wilson.

Aitken, J. (1880–81). On dust, fogs, and clouds. *Transactions of The Royal Society of Edinburgh*, **30**; reprinted in Aitken (1923), paper 7, pp. 34–64.

Aitken, J. (1883–84a). On the Formation of small clear spaces in dusty air, *Transactions of the Royal Society of Edinburgh*, **32**, reprinted in Aitken (1923), paper 11, pp. 84–113.

Aitken, J. (1883–84b). Second note on the remarkable sunsets. *Proceedings of the Royal Society of Edinburgh*, **12**; reprinted in Aitken (1923), paper 14, pp. 123–33.

Aitken, J. (1888). On the number of dust particles in the atmosphere. *Transactions of the Royal Society of Edinburgh*, **35**; reprinted in Aitken (1923), paper 16, pp. 187–206.

Aitken, J. (1889–90, 1892a, 1894). On the number of dust particles in the atmosphere of certain places in Great Britain and on the Continent, with remarks on the relation between the amount of dust and meteorological phenomena, Part I, *Proceedings of the Royal Society of Edinburgh*, (1889–90) **17**; Part II, *Transactions of the Royal Society of Edinburgh*, (1892a), **37**; Part III, *Transactions of the Royal Society of Edinburgh*, (1894) **37**; reprinted in Aitken (1923), papers 23–25, 297–434.

Aitken, J. (1892b). On some phenomena connected with cloudy condensation, *Proceedings of the Royal Society of London*, **51**; reprinted in Aitken (1923), paper 20, pp. 255–83.

Aitken, J. (1900–01). Notes on the dynamics of cyclones and anticyclones. *Transactions of the Royal Society of Edinburgh*, **40**, reprinted in Aitken (1923), paper 27, pp. 438–58.

Aitken, J., (1915). The dynamics of cyclones and anticyclones. *Proceedings of the Royal Society of Edinburgh*, **36**; reprinted in Aitken (1923) paper 28, pp. 459–67.

Aitken, J. (1923). *Collected Scientific Papers*. ed. C.G. Knott. Cambridge: Cambridge University Press.

Badt, K. (1950). *John Constable's Clouds*. London: Routledge & Kegan Paul.

Ben Nevis & Fort William Observatories, Directors (1903). Memorandum by the Directors of the observatories of Ben Nevis and at Fort William in connections with their closures. *Journal of the Scottish Meterorological Society*, **12**, 161–3.

Blackett, P.M.S. (1960). Charles Thomas Rees Wilson 1869–1959. *Biographical Memoires of Fellows of the Royal Society*, **6**, 269–95.

Boas, F. (1887). The study of geography. *Science*, **9**, 137–41.

Brush, S. (1978). Planetary Science: from underground to underdog. *Scientia*, **113**, 771–87.

Brush, S. & Landsberg, H. (1985). *The History of Geophysics and Meteorology: An Annotated Bibliography*. New York and London: Garland.

Buchan, A. & Stewart, B. (1895). Meteorology. In *Encyclopedia Britannica*, vol. 16, pp. 114–84. Chicago: Werner Company.

Cannon, S. (1978). *Science in Culture: The Early Victorian Period*. New York: Dawson and Science History Publications.

Crowther, J.G. (1968). *Scientific Types*. London: Cresset Press.

Crowther, J.G. (1974). *The Cavendish Laboratory 1874–1974*. New York: Science History Publications.

Coulier, J.P. (1875). Note sur une nouvelle propriété de l'air. *Journale de Pharmacie et de Chimie*, **22**, 165–73, 254–5.

Dee, P.I. & Wormell, T.W. (1963). An index to C.T.R. Wilson's laboratory records and notebooks in the Library of the Royal Society. *Notes and Records of the Royal Society of London*, **18**, 54–66.

Feynman, R.P. (1964). *The Feynman Lectures on Physics*. Reading: Addison-Wesley.

Franklin, A. (1981). Millikan's published and unpublished data on oil drops. *Historical Studies of Physical Sciences*, **11**, 185–201.

Geitel, H. (1900–01). Uber die elektrizitätszerstreuung in abgeschlossenen luftmengen, *Zeitschrift für Physik*, **2**, 116–19.

Gentner, W., Maier-Leibnitz, H. & Bothe, W. (1940). *Atlas Typischer Nebelkammerbilder mit Einführung in die Wilsonsche Methode*. Berlin: Julius Springer.

Gentner, W., Maier-Leibnitz, H. & Bothe, W. (1954). *An Atlas of Typical Expansion Chamber Photographs*. New York: Interscience Publishers.

Gillispie, C.C. (ed.) (1981). *Dictionary of Scientific Biography*. New York: Scribner's Sons.

Glazebrook, R. (ed.) (1923). *A Dictionary of Applied Physics*. London: Macmillan.

Goethe, J.G. (1960). *Schriften zur Geologie und Mineralogie*, vol. 20 of the *Gesamtausgabe der Werke und Schriften in Zweiundzwanzig Bänden*. Stuttgart: J.G. Cotta'sche Buchhandlung Nachfolger.

Greene, M. (1982). *Geology in the Nineteenth Century*. Ithaca: Cornell University Press.

Halliday, E.C. (1970). Some memories of Prof. C.T.R. Wilson, English pioneer in work on thunderstorms and lightning. *Bulletin of the American Meteorological Society*, **51**, 1133–5.

Harman, P.M. (1982). *Energy, Force, and Matter. The Conceptual Development of Nineteenth-Century Physics*. Cambridge: Cambridge University Press.

Heilbron, J. (1964). *A history of the problem of atomic structure from the discovery of the electron to the beginning of quantum mechanics*. PhD thesis, University of California at Berkeley.

Heilbron, J. (1977). Lectures on atomic physics 1900–1922, In *History of 20th Century Physics Proc. of International School of Physics Enrico Fermi*, course 57, NY, pp. 40–108.

Helmholtz, R.V. (1886). Untersuchungen über Dampfe und Nebel, besonders über solche von Lösungen. *Ann. Phys. Chem.*, **27**, 508–43.

A History of the Cavendish Laboratory 1871–1910 (1910). London: Longmans, Green, and Company.

Holton, G. (1978). *The Scientific Imagination*. Cambridge: Cambridge University Press.

Howard, L. (1803). On the modifications of clouds, and on the principles of their production, suspension, and destruction; being the substance of an essay read before the Askesian Society in the Session 1802–3. Part I, II; *Philosophical Magazine*, 16, 97–107; 344–57; Part III; *Philosophical Magazine*, 17, 5–11.

Keller, A. (1983). *The Infancy of Atomic Physics*. Oxford: Clarendon Press.

Knoepflmacher, U.C. & Tennyson, G.B. (ed.) (1977). *Nature and the Victorian Imagination*. Berkeley: University of California Press.

Maxwell, J.C. (1885). Physical sciences. In Encyclopedia Britannica, 9th edn, vol. 19. pp. 1–3. Edinburgh (1885) and Chicago (1895): The Werner Company.

Mayr, E. (1982). *Growth of Biological Thought*. Cambridge Mass.: Harvard University Press.

Merz, J.T. (1965). *A History of European Thought in the Nineteenth Century* (first published between 1904 and 1912). New York: Dover.

Paradis, J. & Postlewait, T. (1981). *Victorian Science and Victorian Values: Literary Perspectives*. New York: New York Academy of Sciences.

Paton, J. (1954). Ben Nevis Observatory 1883–1904. *Weather*, 9, 291–308.

Rankin, A. (1891). Preliminary notes on the observations of dust particles at Ben Nevis Observatory. *Journal of the Scottish Meteorological Society*, 9, 125–32.

Reyer, E. (1892). *Geologische und geographische Experimente. Heft I, Deformation und Gebirgsbildung*. Leipzig: Wilhelm Engelmann.

Reynolds, O. (1879). On the manner in which raindrops and hailstones are formed. *Memoirs of the Literary and Philosophical Society of Manchester*, 6, 48–60.

Rochester, G.D. & Wilson, J.G. (1952). *Cloud Chamber Photographs of the Cosmic Radiation*. New York: Academic Press.

Rudwick, M. (1985). *The Great Devonian Controversy*. Chicago: The University of Chicago Press.

Scottish Meteorological Society (1884). Report of the Council. *Journal of the Scottish Meteorological Society*, 7.

Sekido, Y. & Elliot, H. (ed.) (1985). *Early History of Cosmic Ray Studies*. Dordrecht: Reidel.

Shaw, W.N. (1931). A century of meteorology. *Nature*, 128, 925–6.

Sviedrys, R. (1970). The rise of physical science at Victorian Cambridge. With commentary by A. Thackray and reply by Sviedrys. *Historical Studies of Physical Sciences*, 2, 127–51.

Symons, G.J. (ed.) (1888). *The Eruption of Krakatoa and Subsequent Phenomena*. Report of the Krakatoa Committee of the Royal Society. London: Trübner.

Thomson, J.J. (1886). Some experiments on the electric discharge in a uniform electric field, with some theoretical considerations about the passage of electricity through gases. *Proceedings of the Cambridge Philosophical Society*, 5, 391–409.

Thomson, J.J. (1888). *Applications of Dynamics to Physics and Chemistry*. London: Macmillan.

Thomson, J.J. (1893). On the effect of electrification and chemical action on a steam-jet, and of water-vapour on the discharge of electricity through gases. *Philosophical Magazine V*, 36, 313–27.

Thomson, J.J. (1897). Cathode rays. *Philosophical Magazine V,* **44**, 293–316.

Thomson, J.J. (1898). On the charge of electricity carried by the ions produced by Röntgen rays. *Philosophical Magazine,* **46**, 528–45.

Thomson, J.J. (1937). *Recollections and Reflections.* New York: Macmillan.

Thomson, J.J. & Rutherford, E. (1896). On the passage of electricity through gases exposed to Röntgen rays. *Philosophical Magazine, V,* **42**, 392–407.

Tomas, D.G. (1979). Tradition, context of use, style and function; expansion apparatuses used at the Cavendish Laboratory during the period 1895–1912. Unpublished M.Sc. thesis, University of Montreal.

Wedderburn, Sir E. (1948). The Scottish Meteorological Society. *Quarterly Journal of the Royal Meteorological Society,* **74**, 233–42.

Wheaton, B.R. (1983). *The Tiger and the Shark.* Cambridge: Cambridge University Press.

Wilson, C.T.R. (1895). On the formation of cloud in the absence of dust. *Proceedings of the Cambridge Philosophical Society,* **8**, 306.

Wilson, C.T.R. (1896). The effect of Röntgen's rays on cloudy condensation. *Proceedings of the Royal Society of London,* **59**, 338–9.

Wilson, C.T.R. (1897a). Condensation of water vapour in the presence of dust-free air and other gases. *Philosophical Transactions A,* **189**, 265–307.

Wilson, C.T.R. (1897b). On the action of uranium rays on the condensation of water vapour. *Proceedings of the Cambridge Philosophical Society,* **9**, 333–8.

Wilson, C.T.R. (1899a). On the condensation nuclei produced in gases by Röntgen rays, uranium rays, ultra-violet light, and other agents. (abst.). *Proceedings of the Royal Society of London,* **64**, 127–9.

Wilson, C.T.R. (1899b). On the comparative efficiency as condensation nuclei of positively and negatively charged ions. *Philosophical Transactions A,* **193**, 289–308.

Wilson, C.T.R. (1901). On the ionisation of atmospheric air. *Proceedings of the Royal Society of London,* **68**, 151–161.

Wilson, C.T.R. (1902a). On radio-active rain. *Proceedings of the Cambridge Philosophical Society,* **11**, 428–30.

Wilson, C.T.R. (1902b). Further experiments on radio-activity from rain, *Proceedings of the Cambridge Philosophical Society,* **12**, 17.

Wilson, C.T.R. (1903a). On radio-activity from snow. *Proceedings of the Cambridge Philosophical Society,* **12**, 85.

Wilson, C.T.R. (1903b). Atmospheric electricity. *Nature,* **68**, 102–4.

Wilson, C.T.R. (1911). On a method of making visible the paths of ionising particles through a gas. *Proceedings of the Royal Society of London.* **85**, 285–8.

Wilson, C.T.R. (1954). Ben Nevis sixty years ago. *Weather,* **9**, 309–11.

Wilson, C.T.R. (1960). Reminiscences of my early years. *Notes and Records of the Royal Society of London,* **14**, 163–73.

Wilson, C.T.R. (1965). On the cloud method of making visible ions and the tracks of ionizing particles. *Nobel Lectures, Physics, 1922–1941,* pp. 194–215. Amsterdam: Elsevier.

Wilson, D.B. (1982). Experimentalists among the mathematicians: physics in the Cambridge natural sciences tripos, 1851–1900. *Historical Studies of Physical Sciences*, **12**, 325–71.

Wilson, J.G. (1951) *The Principles of Cloud-Chamber Technique*. Cambridge: Cambridge University Press.

Wise, N. & Smith, C. (1988). *Energy and Empire*. Cambridge: Cambridge University Press. (in press).

Worthington, A.M. (1908). *A Study of Splashes*. London: Longmans, Green and Company.

9

LIVING IN THE MATERIAL WORLD: ON REALISM AND EXPERIMENTAL PRACTICE

ANDY PICKERING

He's found his members, said Kleinzeit. He's remembered
himself.
What is harmony, said Hospital, but a fitting together? . . .

What does it all mean? said Kleinzeit.
How can there be meaning? said Hospital. Meaning is a limit.
There are no limits . . .

Place of dismemberment? said Kleinzeit.
Everywhere, all the time, said Hall of Records.

'Rather nicely stabilized, I should say,' said Dr Pink.
'I'm quite pleased with you actually.'
Kleinzeit, *Russell Hoban*

REALISM AND EXPERIMENTAL PRACTICE

The uses of experiment in natural science are many, but one comes immediately to mind: finding out about, and making sense of, the material world. My concern here is to explore what can be said about the processes of finding out and making sense. More explicitly, my aim is to enquire into the relation between articulated scientific knowledge and its object, the material world. I want, that is, to investigate a set of issues conventionally discussed under the heading of realism.

Realism is an old topic but, from the perspective adopted here, traditional discourse on realism has evolved in strange ways. It might seem evident that to say anything for or against realism one should pay close attention to the relation between conceptual and material practices, between performances in the material world and our understanding of them. But this is precisely what is not done in traditional discourse. Philosophical realists clash with antirealists over the status of theoretical

275

entities appearing within well confirmed theories, and the notion of a well confirmed theory takes for granted a particular relation between two sets of *statements*: statements of theoretical prediction are taken to agree with statements of experimental fact. The *constitution* of statements of fact in material practice is left unexamined.[1] Outside this mainstream debate more attention is sometimes paid to the extralinguistic correlates of scientific knowledge, but the typical model of nonlinguistic practice is then *ostension*. Ostension is, to say the least, an impoverished model of experimental practice. A glance into any scientific laboratory reveals a range of engagements with the material world of which ostension is the least intimate: scientists are continually handling material objects, forming and reforming them, arranging and rearranging them: pointing and looking would seem to be a small and not necessarily very revealing part of what goes on.[2]

The conviction which informs this essay is that much remains to be said on the issue of realism, just because the relation of articulated knowledge to practice in the material world has been allowed to go largely unexamined. That is not to say that there are no studies of experimental practice in science. There are. The contributions to this volume, and the works they cite, are part of a growing body of studies of experiment. But the relevance of such studies for issues connected with realism has so far been underplayed. The main concern of analysts has been, for example, with the constitutively social dimensions of knowledge production and use. Without in any way devaluing what has been thus achieved, I want to foreground a different aspect of what can be learned from studies of experiment; I focus upon what they can tell us about realism. I begin with some uncontentious generalities, before sketching out and exemplifying a tenably realistic position in an historical example.

Three elements are conjoined in the production of any experimental fact: a *material procedure*, an *instrumental model* and a *phenomenal model*. By material procedure, I refer to experimental action in the material world: setting up the apparatus, running and monitoring it in the

---

1. See the essays collected in Leplin (1984).
2. Ostension serves as a model for the extralinguistic dimension of practice in, for example, Barnes (1982) and Bloor (1983). Reference to ostension counteracts the philosophical tendency to take 'facts' for granted: it seems clear that by pointing and looking one can divide up the natural world in any number of ways. The question then arises of why some classification systems are preferred to others. Typically, this question is answered in terms of the nonostensive uses to which such systems are put – but this move returns us to the need for a more general analysis of the extralinguistic dimensions of practice, of doing as well as seeing.

laboratory.[3] The instrumental model expresses the experimenter's conceptual understanding of how the apparatus functions, and is central to the design, performance and interpretation of the experiment. It is tempting to imagine that these two elements are the only ingredients of any experiment. Material procedures, interpreted through instrumental models, yield facts directly. But this view is hard to sustain. Experimental facts are produced as *meaningful* facts about the world of natural phenomena, and this is why I suggest that there is a third element intrinsic to any experiment. The third element, which endows experimental findings with meaning and significance, is a phenomenal model, a conceptual understanding of whatever aspect of the phenomenal world is under investigation. Experiments typically aim at, say, the determination of some parameter which is initially unspecified in some phenomenal model. Thus, phenomenal models are both an input to experimental practice – the experiment is designed with a phenomenal model or models in mind – and, more closely specified, an output. The output model carries whatever fact has been determined in the course of the experiment. With this understood, I shall tend to speak either about phenomenal models or about facts, according to which is more appropriate in particular contexts.[4]

These three elements span the material and conceptual dimensions of experimental practice, and it is in this spanning of the two dimensions that connections between the material world and scientific knowledge of it are to be sought. The first point to note is that, in a typical passage of experimental activity, there is *no* apparent relation between the three elements. Incoherence and uncertainty are the hallmarks of experiment, as reported in ethnographic studies of laboratory life.[5] But, at the moment of fact-production, their relation is one of *coherence*. Material procedures and instrumental and phenomenal models hang together and reinforce one another. Material procedures, to be more

3. Note that this conception of material procedure includes, but is not exhausted by, visual scrutiny of the apparatus.
4. 'Phenomenal models' should not necessarily be equated with 'high theory'. I want only to recognise that all facts are, to some degree, conceptualised independently of their material basis, and that instrumental models make translations between material procedures and such conceptualisations. Thus, for example, the low-level 'construals' which Gooding (1986) argues to have been central to Faraday's work on electromagnetism are phenomenal models in the present sense. Pinch (1985) discusses the shifting levels of theoretical significance claimed for experimental findings in controversial situations.
5. See, for example, Collins (1985), Chapter 3, Knorr-Cetina (1981); Latour & Woolgar (1979) and Lynch (1985).

explicit, when interpreted through an instrumental model, produce facts within the framework of a phenomenal model. This observation of three-way coherence in fact-production seems almost trivial. How else could it be? But, following up my remark that uncertainty is endemic to experimental practice, I want to say that such coherence is itself highly nontrivial. And I want further to argue that an exploration of how coherence is achieved can shed important light on the issues surrounding realism.

Explaining coherence

How should the three-way coherence sustaining experimental facts be understood? The simplest answer is that coherence is *natural* and therefore unproblematic. Facts arise when material procedures and conceptual models are all correct, when they correspond to the true state of Nature. Their coherence thus follows automatically: they are all correct within the same world of Nature; Nature is coherent; therefore the three elements of experimental practice must themselves be coherent. The elements are *fixed* for all time by and in Nature itself. One can call this view a *correspondence realism* of facts (and phenomena, and material procedures, and instrumental models). It is, I believe, a fair description of the perspective scientists adopt towards established facts, and it is arguably the perspective underlying the traditional realism debate in philosophy.

It is hard to sustain this kind of correspondence realism. It can be undermined – though I develop a different argument below – by pointing to the possible nonuniqueness of coherent constellations of procedures and models. In *Changing Order* (1985), for example, Collins reviews a range of examples drawn from modern science in which different actors have sustained different coherent sets of phenomenal models, instrumental models and material procedures. Thus, in his classic study of the controversy over gravitational radiation, Collins demonstrates that one scientist, Joseph Weber, interpreting particular material procedures through particular instrumental models, was able coherently to sustain a phenomenal model in which the cosmos was traversed by an unexpectedly high flux of gravity waves. Other scientists, using other procedures and instrumental models, maintained a quite different phenomenal model, in which there was no detectable flux of gravity waves. There are now many studies that point to the same conclusion: coherence between material procedures and conceptual models is no guarantee of naturalness and uniqueness.[6] Correspondence realism at the level of facts is correspondingly thrown into doubt.

This observation opens up the question of realism, and has been argued to point to antirealist conclusions. The nonuniqueness of coherence, it has been suggested, excludes a constitutive role for interaction with the material world in the articulation of scientific knowledge. But I believe this latter conclusion to be mistaken. In what follows, I seek to articulate a more adequate alternative to correspondence realism while avoiding antirealism.[7] I argue that coherences between material procedures and conceptual models should be seen as *made* things, as actors' achievements, and not as arising naturally and uniquely from the material world itself. But I maintain a noncorrespondence realist perspective on the making of coherence. The thrust of my analysis is that, if one examines the evolution of passages of experimental practice in time, it becomes clear that scientific knowledge is articulated in accommodation to *resistances* arising in the material world. There is a direct and analysable relation between scientific knowledge and the material world, though it is one of made coherence, not natural correspondence.

Interactive stabilisation

To discuss this relation further, one needs an analysis of how coherence is made. I suggest that material procedures and conceptual models should be seen as *plastic resources for practice*. And I further suggest that experimental practice should be understood in terms of the deformation and moulding of such resources, with the *objective* of achieving the three-way coherence in which facts are sustained. Thus where the correspondence realist sees naturalness and fixity in all three elements and in bright, unrevisable facts, I suggest that the appearance of naturalness itself arises from the achievement of coherence. It is true, I think, that scientists feel they have made sense of the world and their practice in it when they have achieved such coherence. But that intuition does not legitimate the notion that any particular coherent set of procedures

6. Collins (1985); the gravity-wave controversy is discussed in Chapter 4. For more studies of controversies in experimental science, see Holton (1978); Shapin (1979, 1982); Pickering (1981a, 1984a); Collins & Pinch (1982); Shapin & Schaffer (1985); Pinch (1986), and contributions to Wallis (1979), Collins (1981) and Knorr, Krohn & Whitley (1981).

7. Collins (1985) argues an antirealist perspective on the basis of controversy studies; see Pickering (1987a) for an analysis of the strengths and weaknesses of his position.

and models is *the* natural and predetermined end point of a particular passage of practice.

To emphasise the preceding remarks, I will replace the correspondence metaphors of naturalness andd fixity with one of *interactive stabilisation*. Both terms of the metaphor are important. The reference to 'interaction' is intended to stress the interdependence of the three elements of practice, the mutual credibility that each element bestows on the others when coherence is achieved. The output phenomenal model, and the fact it carries, must be right, it seems, because it is implied by the material procedure and the instrumental model which lie behind it. Of course, as a plastic resource, there is no guarantee that this particular instrumental model is right, but it fits in so nicely between the fact and the material procedure that it is hard to doubt. On the other hand, there is no guarantee that the material procedure is the correct one, material procedures are plastic, too, but . . . and so on.[8] Reference to 'stabilisation', in turn, is intended to draw attention to the potential *instability* of achieved coherences. Coherences are the end points of struggles against incoherences in experimental practice, and they are always liable to come apart in future practice. The achievement of coherence, and the concomitant production of an empirical fact, is thus to be seen as a particular, contingent and potentially temporary *limit to practice*.

Pragmatic realism

Now that the concept of interactive stabilisation has been introduced, it is possible to spell out my position more explicitly. First, as stated, it is a realist position. In the study to follow, as in all available studies of the development of experimental practice in time, it is clear that material practice – interaction with the material world – can play a constitutive role in knowledge production: plastic deformations of material procedures and conceptual models have to be understood as, in part, an accommodation to destabilising resistances in the material world.[9] This is the sense in which a realistic perspective on scientific knowledge

8. Collins (1985), p. 84, discusses some consequences of this kind of circular reinforcement under the heading of 'the experimenter's regress'.
9. Detailed studies illuminating the temporal interaction of the material and conceptual dimensions of practice include Kuhn (1970); Holton (1978); Fleck (1979); Latour & Woolgar (1979); Knorr-Cetina (1981); Pickering (1981a, b, 1984a, b); Gooding (1982, 1986, Chapter 7 this volume); Galison (1983, 1985); Collins (1985), Chapter 3; Lynch (1985).

is appropriate. But it is important to note that material resistances are only manifest *relative to prior expectations*: they have no existence in the absence of such expectations. This situated character of material resistance precludes any move back towards correspondence realism. The relation between knowledge and the material world has therefore to be understood not in terms of fixed correspondence but rather in terms of local, potentially unstable, coherences achieved between material procedures and conceptual models. This is the perspective – let me call it *pragmatic realism* – which I exemplify below in a historical study.[10]

My study concerns a programme of experimental searches for 'quarks' conducted by the Italian physicist Giacomo Morpurgo over a period of more than 15 years. In discussing Morpurgo's experiments I will focus upon what I take to be generally significant features, so it is appropriate that I signal in advance one particular idiosyncracy of Morpurgo's programme. I have spoken of material procedures and conceptual models as plastic resources for practice. I therefore need to comment immediately upon the fact that Morpurgo displayed remarkably little flexibility with respect to phenomenal models: plasticity of phenomenal models was reduced, in Morpurgo's practice, to a choice between two alternatives. I will fill in the technical details and seek to explain this reduction in plasticity below. Here, let me explain why I find it useful to discuss a case which seems so idiosyncratic. The reason is, in part, that the plasticity of phenomenal models is so readily imaginable that it needs little exemplification. Accommodation of phenomenal models to experimental findings, as in the determination of a hitherto unknown

10. The point of departure of the present analysis is the pragmatist philosophy of William James (1978). James saw knowledge as amongst the *means* employed in the pursuit of particular *goals*. He stressed the revisability of knowledge in response to situated obstacles to the achievement of such goals, and took it for granted that obstacles could arise in material practice. But James did not concern himself with the limits to practice. In particular, he did not consider those situations where fact-production is itself the goal of practice. And he did not, therefore, offer an analysis of the conditions under which such goals are said to be achieved. My discussion of the interactive stabilisation of plastic resources is intended to illuminate those conditions. Galison (1983) likewise emphasises the importance of understanding how experiments end. Another aspect of pragmatic realism I need to stress is that the coherence in question is a local, 'vertical', coherence of material procedures and conceptual models. It is not any kind of 'horizontal' coherence across a unified theoretical network. In philosophy, the idea of vertical coherence (and its instability) is typified in the work of Pierre Duhem (1954); the idea of horizontal coherence (and its tenacity) has received more attention, and is usually associated with Quine's (1980) idea of a 'conceptual scheme' and the discussions of incommensurability given by Kuhn (1970) and Feyerabend (1975, 1978).

theoretical parameter, is routinely seen as the point of experiment in general. Conversely, naturalness and fixity are routinely, if tacitly, attributed to material procedures and instrumental models. And exploration of the achievement of interactive stabilisations in a case where phenomenal models are fixed is an excellent way to bring out the plasticity of these latter elements. The oddity of this historical example is, then, the source of its utility in the present context.[11]

We can now turn to the study itself. I will first seek to bring out the related themes of plasticity, material resistance and interactive stabilisation in experimental practice. At the end of the essay I will turn from plasticity to its inverse, in an examination of the *fixed points* of practice. This will make possible an explanation of Morpurgo's phenomenal rigidity that will lead us from the level of the individual to the level of social practice. For the moment, though, the emphasis is on solitary practices, upon Morpurgo himself.

THE HUNTING OF THE QUARK

In the early 1910s, the work of the American physicist Robert Millikan did much to convince the scientific world of the remarkable fact that all of the electric charge in the universe was quantised in terms of a fundamental unit, e, the charge of the electron. Charge, that is, came only in discrete quantities: the charge of any body had to be an integral multiple of e; there was no possibility of finding that some portion of matter carried, say, a charge of $1/4$ or $.645\ e$. In 1964, however, this assumption was questioned in elementary-particle theory. In response to developments which need not concern us here, Murray Gell-Mann and George Zweig independently proposed that the building blocks of the universe might be hitherto unobserved particles called *quarks*. The most unusual property attributed to quarks was that they were supposed to carry third-integral electric charges, either $1/3$ or $2/3\ e$.[12] Immediately following this proposal, several physicists set out to look for quarks experimentally. We will follow the programme of one of them, Giacomo Morpurgo of the University of Genoa, through his published and unpublished writings. In particular, much of what follows

---

11. An emphasis on the plastic moulding of resources does not point to a necessarily conservative image of science. Discussion of plasticity serves to bring out important issues concerning realism, but discontinuous substitution of resources (and goals) is a common alternative response to resistances: see Pickering (1984a) for some examples.
12. For the history of quarks, see Pickering (1984b).

takes the form of a commmentary upon his unusually detailed narrative recollections of the earliest phases of his work.[13]

Morpurgo's search for quarks began not in the material world but in the realm of thought, in *conceptual practice*. He had a goal – the search for fractional (third-integral) charges in his laboratory – and he set out to design an apparatus appropriate to the achievement of that goal. He started from a crude sketch of an *instrumental model*, a conceptualisation of how his apparatus should work. This was inherited from Millikan. According to the unquestioned laws of classical electrostatics, the force on any macroscopic body placed in an electric field is qE, the product of the charge on the body (q) and the magnitude of the applied field (E). Therefore, Morpurgo reasoned, if one could measure the force on a body suspended in a known electric field one could deduce its electric charge. This was just how Millikan had proceded, working with tiny oil droplets. But Morpurgo wanted to measure the charges on much larger samples of matter. In that way he would stand more chance of finding quarks which, if they existed at all, had to be comparatively rare to have gone undetected. Morpurgo then reasoned as follows:[14]

In Millikan's original experiment the same electric field used to measure the charge provides the force which prevents the droplets from falling under the action of gravity. If the droplets are too heavy the electric field needed becomes very large and the difficulties which this produces in Millikan's methods are obvious. Of course if we could use to sustain the grain of matter a force different

13. I should note that Morpurgo collaborated with various individuals in the course of his experiments: 'Morpurgo' is occasionally a convenient abbreviation for a composite entity. One virtue of studying Morpurgo's work is that his experiments are very simple to understand: they are classic benchtop physics and should not be confused with the complex experiments performed by elementary-particle experimenters at accelerator laboratories (I stress this because of possible confusions stemming from the origins of the quark concept in particle physics). Pickering (1981a) presents documentation and analysis of Morpurgo's programme from the time of his first publication; Marinelli & Morpurgo (1982) is an extended technical exposition. The text quoted below on the earlier phases of the programme is Morpurgo (1972). I am indebted to Professor Morpurgo for making this account available to me. It was intended for the series *Adventures in Experimental Physics*, which ceased publication before the article could appear. The typical failings of actors' retrospective accounts are largely absent from Morpurgo's narrative (though see note 22 below). It avoids the editing of 'false starts' and mundane problems which characterises many scientists' accounts of their work, and therefore succeeds in presenting in an economical and comprehensible fashion many typical features of experimental practice otherwise accessible only through ethnographic observation, the reconstruction of laboratory notebooks and comparable techniques. Instances of all of the features discussed here could be documented in, for example, the works cited in note 9 above. On editing and the enforcement of correspondence realism, see Pickering (1988a).
14. Morpurgo (1972), p. 2.

from the electric field which is used to ascertain its charge it would be much better. But how to sustain the grain of matter? We first thought of having our grains floating in a dielectric liquid . . .

These sentences introduce the first *interactive stabilisation* achieved by Morpurgo, though at this stage it remained a two-way stabilisation in conceptual practice rather than the kind of three-way stabilisation of material and conceptual practices discussed above. In formulating his plan to suspend samples in a liquid, Morpurgo had contrived an inter-active stabilisation of *means and ends*.[15] Morpurgo's refined instrumental model, in which the electric field would be applied to a particle sus-pended in a liquid, made his goal of measuring charges on large samples of matter appear attainable. Conversely, his goal only made sense once such an instrumental model had been devised. Without such an inter-active stabilisation Morpurgo's programme could not have gone ahead. As I have said, reference to 'stabilisation' is intended to signal the contingency and incipient instability of achieved coherence: it is impor-tant to recognise that Morpurgo had not brought means and ends into unbreakable alignment for all time. He had a plan for his experiment, but it could fail when he attempted to translate it from conceptual to material practice. And it did:[16]

for a few days we were looking at the microscope to graphite grains floating in a mixture of C Cl₄ [carbon tetrachloride] and something else (I don't remember any more) in the presence of a strong transverse electric field. Had we been a little more wise we could have spared this attempt at all. In fact we saw our grains moving back and forth from one side of the container to the other (we understood after a while that this motion was due to irregular and conspicuous changes in charge of the grains in contact with the liquid). However this attempt was not useless. I was so irritated − perhaps by the inhalation of a little too much C Cl₄ − that I was stimulated to think more on what we were doing. I realized that the floating idea was good; only we should have our grains levitating in vacuum, not in a liquid.

This passage of mundane unproductive laboratory life illustrates some fundamental aspects of the relation between articulated knowledge and the material world. Note that Morpurgo had no reason to expect his liquid suspension system to fail: it worked perfectly in conceptual practice. That it failed in the laboratory points to the potential *resistance* of the material world to conceptualisation: Morpurgo's conceptual

15. The discussion of the means and ends of scientific practice leads back to the pragmatist analysis sketched out in note 10 above. See also Pickering (1987b).
16. Morpurgo (1972), pp. 2–3.

coherence of means and ends was *destabilised in material practice*. Note also that Morpurgo reformed his conceptual practice in response to this resistance in two ways. He *moulded* his instrumental model to make sense of what he had found, by including in the model a possibility he had hitherto ignored: that his samples might exchange charges with the suspending liquid. And he abandoned liquid suspension as the means to his chosen end. This sequence – conceptual practice; resistance in the material world; reformed conceptual practice – points to the need for a realistic understanding of scientific knowledge and practice. Morpurgo's understanding of his current material practice and of how he should proceed were reformed on the basis of his specific, nonverbal interaction with the material world. In this sense, Morpurgo's material practice was constitutive of his conceptual practice.

If we consider just how Morpurgo made sense of his experimental findings, it becomes clear that he achieved a new kind of interactive stabilisation, this time a synchronic stabilisation between his *material procedure* and his *instrumental model*. Morpurgo's material performance in the laboratory made sense in terms of his new instrumental model; the new model, which was not specified in advance, was sustained by his material practice. But this synchronic stabilisation cut destructively across Morpurgo's temporal stabilisation of ends and means. According to his new instrumental model, liquid suspension was not an appropriate route to his goal. In the face of this conclusion, as indicated in his narrative, Morpurgo retreated from material to conceptual practice. He began to design an alternative, magnetic, suspension system. There is no need to go into the physical principles of this system here, but some features of its initial implementation are relevant. As in the case of the liquid suspension system, Morpurgo's attempts to materialise his magnetic suspension system did not go smoothly. He continually encountered resistances which destabilised his instrumental models and threatened his alignment of means and ends. But he now displayed greater perseverance. Resistances were met with trial-and-error adjustments to material procedures and instrumental models, and interactive stabilisations were repeatedly restored.[17]

Producing the first fact

Eventually, by the summer of 1965, Morpurgo had before him a 'working' detector: he was able to suspend particles of graphite between two metal plates, and to observe measurable responses when known vol-

17. Ibid., pp. 3–5.

tages were applied to the plates.[18] All that remained was to make the measurements. We can return to Morpurgo's narrative:[19]

I recall clearly the measurements on the first grain . . . . We had a grain which, on applying the electric field, moved by seven divisions of our graduated scale. We had lunch and after, maybe, two hours we came back. The grain was of course still there. Only on applying the same electric field it did not move anymore; or better it did move by only one division, and the *same* movement, in the *same* direction it had when the electric field was reversed. (This meant, of course that there was a gradient in the electric field, since the force due to such a gradient $E\partial E/\partial x$ is unchanged if the field is reversed). Clearly the grain in these conditions was neutral; during the lunch time it had captured an ion from the residual gas and the charge had probably passed from one to zero. Because I had to go to a faculty meeting we postponed the beginning of the measurements to the end of the meeting; when, after three hours, I came back the grain was in the same charge state as before; and in about one hour we could perform the measurements which are reported in the first column of the table 1 of [our first publication] . . . .

In the next few days we performed the five measurements reported in the table; there was no indication of fractionary charges, so far, and we had already performed the equivalent of one thousand Millikan measurements (the standard Millikan droplet weighs 10^{-11}g.) We therefore decided to write a letter to Physics Letters.

With Morpurgo's first fact, submitted for publication in November 1966, we can examine his first achievement of a three-way interactive stabilisation of material procedures and conceptual models.[20] Note that this coherence did not follow automatically or naturally, even once Morpurgo's apparatus was 'working'. Thus, in Morpurgo's after-lunch measurements on the first grain, the material world continued to resist him: the grain acted strangely, moving in the same direction when the direction of the electric field was reversed, a result, according to Morpurgo's existing instrumental model, quite unacceptable within his space of phenomenal models. Here we encounter the phenomenal inflexibility I mentioned earlier. Morpurgo was prepared to confirm one of only two phenomenal models; that charge was quantised in either integral or third-integral units. It is not clear what the phenomenal implications of the strange behaviour of the grain might have been when interpreted through Morpurgo's existing instrumental model, but certainly it did not point to the quantisation of charge in

18. Becchi, Gallinaro & Morpurgo (1965).
19. Morpurgo (1972), pp. 5–6.
20. Gallinaro & Morpurgo (1966).

either integral or third-integral units. A situated incoherence, then, emerged in material practice between Morpurgo's material procedure and his instrumental and phenomenal models. Morpurgo responded to this incoherence with a plastic deformation of his instrumental model. And, with a simple elaboration of the model – the inclusion of a term to allow for the interaction between an induced electric dipole moment on the sample and inhomogeneities of the applied field – he at last achieved a three-way interactive stabilisation in material and conceptual practice. His material procedure, interpreted through his new instrumental model, yielded results which confirmed one of his two phenomenal models: he began to find only integral charges on a series of graphite grains, and published the fact that he was unable to find any quarks in his laboratory.

This sequence of events exemplifies the pragmatic realist perspective I outlined above. A situated resistance encountered in material practice was accommodated by a plastic deformation of Morpurgo's instrumental model. This deformation produced a coherent, three-way interactive stabilisation of Morpurgo's material procedure and conceptual models; it marked a (temporary) limit to Morpurgo's practice and was the point of fact-production. But it is, of course, precisely at the point of fact-production that it becomes most difficult to think in terms of plasticity and interactive stabilisation: all seems fixed, and metaphors of correspondence beckon.

Accommodating the material world

Fortunately, the limit to Morpurgo's practice was short-lived. He continued to make measurements, and an examination of subsequent developments is an antidote to correspondence intuitions:[21]

A few days after having submitted it [the report for publication in *Physics Letters*] we found the first 'anomalous' event. The grain number 7 looked as if it contained a quark. On reversing the electric field when the grain was in its state of minimum charge it had a displacement, in the same direction, but not equal to the previous one; the difference in these displacements corresponded to one fourth (or with 'some goodwill' to one third [i.e. a quark]) of the difference in displacements when the object had captured an electron.

For a few days we were rather excited; the excitement decreased however, when, after a while, we saw that several other grains had a similar, but not quantitatively identical behaviour. We finally understood what was going on; the clue came when we decided to make measurements, on the same 'anomal-

21. Morpurgo (1972), pp. 6–7.

ous' grain, at different increasing separation of the platelets. In fact the first of these measurements gave the following result: we had a grain showing a residual charge of 1/9 when the platelets had been placed at a separation of 1.6 mm.; this residual charge decreased to less than 1/32 when the separation between the platelets was increased to 2.7 mm. (the applied voltage was also increased to keep the electric field the same). We concluded that we had been observing a spurious charge effect.

Its explanation is the following: . . . the total field E acting on the grain cannot really be identified with the field E_a we apply. In between the platelets there exists, in addition, a very small electric field E_v due to Volta effects which we cannot reverse (also if the platelets are fabricated of the same metal, they are not monocrystals, and moreover there are always irregular deposits of graphite on them). . . . [Effects due to E_v] cannot be distinguished from a charge force on reversing E_a. They simulate a spurious charge; they can be decreased only by increasing the distance between the platelets; indeed this is what we had done.

Here we have an account of a three-way stabilisation breaking down and being repaired in material practice. Morpurgo's further measurements on graphite grains produced apparent charge measurements like 1/4 and 1/9, which pointed rather to the continuous divisibility of electric charge than to either of the phenomenal models he was prepared to entertain. Once more Morpurgo had encountered resistance in the material world. It is important to note that it was a *situated* resistance: it was only because of Morpurgo's rigid phenomenal expectations that it counted as a resistance at all. If, for example, he had been prepared to accept that charge was continuously divisible, there would have been no obstacle to overcome. But within the space of Morpurgo's phenomenal expectations the material world did resist him, and he responded to this resistance along lines which should be becoming familiar.

Morpurgo first made a plastic deformation of his material procedure. He increased the separation between the metal plates ('platelets' in the above quotation) and found that he could obtain consistent measurements of only integral charges in this way. Thus he restabilised a new material procedure against his conventional phenomenal model. A further plastic deformation to his instrumental model restored a full, three-way stabilisation. He found that, if he included in his instrumental model terms to account for surface effects on the plates, he could explain why increasing the plate separation had the observed effect. Once more, then, coherence and interactive stabilisation were achieved between the elements of Morpurgo's practice. His material procedure, interpreted through his instrumental model, pointed to the validity of an acceptable phenomenal model. Once more, in a definitive summary publication in 1970, Morpurgo reported a fact: the absence of quarks on relatively large quantities of graphite.[22]

Again, pragmatic realism offers an appropriate perspective on this sequence of events. The material world had a manifest role in destabilising the prior coherence of Morpurgo's procedures and models; deformation of material procedures had a manifest role in re-establishing coherence; and Morpurgo's new instrumental model was explicitly moulded around the effect of this shift in procedures. A realistic accommodation to the material world is evident. But again, it is precisely at the moment when coherence is rescued and stabilisation achieved that the metaphor of interactive stabilisation seems least appropriate, and one is tempted to slide from pragmatic to correspondence realism. As a further corrective, I close this section with some remarks on the subsequent history of Morpurgo's quark search.[23]

22. Morpurgo, Gallinaro & Palmieri (1970). Two remarks are appropriate here. First, there is a sense in which Morpurgo's retrospective account makes his work appear too 'easy'. The preceding quotation, for example, almost certainly underplays the element of open-ended *trial and error* involved in Morpurgo's restabilisation of his practice. Ethnographic studies of science reveal that trial-and-error tinkering with procedures and models is an endemic feature of experiment (Latour & Woolgar 1979; Knorr-Cetina 1981; Lynch 1985) and one should think of Morpurgo's response to destabilisation as that of tinkering with his material procedure and instrumental model with no guarantee of future success. Trial and error is a central element of the pragmatist analysis of knowledge and practice. The second remark is this. I have noted that, suitably elaborated, Morpurgo's instrumental model explained why increasing the plate separation had the observed effect. But this explanation had some interesting features. It was essentially an *explaining away* of the early anomalous findings: as Morpurgo himself wrote, 'it is not clear that the estimates . . . have much meaning' when applied to the small plate separations at which the anomalous findings were obtained (Morpurgo, Gallinaro & Palmieri, 1970, p. 113). And the elaborated instrumental model was not used in interpreting the results obtained at increased plate separation: with the anomalous findings explained away the previous, and simpler, model remained in use. Thus Morpurgo's new material procedure was stabilised rather directly against the no-quarks phenomenal model, with the new instrumental model playing a *legitimating* rather than a *constitutive* role in making the bridge between them. This observation connects up, I think, with discussions initiated in philosophy of science by Ian Hacking, who noted the importance of two-way stabilisations of material procedures and phenomenal models in the history of experimental science. A knowledge that, rather than how, the apparatus works seems often sufficient in science, as in the case of microscopes and bubble chambers. See Hacking (1983); Ackermann (1985); Baird (1988). Zandvoort's (1986) analysis of the development of the theory of nuclear magnetic resonance is also relevant here.
23. See Pickering (1981a) for details and documentation.

Resisting realism

Following each production of fact, Morpurgo revised his goal upward, seeking to examine ever larger quantities of matter. These revisions were accompanied by innovations in experimental technique designed to make the newly defined goals attainable. Thus, for example, in the late 1960s Morpurgo began to work with oscillating instead of static electric fields, and in the 1970s he switched from a diamagnetic to a more powerful ferromagnetic suspension system. And as he experimented on ever more massive samples, the familiar sequence of breakdowns in, and repairs to, interactive stabilisations of material and conceptual practice was continually repeated. I will note two examples.

In the ferromagnetic suspension experiments of the mid-1970s, Morpurgo found at first that the charges of iron cylinders seemed to drift overnight – from zero to 1/10 e, for example. His three-way stabilisation fell apart once more in material practice. As before, Morpurgo began his repairs in a plastic deformation of his material procedures: he found that if he made measurements on cylinders which were spinning relatively rapidly in space (rather than stationary, as in the earlier version of the experiment) he could achieve stable measurements of integral charge. He thus realigned his material practice with his phenomenal model. Further tinkering with his instrumental model, he found that he could make sense of this outcome in terms of 'small torques [which] can arise under the action of the oscillating electric field'. At this point, as before, Morpurgo reported his inability to find evidence for isolated quarks.[24]

Continuing to make measurements using the same apparatus, Morpurgo yet again started finding unacceptable results: charges which were neither integral nor third-integral. This provoked another elaboration of Morpurgo's instrumental model. In 1980, after thinking for a decade and a half about a simple classical electromagnetic model of a simple piece of apparatus, he announced the discovery of a new and unexpected force at work in his apparatus – a magnetoelectric force, capable of mimicking charges up to four times that of the electron. When this force was included in Morpurgo's instrumental model, three-way interactive stabilisation was restored once more, and the absence of quarks on a total mass of 3.7 mg of steel was reported. (This, inciden-

---

24. Gallinaro, Marinelli & Morpurgo (1977), the quotation is from p. 1257.

tally, corresponded to a gain in sensitivity of the order of 10^8 over the standard Millikan-type experiment.)[25]

At this point Morpurgo decided that he had looked for quarks long enough, and discontinued his experimental programme. At this point, too, enough has been said to undermine the correspondence metaphors of naturalness and fixity. The history of Morpurgo's experiments was one of continual deformations of plastic resources – material procedures and instrumental models – in pursuit of interactive stabilisation. The manifest role of material resistances in destabilising existing coherences, and of deformations in material procedures in the re-establishment of coherence, points to a realistic understanding of articulated scientific knowledge, but it is pragmatic realism not correspondence realism that recommends itself.

PLASTIC RESOURCES AND FIXED POINTS

To conclude this essay I want to widen its frame of reference. Though I continue to take Morpurgo's programme as my example, I want to move from the discussion of individual practice towards the social. I have emphasised the plasticity of the resources with which Morpurgo worked. But not everything was plastic in Morpurgo's practice. As we have seen, Morpurgo worked in a fixed space of phenomenal models. He would accept only one of two candidates, when many more would seem to be conceivable: why not accept that charges can be neither integral or third-integral, for example? Morpurgo's instrumental models, though evidently plastic, were all built around a fixed skeleton given by classical electrostatic theory. Enough has been said to illustrate the theme of plasticity; I turn now to its opposite, in a discussion of the *fixed points of practice*.

The single feature of Morpurgo's practice untypical of experiment in general was the fixity of the phenomenal space in which he worked. To understand this fixity, it is necessary to look beyond Morpurgo to the technical culture of his field. One then sees that, in choosing to operate within a restricted space of possibilities, Morpurgo aligned his conceptual practice with that of his colleagues. It was not so much Morpurgo as his commmunity that refused to countenance all but two phenomenal models. The idea that charge was quantised in units of

25. Marinelli & Morpurgo (1980, 1984).

the charge of the electron had been institutionalised in physics since the 1910s. From 1964 onwards the quark proposal was increasingly embedded in the practice of elementary-particle physicists; but beyond the quark proposal, no conceptual models of nonintegral electric charges commanded more than individual support in any sector of the scientific community. Thus, in working within a binary space of phenomenal possibilities, Morpurgo laid the foundations for a *social stabilisation* of his practices and findings. Within the restricted space, whatever findings he reported would, to some extent, interactively support and be supported by the practice of his colleagues and predecessors. This dimension of social stability would have been trivially absent if Morpurgo had worked outside this space.

Despite this degree of fixity in Morpurgo's practice, the notion of interactive stabilisation remains relevant here. If Morpurgo had chosen to move outside the binary space – if, for example, he had found it impossible to stabilise his practice within it and had, instead, reported the existence of a continuum of nonintegral charges – then both his own findings and institutionalised beliefs would have been destabilised. The theoretical work of Gell-Mann and Zweig had already destabilised existing notions of charge quantisation by opening up a space in which third-integrally charged quarks became conceivable. In this respect, I repeat that I do not insist that practice in the material world is the only route to the destabilisation of material and conceptual practices: the impact of the work of Gell-Mann and Zweig would be a counterexample to this. My intention here is simply to thematise the material dimension of practice as a means of exploring the realistic aspects of scientific knowledge.

Similar comments apply to the other fixed point of Morpurgo's conceptual practice, his adherence to a skeleton of classical electrostatic theory in the construction of instrumental models. Like the belief in charge quantisation, classical electrostatics had been institutionalised in physics for many years before Morpurgo began his experiments. I know of no alternative schemes to which Morpurgo might have appealed in the construction of alternative models. Dependence upon classical electrostatics was, in effect, compulsory if Morpurgo's findings were to achieve any kind of social stability. But the most interesting observation to be made is that, despite the taken-for-granted status of classical electrostatics, the theory was insufficient to specify an instrumental model for Morpurgo's experiment in advance. We have seen that, at all stages of the experiment, space remained for further elaboration and modification of the model. The fixity of the skeleton did not efface the plasticity of the models built around it.[26]

Having discussed these fixed points in Morpurgo's conceptual practice, his material practice deserves some attention. Can one discern any fixed points there? One would expect to be able to, although I have not sought to bring this out in the above account. Apprenticeship in physics transmits an immense amount of lore in the construction of, say, electrical circuits and optical systems. This lore is embedded in material practice and is only partially articulated at most. It informs the construction of any particular apparatus, and there seems no reason to doubt that the language of interactive stabilisation would prove applicable in this domain, too.[27] But it is useful to bear in mind that apparatus always remains distinguishable from its conceptualisation and irreducible in its particularity. Two pieces of apparatus might be built respecting the same lore, and yet still remain distinct in their material existence: think of Morpurgo's apparatus before and after he increased the plate separation, for example. And it is because of this particularity that plasticity inheres in material practice. Like fixed points in conceptual practice, fixed points in material procedure do not efface plasticity; they do not determine future practice.[28]

To sum up, my analysis of the relation between articulated scientific knowledge and the material world has been this. I have argued for a realist perspective by pointing to the constitutive role in knowledge production of situated resistances encountered and overcome in material practice. At the same time, I have opposed the correspondence metaphors of naturalness and fixity. I have argued that material procedures and conceptual models are plastic resources for practice, and that fact-production rests upon the contingent achievement of coherence

26. All of Morpurgo's instrumental models were constructed in accordance with accepted theory. Morpurgo found space for plastic deformation in the specification of 'significant' aspects of his apparatus which had been assumed to be 'insignificant' in prior models. The upshot of such deformations was to weaken the social stabilisation of Morpurgo's instrumental models: they referred more and more to the specifics of his apparatus, and relatively less to the agreed laws of electrostatics.

27. One difference between the fixed points of conceptual and material practices is that the former domain is, in physics at least, much more highly codified than the latter. Baird (1987) suggests that one should think of regularities in material practice in terms of 'instrument cookbooks'.

28. For more examples of fixed points and their inadequacy to determine practice, see Pickering (1981b) on the use of phenomenal models as 'benchmarks' and on the role of prior agreements concerning material procedures, and Collins (1985), pp. 100–106, on calibration.

amongst these elements. I have used the notion of interactive stabilisation to emphasise both the reinforcement which each element bestows on the others when coherence is achieved, and the incipient instability of achieved coherence. In the last section of this essay, the move from plasticity to a concern with fixed points has led into a discussion of the social stabilisation of practice. As far as my historical example has been concerned, this social dimension has taken a very simple form: the fixed points of Morpurgo's practice were institutionally defined and unreflectively respected. This makes Morpurgo's work a convenient site for discussions of plasticity and realism, but I do not want to suggest that simplicity in the social stabilisation of practice should be seen as typical.[29] To the contrary, I believe that an analysis of social stabilisation is one of the most important topics which can be addressed via detailed case studies.[30] But analysis of social axes of stabilisation and destabilisation will not, I think, change the conclusions of this essay. However we come to conceptualise the full network of interconnections within which empirical knowledge is stabilised, it seems clear that one terminus of the network will be situated in material practice, and that the pragmatic realism sketched out above will remain an appropriate perspective.

ACKNOWLEDGEMENTS

This paper was written while I was a member of the Institute for Advanced Study, Princeton; I gratefully acknowledge its support. Early versions of related papers were presented at the 'Uses of Experiment' conference, Bath, September 1985, the annual meeting of the Philosophy of Science Association, Pittsburgh, October 1986 (Pickering 1988b), the Institute for Advanced Study, January and March 1987, and the Philosophy Department of the University of Western Ontario, March 1987. I thank all of these audiences for valuable comments. I owe particular debts to Bill Connolly, James Cushing, Paul Feyerabend, David Gooding, Ian Hacking, Thomas Kuhn, Barbara Herrnstein Smith

29. In quark-search experiments, social stabilisation became problematic when LaRue, Fairbank & Hebard (1977) announced that they had found evidence for fractional charges in an experiment similar to Morpurgo's. Subsequent developments further exemplify and elaborate the perspectives of the present essay, and point beyond it to the need for an extended pragmatist analysis of the dynamics of scientific practice: see Pickering (1981a).
30. For some general analyses, see Latour & Woolgar (1979); Knorr-Cetina (1981); Barnes (1982); Bloor (1983); Pickering (1984b, 1987b); Callon & Law (1988).

and, especially, to Yves Gingras & Silvan Schweber (1986) for their provocative essay review of my book, *Constructing Quarks* (1984).

References

Ackermann, R.J. (1985). *Data, Instruments, and Theory: A Dialectical Approach to Understanding Science*. Princeton: Princeton University Press.

Achinstein, P. & Hannaway, O. (ed.) (1985). *Observation, Experiment, and Hypothesis in Modern Physical Science*. Cambridge, Mass: MIT Press.

Baird, D. (1987). *Pragmatic Tinkering and Instrumental Realism*. Mimeo, University of South Carolina at Columbia.

Baird, D. (1988). 'Instruments on the Cusp of Science and Technology: The Indicator Diagram'. In ed. L. Hargens, R.A. Jones & A. Pickering, Greenwich, Conn.: JAI Press (in press).

Barnes, B. (1982). *T.S. Kuhn and Social Science*. London: Macmillan.

Becchi, C., Gallinaro, G & Morpurgo, G. (1965). Measurement of small charges in macroscopic amounts of matter: discussion of a proposed experiment and description of some preliminary observations. *Nuovo Cimento,*39, 409–12.

Bloor, D. (1983). *Wittgenstein: A Social Theory of Knowledge*. London: Macmillan.

Callon, M. & Law, J. (1988). On the construction of sociotechnical networks: content and context revisited. In ed. L. Hargens, R.A. Jones & A. Pickering. Greenwich, Conn.: JAI Press (in press).

Collins, H.M. (ed.) (1981) Knowledge and controversy: studies of modern natural science. *Social Studies of Science*, 11(1).

Collins, H.M. (1985). *Changing Order: Replication and Induction in Scientific Practice*. Beverley Hills: Sage.

Collins, H.M. & Pinch, T.J. (1982). *Frames of Meaning: The Social Construction of Extraordinary Science*. London: Routledge and Kegan Paul.

Duhem, P. (1954). *The Aim and Structure of Physical Theory*. Princeton: Princeton University Press.

Feyerabend, P.K. (1975). *Against Method*. London: New Left Books.

Feyerabend, P.K. (1978). *Science in a Free Society*. London: New Left Books.

Fleck, L. (1979). *Genesis and Development of a Scientific Fact*. Chicago: University of Chicago Press.

Galison, P. (1983). How the first neutral current experiments ended. *Reviews of Modern Physics,*55, 477–509.

Galison, P. (1985). Bubble chambers and the experimental workplace. In ed. P. Achinstein & O. Hannaway (1985). pp. 309–73.

Gallinaro, G., Marinelli, M. & Morpurgo, G. (1977). Electric neutrality of matter. *Physical Review Letters*, 38, 1255–8.

Gallinaro, G. & Morpurgo, G. (1966). Preliminary results in the search for fractionally charged particles by the magnetic levitation electrometer. *Physics Letters*, 23, 609–13.

Gingras, Y. & Schweber, S.S. (1986). Constraints on construction. *Social Studies of Science*, 16, 372–83.

Gooding, D. (1982). Empiricism in practice: teleology, economy and observation in Faraday's physics. *Isis*, 73, 46–67.

Gooding, D. (1986). How do scientists reach agreement about novel observations? *Studies in History and Philosophy of Science*, 17, 205–30.

Hacking, I. (1983). *Representing and Intervening*. Cambridge: Cambridge University Press.

Hargens, L., Jones R.A. & Pickering, A. (ed.) (1988). *Knowledge and Society: Studies in the Sociology of Science, Past and Present, Vol 8*. Greenwich, Conn.: JAI Press. (in press).

Holton, G. (1978). Subelectrons, presuppositions and the Millikan – Ehrenhaft dispute. In Holton, *The Scientific Imagination: Case Studies*, pp. 25–83. Cambridge: Cambridge University Press.

James. W. (1978). *Pragmatism and The Meaning of Truth*. Cambridge, Mass. and London: Harvard University Press.

Knorr-Cetina, K. (1981) *The Manufacture of Knowledge: An Essay on the Constructivist and Contextual Nature of Science*. Oxford and New York: Pergamon.

Knorr, K.D., Krohn, R. & Whitley, R.D. (ed.) (1981). *The Social Process of Scientific Investigation. Sociology of the Sciences, Volume IV, 1980*. Dordrecht: Reidel.

Kuhn, T.S. (1970). *The Structure of Scientific Revolutions*, 2nd edn. Chicago: University of Chicago Press.

LaRue, G.S., Fairbank, W.M. & Hebard, A.F. (1977). Evidence for the existence of fractional charge on matter. *Physical Review Letters*, 38, 1011–4.

Latour, B. (1987). *Science in Action*. Cambridge, Mass: Harvard University Press.

Latour, B. & Woolgar, S. (1979). *Laboratory Life: The Social Construction of Scientific Facts*. Beverley Hills: Sage.

Leplin, J. (ed.) (1984). *Scientific Realism*. Berkeley: University of California Press.

Lynch, M. (1985). *Art and Artifact in Laboratory Science: A Study of Shop Work and Shop Talk in a Research Laboratory*. London: Routledge and Kegan Paul.

Marinelli, M. & Morpurgo, G. (1980). New results in the search of quarks in matter by the magnetic levitation electrometer. *Physics Letters*, 94B, 427–32.

Marinelli, M. & Morpurgo, G. (1982). Searches of fractionally charged particles in matter with the magnetic levitation technique. *Physics Reports*, 85, 161–258.

Marinelli, M. & Morpurgo, G. (1984). The electric neutrality of matter: a summary. *Physics Letters*, 137B, 439–42.

Morpurgo, G., (1972). A Search for Quarks (a Modern Version of the Millikan Experiment). One Researcher's Personal Account. Mimeo, Genoa.

Morpurgo, G., Gallinaro, G. & Palmieri, G. (1970). The magnetic levitation electrometer and its use in the search for fractionally charged particles.

Pickering, A. (1981a). The hunting of the quark. *Isis*, 72, 216–36.

Pickering, A. (1981b). Constraints on controversy: the case of the magnetic monopole. In ed. H.M. Collins (1981), pp. 63–93.

Pickering, A. (1984a). Against putting the phenomena first: the discovery of the weak neutral current. *Studies in History and Philosophy of Science*, 15, 85–117.

Pickering, A. (1984b). *Constructing Quarks: A Sociological History of Particle Physics*. Chicago: University of Chicago Press; Edinburgh: Edinburgh University Press.

Pickering, A. (1987a). Forms of life: science, contingency and Harry Collins. *British Journal for History of Science*, 20, 213–21.

Pickering, A. (1987b). Models in/of scientific practice. *Philosophy and Social Action*, 13, 69–77.

Pickering, A. (1988a). Editing and epistemology: three accounts of the discovery of the weak neutral current. In ed. L. Hargens, R.A. Jones & A. Pickering. Greenwich, Conn.: JAI Press (in press).

Pickering, A. (1988b). Against correspondence: a constructivist view of experiment and the real. In ed. A. Fine & P. Machamer, *PSA 1986: Proceedings of the Biennial Meeting of the Philosophy of Science Association*, vol. 2. East Lansing: PSA., 196–208.

Pinch, T.J. (1985). Towards an analysis of scientific observation: the externality and evidential significance of observational reports in physics. *Social Studies of Science*, 15, 3–36.

Pinch, T.J. (1986). *Confronting Nature*. Dordrecht: Reidel.

Quine, W.V.O. (1980). *From a Logical Point of View: Nine Logico-Philosophical Essays*, 2nd edn. Cambridge, Mass. and London: Harvard University Press.

Shapin, S. (1979). The politics of observation: cerebral anatomy and social interests in the Edinburgh phrenology disputes. In ed. R. Wallis (1979), pp. 139–78

Shapin, S. (1982). History of science and its sociological reconstructions. *History of Science*, 20, 157–211.

Shapin, S. & Schaffer, S. (1985). *Leviathan and the Air Pump: Hobbes, Boyle and the Experimental Life*. Princeton: Princeton University Press.

Wallis, R. (ed.) (1979). *On the Margins of Science: The Social Construction of Rejected Knowledge*. Sociological Review Monograph 27. Keele: University of Keele.

Zandvoort, H. (1986). *Models of Scientific Development and the Case of Nuclear Magnetic Resonance* Dordrecht: Reidel.

10

JUSTIFICATION AND EXPERIMENT

THOMAS NICKLES

GENERATIVE JUSTIFICATION

The rejection of simple inductivist methodologies of science by twentieth-century philosophers and historians has led to a neglect of experimental research and to an unwarranted dismissal of generative methodology in favor of an overly hypotheticalist view of scientific research. Generativists attempt to justify scientific claims by deriving them as conclusions from already established premises. Hypotheticalists employ the claims themselves as conjectural premises of predictive arguments and identify successful predictions as the source of all empirical support. Roughly speaking, generativists believe that theoretical claims can sometimes be 'deduced from the phenomena', while hypotheticalists deny this and reduce experimental research to the testing of theoretical claims. This volume manifests the current revival of interest in both generativism and experimental research. Its contributors wish to correct the overemphasis on theory at the expense of experiment, especially in philosophers' accounts of research.

One topic that emerges from recent discussion is surprisingly new: *experimental reasoning*. According to received accounts, there can be no specially interesting forms of reasoning in experimental contexts. After all, what do experimenters do except build and manipulate instruments and then observe the results? Philosophers have considered observation in turn to be a species of perception involving no inferences, at least none of relevance to methodological accounts.[1] Consequently, many philosophers of science have considered experimental contexts to be devoid of resources sufficient to support reasoning of interesting kinds. By default, theory-dependent reasoning (consisting of deductions of experimental predictions from theory) exhausts the subject of 'experi-

1. Shapere (1982) criticizes this view and provides references.

mental' reasoning. This story may have a familiar ring, for a parallel tale held that so-called 'context of discovery' was too impoverished to support reasoning of methodologically significant kinds.

Contrary to the empiricist tradition, which viewed our observational access to Nature as the least problematic cognitive relation (and hence as a basis for all the rest), Bogen & Woodward, Collins, Franklin, Galison, Gooding, Hacking, Pinch, and others have shown just how problematic observation-experimentation processes are and how intricate are their linkages to theory. Given the highly formal treatment of reasoning by many philosophers, however, even to speak of experimental *reasoning* already threatens to impose an overly verbal, rule-based, indeed theoretical, perspective and to ignore the skilled practice and judgemental behaviour which characterise experimental work. While philosophers and other students of science have long debated the theory-ladenness of observation, Gooding points out what we might call the *technique-* or *skill-ladenness* of observation.[2] He reminds us that at the frontier experimentalists are, in some respects, novices rather than experts, and he is thereby able to backlight the surprisingly large gaps between the initial detection of observational novelty, its eventual cognitive organisation in the work of an individual, and its later articulation as a finished scientific communication. Coming to recognise and characterise the 'data' of an experiment is a complex task, and the path from communicable experimental data to phenomenal claims about the world can be long and difficult. By the time the experimentation process has become so refined that its practitioners are experts, Gooding argues, the skilled activity components of experimental meaning become invisible, making it appear that the experimentalists are simply reading their phenomenal claims off the natural world.

Accounts of science typically start from this point. By failing to notice the considerable amount of socio-cognitive construction and reconstruction already completed, they provide a distorted picture of scientific learning in which (relative) novices are not distinguished from experts. Ironically, the same confusion occurs in standard hypothetico-deductive (H-D) accounts of the justification of theory, only in reverse historical order: scientific experts are treated as perpetual novices who can never employ stronger methods than that of blind generation of conjectural problem solutions followed by predictive testing of those conjectures. In fact, many case studies show that proposing conjectures and testing their consequences is most conspicuous during the early work on a problem. Testing retains some importance later in the game, but stronger

2. Gooding (1982, 1986).

forms of justification often emerge as we enter the 'expert' phase in which the problems become so well defined and heavily constrained that, in the most successful cases, theoretical claims can be derived from what scientists now 'know' about the world.[3]

I emphasise that 'expert' and 'novice' are relative terms.[4] I do not mean that, at the frontier of research, trained scientists are no better than rank amateurs, only that, *ex hypothesi*, they have not yet brought new phenomena, ideas, techniques, etc., under complete intellectual and practical control. Scientist novices-at-the-frontier are normally much better off than complete lay novices would be, since their training provides a large stock of resources and previous experience to apply to the new problems. Sometimes they are experts at being frontier-novices! Also, scientists who regularly work at the frontier tend to have exceptional ability and to possess great confidence in their ability to make sense of new and puzzling material.

One way of restating my opening remark is to say that previous accounts of science have failed to recognise the remarkable extent to which mature results of both experimental and theoretical science are the products of construction and reconstruction. (I use these latter terms in a neutral sense which does not prejudge the 'objectivity' of scientific claims, as 'mere' constructions *versus* discoveries.) Gooding, Galison and others focus on experiment.[5] They show that experiment is a source of conceptually deep discovery and of theoretical meaning, that experiment undergoes conceptual and technical (i.e., technique-al, practical) transformations, and that experimental research falls into traditions and programmes just as theoretical work does. Their studies can be read as indictments of standard hypotheticalist views according to which most experimental work is dictated by theory and is not conceptually and technically problematic.

In effect, I shall extend Gooding's point about transformation to the reconstruction of published and accepted experimental and theoretical results. Like too many of my philosophical predecessors, I shall begin where an account such as Gooding's ends, taking as 'given' well established claims about natural phenomena. But unlike most of my predecessors, I shall not employ this information in a fixed, hypotheticalist manner. On the contrary, I shall defend a generativist position on scientific justification. Generative justification is typically the upshot of a process which transforms both experimental and theoretical results.

3. See also Schaffer's discussion of 'transparency' in this volume.
4. On experts and novices, see Larkin *et al.* (1980).
5. Gooding (1986); Galison (1987).

This reconstruction is a social process – a sort of continuation of Gooding and company's constructivist account of experimental inquiry.

Other defenders of generative justification neglect the social element, presumably because they have failed to see any place for it, or any epistemic interest in it. This oversight is understandable, for, to be fully successful, the social process of justification must render itself 'transparent' and hence invisible *in order to* render the resulting claims unproblematic – as obvious facts read off Nature in one sort of case and as purely logical proofs in another. Paradoxically, one second-order 'function' of social reconstruction is to render itself invisible! Reconstruction succeeds in making the reconstructed experimental and theoretical claims unproblematic only in so far as it succeeds in rendering itself invisible. This is just another way of talking about closure in science.

THE PROBLEM: LOGIC AND SOCIETY

Philosophers have done justice neither to the strength of scientific justification nor to its social nature. They have assumed that experimental information can justify claims only by testing their logical consequences. They have ignored generative justification – the derivation of a claim from what scientists already 'know – in favour of purely consequential theories of justification. Naturally, experimentally established phenomena can play an important role in generative justification, indeed, a stronger role than in standard, consequential theories. 'Deduction from the phenomena' is a special case of generative justification.

Against the background of philosophers' criticisms of sociologists of scientific knowledge, my first claim is puzzling. For are not social accounts of justification 'softer' than logical accounts? How can emphasising the social nature of justification *strengthen* the logical warrant available? In particular, how is direct justification by logical 'proof' compatible with a social account of justification? My aim is to resolve one dimension of this perplexity. To call attention to the social basis of logical proof means that I must swim against the current of the above claim that construction and reconstruction succeed only in so far as they render themselves 'transparent'.[6] Of course, when successful, they do eventually become just as 'transparent' in the case of logical proof as in the case of experimental claims.

6. Calling attention to the social basis of proof claims is not at all original with me. See, e.g., Pinch (1977, 1985b) and especially Bloor (1976, 1983), who attempts a far more comprehensive account than I shall.

This chapter thus deals with the surprising intersection of two central themes. The first theme is that, even within scientific justification, experimental observation plays roles other than the overly familiar one of consequential testing. The second theme is the social nature of inquiry. In both cases my emphasis will be on justification. Many philosophers and sociologists have held that any bearing of social factors on epistemology is negative – undermining rather than supporting the justificatory process.[7] My own view is that if inquiry really is a social process, as even philosophers admit, then a social account of inquiry ought to be possible in which the social considerations contribute positively rather than only negatively to the possibility of reasoned judgment. What counts as sound, reasoned judgment is itself socially defined.

I shall examine the following connections between 'logic' and 'society', and only these. The existence of the right sort of social consensus of experts answers a major logical objection to generative theories of justification, namely that such theories are caught in a vicious regress of logical justification. Given the imperfect information and fallible modes of inquiry available to us, justification is unavoidably local and social. I claim that methodology is not a purely logical subject but must also satisfy cognitive and social constraints. These constraints I term the *cognitive* and the *social realisability requirements*. A viable 'logic' of science must be realisable by communities of human investigators and must closely resemble science as we know it. Conversely, any psychological and/or sociological treatment of scientific activity that can account for the demonstrated success of much scientific work must arguably satisfy a *methodological tenability requirement*. Here I disagree with those who say that the success of modern science is entirely socially defined.[8]

My terminological distinction implies no opposition between the 'cognitive' and the 'social'. 'Cognitive' here refers rather narrowly to the perceptual and memory capacities, computational speed, etc. of individuals – topics studied by cognitive psychologists. Social processes can, of course, be cognitive in a wider sense. I also emphasise that no particular social or logical account is built *a priori* into my realisability requirements. Rather, these and other constraints will guide an empirical search for adequate 'theories' of science. Work in philosophy and the science studies disciplines will be both mutually informative and

7. This view pervades positivist and Popperian writings. More recently, see Laudan (1977), various papers in Brown (1984), and Franklin (Chapter 14, this volume).
8. Consult Brown (1984) for references. Brannigan (1981) provides a sophisticated account of scientific discovery as social attribution.

mutually demanding.[9] If we succeed, we shall find accounts of various pieces and kinds of research which robustly satisfy logical, cognitive psychological, historical, and sociological demands.

GENERATIVE VERSUS CONSEQUENTIAL LOGICS OF EXPERIMENT

Classical *generativists* like Bacon and Newton held that the best theories are those derived from what we already know about the world – the phenomena plus other previously established results.[10] The basic, seventeenth-century methodological conflict set generativists against *hypotheticalists* or *consequentialists* (as I prefer to call them), who held that empirical support results from successfully testing the predictive *consequences* of theories, no matter how these theories were obtained. Today most philosophers of science believe that the consequentialists won the battle: only the consequential testing of theories provides empirical support.[11] We may express the 'central dogma' of H-D methodology as follows: All empirical support = empirical evidence = empirical data = successful test results = successful predictions = true empirical consequences of the claim (plus auxiliary assumptions). This common formulation conflates experimental data with natural phenomena.[12] It means that all scientific justification reduces to a claim's record of predictive success. For example, consistency with theoretical constraints, derivability from other claims, and reproducibility of a finding by means of standard techniques all count for nothing. In my view this is a narrow, single-track view of scientific research, one which sacrifices the contextual richness of scientific work to a logical schema.

Popperian and Lakatosian methodologists constitute an extreme but important example of modern day consequentialism. On their view only 'novel' predictions provide support (or corroboration) – a view that we may call *novel* consequentialism. The meaning of 'novel' is given by their requirement that no information used to construct a

---

9. Among the 'science studies' disciplines I include history, psychology, and sociology of science. In a wider sense, my paper is an argument for transforming philosophy of science into an empirical, science-studies discipline – and also for making the latter more 'philosophical', i.e., more epistemically oriented.

10. The term generativists is my own and a modification of Laudan's (1980) 'generators'.

11. Bayesian confirmation theorists, with their emphasis on the importance of prior probabilities, represent a growing body of opposition to the strict consequentialist view. I am not Bayesian, but I agree that their analyses are useful in local, well defined contexts.

12. Distinguishing these has important ramifications throughout history and philosophy and science: Bogen & Woodward (1988).

theory may count in its support.[13] If a novel prediction is also suprising, so much the better. The novel consequentialist view of experimental justification is precisely the opposite of the most extreme (anti-novelty) form of the generativist view-point, according to which *only* the evidence used to construct the theory counts in its support. The Popperian caricature of classical inductivism amounts to this position: the only admissible scientific theories are those which are derived from the known facts. According to this stereotype, inductivists attach no importance to predictive consequences, whereas novel consequentialists deny that inductive information provides any support at all. Each of the two traditionally great methodologies of science stands the other on its head. I term the change from predominantly generativist to consequentialist methodologies *the great logical inversion*. This inversion is supposed to have occurred in nineteenth-century science.

Was generative methodology abandoned?

On my reading, the history of science (in particular, the history of physics) does not reveal an abandonment of generative methodology by *working scientists*. The methodological sea-change is evident in nineteenth-century *philosophical* writing about science, but recent commentators have exaggerated even the philosophical shift.[14] It is true that inductivism was largely abandoned as a general methodology, but inductivism is only one type of generative methodology.

John Herschel, for instance, is often quoted as a source of modern consequentialism. It does not matter how a theory is discovered (generated), he says, only that it be tested as thoroughly as possible.[15] The larger context, however, reveals that Herschel is hardly a precursor of Popper. His point is the very opposite! It is precisely because logical generatability is what ultimately matters to justification (i.e., logical reconstructability late in the game in contrast to original construction) that the *original*, historical mode of discovery is of little consequence and, accordingly, that hypotheses are permissible and useful in the early stages of research. If sufficiently careful testing is conducted across

13. Worrall (1978). Worrall has since altered his position.
14. The 'mystery' of the great logical inversion is the problem of explaining why it occurred. According to Laudan (1980), the main explanatory factors are the onset of fallibilism together with an increasing interest in deep-structural theorising, in the early nineteeth century. He identifies Herschel and Whewell as the major philosophical figures. For my reply and a fuller discussion of these major historical and logical issues, see Nickles (1985, 1987a).
15. Herschel (1830), Pt. 2 chap. 7.

the entire range of a claim, he thinks, then we may eventually acquire sufficient empirical information, in conjunction with previously established theoretical results, to determine (i.e., to derive) the theory in question. For Herschel, use of hypotheses is an economical, heuristic, preliminary stage of investigation which eventually gives way, in successful cases, to a quite different logic of experimental justification. Part of Herschel's point, then, is that it is not the original, historical manner of thinking up a theory that is important to justification; rather, it is the logical derivability conditions that we may be able to establish later. To employ terms I introduced elsewhere,[16] it is discover*ability* (generatability, derivability) which is important, not the original, historical *discovery* (original mode of generation). Although Herschel gives original discovery a back seat in his theory of justification, it is hardly H-D consequentialism which takes its place.

Generative justification: a defence

Since discoverability arguments are arguments, generative justification is a logical rather than an historical conception of justification.[17] It thus avoids the many problems with historical conceptions of justification – problems which require answers to such questions as, 'What did Einstein know and what was in his mind when he . . . ? However, this does not mean that generative justification is context-free.

In recent papers I have defended generative justification as a form of justification which is complementary to consequential testing and is sometimes logically stronger than the latter. Generative justification furnishes sufficient conditions for knowledge claims rather than the necessary conditions that H-D justification provides. A theory can be justified directly by reasoning *to* the theory from what we already 'know' about the world as well as indirectly by reasoning *from* the theory plus auxiliary assumptions to predictive consequences. Generative justification is constructive; consequential testing is purely eliminative.

While a full discussion is impossible here, a series of objections with brief replies will help to explain what I have in mind and to set the agenda for the rest of the paper.

Objection 1. Deduction from the phenomena is strictly impossible. Since there are no valid ampliative inference rules, one cannot reason from experimental claims to a general law or theory, nor can one derive deep-structural theories from phenomenal laws.

16. Nickles (1985).   17. Musgrave (1974).

Reply. Instances of (partial) deduction from the phenomena are more common than usually recognised. The derivation is from phenomenal claims, not literally from experimental data; and often the derivations involve approximations and simplifications and so are not purely deductive. The derivations make use of additional premises, among which are previously established laws, principles, and theoretical results. A phenomenal claim, in conjunction with well-established theoretical premises, can have revolutionary consequences. An example is that from the constancy of the velocity of light together with some innocuous assumptions about the transformation of co-ordinates, one can derive the Lorentzian instead of the Galilean transformation.[18] A second example is that from Planck's *empirical* radiation formula, together with generalisations (abstractions) of classical statistical-thermodynamical principles, Ehrenfest was able to derive the discreteness (quantisation) of oscillator energy levels in 1911, about six years after it was probably first assumed by Einstein.[19]

Contrary to the usual objections to 'Newtonian' methodology, then, generatability need not be at all inductive. Nor is it an 'all or nothing' affair of complete logical deducibility from phenomenal claims alone. Partial generative support (partial derivability) is only one reason why instances of 'deduction from the phenomena' are more common than usually recognised.[20] Another reason is that tight 'discoverability' arguments are one device for expanding the sphere of the observable. What the scientific community considers to be observable phenomena expands with the closure of debate about experimental and theoretical questions.[21] When scientists say that neutrino detection gives them a way of directly observing the interior of the sun, we have the ironic situation, relative to the knowledge of 30 years ago, of directly observing an unobservable object by means of undetectable particles!

Objection 2. The answer to the first objection commits the generativist to an overly optimistic epistemology that is incompatible with the fallibilism which characterises twentieth-century thinking. For in permitting lawful and theoretical premises in its justificatory derivations, generative justification assumes that previously established results have been shown to be true.

Reply. No, the generativist must hold only that previous results are justified, that they are *reliable* enough to provide support for the claim in question. Generative theories of justification need not be infallibilistic

18. For an accessible presentation see Spector (1972), chapter 10.
19. Ehrenfest (1911).   20. Many 'big' cases are nicely treated by Dorling (1973).
21. See Shapere (1982); Pinch (1985a).

anymore than consequential theories are. Derivability remains important within a fallibilistic methodology; indeed, it is more important than ever. An infallibilist needs only a single proof. Fallibilists need all the justification they can get!

Objection 3. No such reliable justification is possible. Generative justification by derivation of conclusions from premises begins a vicious, linear regress of justification. At any given stage, the argument begins from unproved (ungenerated) premises; hence, nothing is justified.

Reply. No sane generativist demands that *all* justification be generative, just as no reasonable consequentialist demands that all justification be consequential. (The regress objection can be launched against consequentialists also and, indeed, against any position which makes essential use of arguments!) In any case, generative justification does not entail a linear, regressive theory of justification.[22] Reliability is enhanced when a result can be derived in more than one way, i.e., from different combinations of premises, and when this result can be used as a premise in other derivations, including some which yield the former premises as conclusions. A single 'proof' is nice, but multiple derivations are naturally preferred in science because they are robust. They amount to what we might term *theoretical replication* – the replication of a theoretical result – by 'locking in' the claim as an unproblematic node of the accepted body of theory and practice. A robust result is a constancy against a changing background, an invariance, something that different, even competing, methods or techniques produce in common.[23] For example, if very different theoretical models or research techniques yield the same result, we have robustness. This robustness can be of several kinds. In the present context, direct derivability plus consequential testing provides one sort of robustness. In the form of multiple derivability, discoverability itself can be robust. The failure of a single premise does not spoil a multiply derivable result. Such multiple derivations are also indispensable in crystallising out the essential scientific (e.g., physical) meaning of theoretical principles – an aspect of scientific thinking overlooked by most philosophers and one related to robustness. I shall return to both meaning and robustness below.

Objection 4. Invoking generative justification is a move in the direction of a formalist, universal, and purely logical conception of justification and away from social, constructivist, localistic, and 'finitist' accounts of justification.

Reply. We shall see.

---

22. See Nickles (1987b).   23. Wimsatt (1981).

RECONSTRUCTION IN SCIENCE

The derivability of a claim from previously established results typically can be established (if ever) only long after the claim is first formulated and tested. In the reconstructed versions of the theory, the early heuristic motivation for producing a particular theory is often lost or tranformed.[24] The very character of the theory may change in the process. The same can be said for the experimental work from which the phenomenal claims were originally constructed and which eventually are reconstructed as key premises for such theoretical derivations. Just as deductions from the phenomena are a special case of discoverability arguments, the assembly of discoverability arguments is only part of a more general process of conceptual and technical *reconstruction* in science. I emphasise that reconstruction is part and parcel of scientific work itself, not an idealised characterisation of scientific activity imposed externally by philosophers. I do not mean the notorious 'rational reconstruction' of philosophers. 'Reconstruction' in my sense is closer to the 'construction' studied by sociologists and historians; indeed, it is a continuation of that process.[25] It consists of second-order work on problem solutions and other results already gained rather than the work of constructing those results in the first place. When successful, reconstruction results in the replacement of poorly defined problems, weak problem-solving techniques (including experimental techniques) and conjectural solutions by well structured problems, solutions, and solution techniques (which have been reduced to routine and, in some cases, even automated).

A body of scientific problems and solutions may be reconstructed more than once and may be transformed eventually into something quite different from the original examplars. One has only to contrast the standard problem-formulations and solution methods of the latest physics and genetics textbooks and chemistry manuals with their ances-

24. This was the case with Einstein's principle of equivalence, the great heuristic idea behind general relativity. It is difficult to find an exact heir to the principle in the 'final' theory at all, let alone one that retains the heuristic power of the original versions (Earman & Glymour 1980, p. 177). Even the clear derivability of predictions may not be established until much later. According to Earman and Glymour, an unproblematic derivation of the red shift formula from general relativity was obtained only a decade after Einstein proposed the theory. Actually, Einstein first announced his prediction in 1907, from his evolving principle of equivalence, before the full theory was even formulated.

25. Sociologists have given most of their attention to what I term 'construction' – work concluding with the writing of scientific papers for publication. In recent years, however, sociologists such as Pickering, Collins, and Pinch have increasingly attended to reconstruction.

tors, the original historical problems and solutions. Consider the evolution of the two-body problem from Kepler to Newton's *Principia* (1687) and thence to the contemporary mechanics text. Reconstruction is a process by which scientists gain expert control of previously announced results. As the preceding examples suggest, the process may be transgenerational. Scientists themselves can be remarkably unaware of the process, as Kuhn and Pickering have noted.[26]

Unlike philosophers, working scientists (at least nowadays) are notoriously uninterested in foundational issues for their own sake. The occasion for reconstructive transformation is typically the application of previous problem-solutions to new problems at the frontier, in line with the latest ideas and techniques. Kuhnian exemplars do not always remain fixed models for further work but may themselves be transformed by that work. The modeling is interactive.

Construction and reconstruction

In my terminology, construction concerns the original manufacture of knowledge claims (usually in the form of published articles), and their immediate interpretation and acceptance (or rejection) by the relevant scientific community. Reconstruction concerns what happens to these results later – how they are streamlined, articulated as nodes of well understood conceptual frameworks, and, in some cases 'black boxed' so that they can be routinely and reliably used *without* full understanding.[27] ('Black boxing' makes possible their reliable use by technicians who lack full understanding.) In historical application, it is somewhat arbitrary where we say that construction ends and reconstruction begins, but the first published article is a useful boundary marker. Indeed, the two processes often operate simultaneously at different levels: partly on the basis of a recasting of older material, an author hazards a new knowledge claim.

The construction–reconstruction distinction may be applied to experimental as well as theoretical work. Gooding points out the surprisingly large gap between the initial detection of observational novelty by Michael Faraday and its eventual cognitive organisation and articulation as a finished publication.[28] Reconstruction concerns what happened after this stage as these particular results were refined experimentally

---

26. Kuhn (1962); Pickering (1984a,b). Since reconstructive transformations are so often unintentional and unrecognised by scientists, it would be dangerous to consider it an 'actors' category'.
27. On 'black boxing', see Pinch (1986).   28. Gooding (1986).

by Faraday himself and other scientists and reconceptualised theoretically by Clerk Maxwell and others.[29] Biochemistry in recent decades provides many examples to illustrate the distinction. A new experimental technique which is technically difficult and not fully understood (by later standards) is announced. Further work yields fuller understanding on the conceptual side and the mechanisation of the technique on the side of practice, so that laboratory technicians can routinely perform the procedure in the future. The procedure therefore becomes 'transparent' or 'black boxed' in two distinct senses: *conceptually* 'transparent' and *instrumentally* 'transparent'. While they are mutually supportive, each of these types of 'transparency' can occur in the absence of the other. Scientists may gain excellent conceptual control without being able to exploit that control technologically; and they may achieve remarkable instrumental control without correctly understanding their instruments, as Hacking's nice discussion of the microscope illustrates.[30]

There are many ways in which the logical status of experimental results can change under reconstruction. For instance, when an experimental claim (which might originally have been a consequential test) comes to impose a condition of adequacy on the form of theories of the domain, the claim can be given a generative role. That is, it can be employed to build the theory in the first place rather than to test it. An experiment originally considered to be crucial may later be reconstructed by scientists themselves as noncrucial. And contrariwise, on a later reconstruction of the domain of knowledge, a once-ordinary experiment may become crucial in deciding among the leading candidates for a problem-solution.[31] Failing to acknowledge these possibilities is an error almost as serious as thinking that an empirical anomaly facing a new theory will forever remain anomalous, that all anomalies are logical refutations.

Good *discoverability* arguments are rarely available when a new hypothesis is first discovered or constructed. Much work is required to obtain theoretical, experimental, and observational 'closure' by using hypotheses and their tests as tools to conceptually reconstruct the original problem, the constraints on the problem, and, ultimately, to reconstruct the solution itself. Establishing the legitimacy (or the reverse) of the various attempts at derivation is a complex social process. A classic example, described in technical but not social detail by historians Klein and Kuhn, is the reconstruction of Planck's work by Einstein, Ehrenfest, Debye and others. Ultimately, Planck's empirical law could be derived

29. Nersessian (1984).   30. Hacking (1983), chapter 11.
31. For further discussion and examples, see Nickles (1988).

in many acceptable and theoretically sophisticated ways, and the existence of quantised energy levels could be derived from it.[32]

Although not razor sharp, the construction–reconstruction distinction is none the less important. Philosophers and science-studies experts have often written as if scientific innovation stops with the publication and acceptance of a 'discovery' or construction. In fact major innovation can occur at all stages of research. Worse, rational reconstruction philosophers have tended to think of all phases of the research process in terms of the logical structure of the final products of research as expounded in the latest textbook. To coin a term, formalists have assumed that history must *pre*capitulate' their logic. Historically-sensitive writers ('historicists') have retorted that any legitimate logic must simply *re*capitulate (report) previous investigation. I exaggerate to make a point. However, some sociologists have treated even the writing up of results for publication as the creation of an artifact (in the pejorative sense), in so far as the paper is not merely an historical report of the laboratory work done. Why not consider the writing process a continuation of 'real' scientific research rather than a fraudulent misrepresentation of it?[33]

Both sides in this debate tend to collapse science into a 'single pass' affair which allows for no essential transformation over time, by successive refinements of previous work. Both are guilty of what I call *the one-pass fallacy.* This is a form of the genetic fallacy, or of its reverse – the mistake of arguing from the present nature of a thing to conclusions about what its origins must have been. Another characterisation of the mistake is that one-pass accounts of science are *essentialist.* They treat science as just one sort of activity throughout. They cannot appreciate the many uses of experiment. Yet another characterisation is that both sides are guilty of (opposite forms of) *presentism,* the historical fallacy of trying to understand other periods in terms of some present (perhaps

32. See Klein (1962, 1970); Kuhn (1978).
33. See (Nickles, 1987c, 1988) for development of this point. E.g., Knorr-Cetina (1981) writes as if a scientific paper is supposed to be merely a historical report of the research one already has done. Medawar's (1964) suggestion that the typical scientific paper is fraudulent derives from the same assumption. In my opinion even Kuhn (1962, chapter 3, 4) in his many pages on normal science tends to treat exemplary problem solutions in too fixed a manner, as historically datable achievements. For Kuhn the limited innovation of normal science consists more in the articulation and application of these exemplars than in their transformation. Accordingly, normal science comes out looking more conservative and less creative than it really is. Kuhn (1962, chapter 11) notes the invisibility of revolutions but arguably misses much of the innovation of normal science because normal reconstruction is also largely invisible.

an historical present). The formalists who attempt to understand the past in terms of the (actual) present are guilty of whig history, while those historicists who require the future to be understood in terms of some historical present are guilty of an overly extreme anti-whiggism. Scientists are notoriously whiggish, for example, in failing to recognise the extent to which they reconstruct previous work. However it would be silly to upbraid them for this and to consider their work fraudulent as long as they are not claiming to *do* history.

Reconstruction in science raises many intriguing questions for science-studies scholars. As previously suggested, some of the issues discussed by rational reconstructionist philosophers gain new legitimacy when examined as part of the scientific research process itself, a perspective ironically foreign to those approaches themselves. Fortunately, historians, sociologists, and historically-sensitive philosophers are now furnishing the resources for undertaking this study. The previous discussion also suggests that problems in accounting for experimental and theoretial closure derive in part from insufficient attention to reconstruction in science. To the sociologist, reconstruction opens up the possibility of a deeper constructive view of science: social construction by reconstruction. I believe that the topic also provides an opportunity to bridge the gap that has hitherto existed between social and logical accounts of scientific work. That is the entrance cue for our third main theme: the unification of logic with a social conception of justification.

GENERATIVE JUSTIFICATION AND SOCIAL CONSENSUS

How can generative justification be compatible with a social conception of inquiry? Does not generative justification make justification a matter of logical proof rather than of social licensing? This section and the next one sketch my answers to these questions.

If the derivation or 'proof' itself is of a novel kind, then this new type of argument must be socially legitimated. Even when the type of reasoning is not controversial, however, the construction of a derivation involves all of the processes of conceptual and technical hygiene alluded to above – cleaning up experimental techniques, purifying the data, debugging arguments to phenomenal conclusions, cutting down problem spaces and restructuring problems. A consequence of tidying up the relevant stock of background knowledge and skills is that the process of doing this becomes invisible and the resulting claims appear to be dictated by logic from premises given by Nature.

Where does justification end?

Here I shall provide only a rough sketch of one part of the story – the part which answers the most serious objection to a generative conception of justification.[34] This is the linear regress objection (Objection 3 above): A proof or weaker derivation is only as strong as its premises, and what is to justify those premises in the case of empirical scientific claims? Nothing in science can be generatively justified, even in principle (the objection continues), for at every stage of the generative argument there will be premises which have not themselves been generatively justified.

An early instance of the use of a generativist methodology will both highlight the difficulty and suggest its solution. In his *Opticks* (1704), Newton presents his theories of the nature of light, refraction, reflection, etc., as 'deductions from the phenomena'. The *Opticks* is written in geometrical style, with theorems and their corollaries derived from axioms, definitions and other propositions 'proved by experiment'. Newton's axioms include the statements that the angles of reflection and refraction lie in the same plane, that the angles of incidence and reflection are equal, and that the sine law holds for refraction. How does Newton justify the axioms? In the *Opticks* he remarks: 'I have now given in Axioms and their Explications the sum of what hath hitherto been treated of in Opticks. For what hath been generally agreed on I content my self to assume under the notion of Principles, in order to what I have farther to write.'[35] Rather than continue the regressive, logical–analytical search for proofs here, Newton makes a dialectical–rhetorical move. He departs from purely formal reasoning to a more practical stance which takes into account his intended audience and its generally accepted judgements. Newton addresses a particular audience rather than attempting the impossible task of satisfying all possible audiences by means of absolute demonstrations of his conclusions.[36] The Euclidean arguments cannot stand alone; the audience must furnish something. The consensus of the relevant community of experts concerning a body of positive scientific results (not to mention the legitimacy of the overall form of reasoning) is, at one level, justification

---

34. See Bloor (1976, 1983) and Pinch (1977, 1985b) for sociological treatments of logical arguments. Although I believe that they go too far, their accounts do reveal the serious deficiences in nonsocialised, purely logical accounts of science.
35. op. ciit., pp. 19–20.
36. For the dialectical–rhetorical dimensions of practical reasoning, see, e.g., Perelman (1982). The sort of move Newton makes is arguably present in the tradition fixed by Euclid, in the status of the postulates and common notions of geometry.

enough, Newton thinks. Where no actual, serious objections remain unanswered, there is no need to supply justification. Where it doesn't itch, don't scratch.

To speak of consensus is admittedly only the beginning and not the end of analysis. Post-Kuhnian philosophers refer glibly to 'the scientific community' as a structureless pudding of roughly equal experts who somehow reach consensus.[37] Here it must suffice to say that such a consensus is no mere coincidence of subjective preferences. Newton's axioms were not simply conjectural assumptions upon which the community (such as it then was) happened to agree. The axioms were themselves products of an extended empirical, theoretical, and critical enterprise among investigators who frequently disagreed about problems, methods and substantive claims.[38] Newton recognised that there was no need to rehash all of this in every new paper. Thus, even a fairly comprehensive text like the *Opticks* could treat justification in a localised manner by appealing to what his primary audience already took for granted.

Social interaction is essential here. Through the process of critical reception, the community in effect 'licenses' the use of a body of previous results. (Such metaphors are today being replaced by hard empirical studies of scientific activities and the mechanisms which govern them.) The determination that objections have been satisfactorily answered and that there is sufficient reason to endorse certain claims ultimately involves human judgment: it is not purely a matter of formal logic. '[T]here is no standard higher than the assent of the relevant community'.[39] What the Newton example suggests, therefore, is that even discoverability arguments are located in a dialectical–rhetorical context. This context becomes largely invisible once the social agreements are sealed. At the very least, then, recognition of the social basis of justification is necessary to stop the logical regress of justification and, specifically, to 'prime' generative justification. As Kuhn noted, a social account is also necessary to explain why a theory framework is

37. See Fuller (1988). See also Collins (1981a) and Pinch (1981) on the 'core-set' as an example of much needed sociological work on the structure of consensus. An objection to my account is that by invoking consensus I presuppose closure rather than explain it. My brief reply is that generative justification can parlay closure at one level into closure at other levels. A special problem for early science and for emerging specialties is that the number of investigators is so few and so scattered that it is not clear in what sense 'a community' exists. On this problem see Gooding (chapter 7 this volume).

38. For example, Newton himself had used mechanics to derive the sine law from assumptions about the nature of light in the *Principia*, (1687), bk. I, §XIV, prop. XCIV, thm. XLVII.

39. Kuhn (1962), p. 94.

protected from logical refutation by one or more anomalies, which are always available. Obviously, consequential as well as generative justification depends on social processes.

A further examination of the Newton case (and of many others) would challenge the apparent *linearity* of generative justification. Newton's geometrical exposition marks him as a strong exponent of linear justification. Working in the Euclidean, axiomatic tradition, he sought the one most convincing mode of presenting his principal claims, as 'deductions from the phenomena'.[40] Yet in writing the *Opticks* and in the years of optical research apparent in his Cambridge lectures (1670–72), Newton had recognised that there were any number of ways in which his experimental results and arguments could be arranged to support one claim or another.

Circularity

Contrary to the implications of axiomatic representations of scientific work, scientific justifications may proceed in more than one direction. Accordingly, a claim does not have a unique position in a context-free logical framework. To put it bluntly, nonlinear justification involves patterns of reasoning that are broadly circular.[41] This being the case, the objection that a particular argument is *trivially* circular and thus question-begging and hence nonsupportive is always available. Social judgements determine whether or not such objections are allowed to stand, just as they determine whether an anomaly should be considered a refuting instance or whether demanding a deeper justification for a claim would be silly. 'Social judgment' can involve many things, including critical discussion and negotiation. Such judgment is rarely arbitrary; on the contrary, it must invoke previously licensed resources, including accepted scientific reasons and goals. The judgment usually issues in the form of a consensus, at least among the dominant authorities, who carry much of the rest of the field with them. The losing party may or may not join this dominant position.

The circularity objection can be damped by the provision of consilient results from different quarters – a situation in which the logical overlap

40. Besides the historical evidence, there is powerful internal evidence in the *Opticks* (1704), and in his famous 1672 optics paper, that not even a Newton could have constructed his experiments and arguments in a single afternoon-or in a matter of months. Even the 1672 paper reveals Newton as a sophisticated expert on these matters.
41. Compare Kuhn (1962), p. 94. The circularity should not be 'blamed' on the social nature of justification.

is reinforced by a social meshing. A nonlinear account of scientific reasoning stresses the importance of multiple derivations, multiple cross-linkages, as a way of increasing the reliability of a body of claims and methods.[42] However, even if justification of the body of 'positive science' (i.e., previously established results) is globally nonlinear, individual derivations are linear and must be primed by acceptable premises.

There is a sense in which multiple derivability makes a result less context-sensitive, especially when the derivations spring from quite different quarters. One source of the reliability of Planck's law was its multiple derivability and its use in the derivation of many other results. Such robustness of a scientific claim usually reflects a convergence of work within several specialty communities. Here my idea ties in with Pickering.[43] Multiple connectedness helps to unite and enlarge the relevant scientific sub-communities and tends to confer reality on the result. Newer problem conceptions and solution techniques are always undergoing standardisation or routinisation at any time, while others are losing their status as standard procedures. Obviously, the logico-mathematical moves of those derivations which are multiply derivable do not critically depend on a single set of premises (and corresponding community commitments); and those with multiple derivations rooted in more than one specialty acquire an authority independent of any single community of experts.

THE REALISABILITY REQUIREMENTS

The old view of methodology as a branch of formal logic or as an application of *a priori* epistemology is no longer tenable. This section goes further than the last in dispensing, at one level, with the idea of an *a priori* method of science that is fixed in advance and into which we merely plug social decisions to stop logical regresses. We must examine more closely the 'social backing' of logic and methodology and also the logical backing of social formations. However, at this level I continue to take for granted the validity of logico-mathematical derivation techniques as discoveries or reconstructions already fixed by the previous history of human inquiry.[44] Although they are subject to revision, logic and method are not reinvented or totally reconstructed by every new generation of scientists. This does not mean that I relegate all of logic and methodology to ancient history, along with Euclidean

42. Wimsatt (1981); Nickles (1987b).   43. Pickering (1980, 1984b).
44. Bloor (1976, 1983) *begins* from the sociological 'side' and develops an account of mathematics and logic in sociological terms.

geometry and Aristotelian logic. Every generation of scientists has developed new problem-solving methods – new mathematical and logical inferential techniques, new experimental procedures, etc. These novelties often coincide with major changes in the substantive claims of science.

In so far as it concerns the governance of human research communities, methodology is a human, social–scientific subject and not a purely logical subject.[45] Most of the recent historical, psychological, and sociological work critical of various philosophies of science attempts to show that the proposed methodologies fail utterly because they are not cognitively and socially *realisable* for human investigators. (Recall my argument that 'cognitive' and 'social' are not terms of opposition.) This criticism has force because science as we know it does not and *cannot* operate according to the methodological rules proposed. Normative philosophies of science presume that 'ought' implies 'can'; it follows that 'cannot' implies 'not ought'. Hence, those methodological rules must be rejected as inapplicable to actual science. Repeatedly, successful kinds of scientific practice depart radically from *a priori* reasonable methodologies and theories of justification. Kuhn's critique of Popper details such a methodological failure.[46] Not only does history fail to fit Popper's methodology but also it is clear that even if Popperian methodology *could* be employed by a community of inquirers, the result would be *nothing like science as we know it*. It would be more like philosophy, perhaps.

If we assume that scientific methodology *is* intended to govern the community of human scientific inquirers, then any viable methodology must be realisable by communities of human individuals and their resources (including super-computers!). Such realisability need not be perfect, of course, but its goals, strategies, and methodical procedures must be practicable, else it will be irrelevant to working science. With this qualification, any method which is cognitively and/or socially impracticable can be rejected.

Of course, what is cognitively and socially feasible depends on many things, including the state of technology, educational and political institutions and the economy. The importance of these factors is revealed by empirical studies of science, not by *a priori* reflection on what science must be like. However, rejection of such methodologies is not merely on the ground of *descriptive* inadequacy in the light of empirical studies; it is also a *normative* rejection, a matter of economic imperative. Thus Herbert Simon rejected certain classical economics approaches for their hopelessly unrealistic requirement of perfectly rational agents.[47] As

45. Will (1981).   46. Kuhn (1962, 1970).   47. Simon (1947).

Simon has seen more clearly than anyone since Charles Peirce, methodology itself is largely a matter of economical organisation. It is only because our intellectual and other resources are limited that methodically-ordered inquiry becomes necessary. Crudely stated, methodology is the managerial science (the economics, broadly speaking) of the scientific community. Leading philosophical accounts of research, including 'theory of justification', have been frightfully uneconomical in their failure to recognise human limitations, both individual and social (e.g., the 'boundedness' of human rationality). These limitations are magnified by the amibitiousness of familiar methodological goals, e.g., to select one comprehensive, true theory from an infinite domain of candidates on the basis of very imperfect information.

Justification in context

Whatever the details will look like, such a conception brings the social and institutional dimensions of science *essentially* into the 'theory of science'. Since such dimensions are not historical–contextual constants, the resulting accounts of scientific research will end up looking still less like a purely logical subject. So far I have been talking about methodology in general. The cognitive and social realisability requirements hold for theories of justification in particular. Any conception of justification as a purely logical affair, a matter of applying a formula, is, to this extent, a global conception of justification, because it requires us to remove justification claims from any contextual complexity that is not explicitly built into the formula. (Even fallibilistic conceptions of justification may be global in this sense.) These global conceptions of justification are not humanly realisable; thus we may reject as untenable all accounts which consider justification nothing more than an individual's satisfaction of timeless logical standards.

This argument has two implications: justification is local and justification is social.[48] It is social for many reasons, not least that it is wasteful

48. Philosophers who question how derivation from theoretical premises can possibly be justificatory miss this. Despite their rejection of global, 'Cartesian' *scepticism* in favour of local, specific-reasons scepticism and their rejection of an 'absolute' observational –theoretical distinction, they stand outside any scientific community and outside of history and judge all theoretical commitments therein to be unwarranted as premises in a deriviation. They fail to see that local scepticism and local justification are opposite sides of the same coin. The fact that all justification is more or less local does not contradict my claim that generative justification is logical rather than historical. The point is that it is the logical order of reasons which is important to justification, not the historical order of events. Generative justification remains historically and socially located in another sense.

and irrational to provide more justification than the relevant community of experts demands.[49] For those who regard the claim that justification is social as a dangerous statement requiring many qualifications, I shall say just that 'social' does not entail 'illogical'. Recall Kuhn's statement that there is no higher court of appeal than the relevant community of experts, left to freely investigate the claims. To be sure, this does not make community decisions automatically correct. Later generations may alter, reinterpret, or even reverse a community's judgments. Later generations may determine that their predecessors unwittingly violated those predecessors' own rules of governance.

Justification is local in the sense of residing within a logico-historical niche of positive scientific attainments and problems. It is also necessarily local because of human limitations, and this locality implies sociality. Few people can be expert about more than one or two things. Thus to deal effectively with large problems, we need communities of people who are experts on the different facets of the problems. Recent empirical studies show that the limitations are still more severe: one can hardly become an expert on anything except by being trained within a community of experts. Hence, *specialty* communities are necessary, not only communities of experts on diverse subjects.

Conversely, the sociality of scientific work – and specifically the cultural nature of the transmission of scientific knowledge – implies its locality. Even within the same field communication among different research groups in different microenvironments can be difficult.[50] On a larger scale historians, sociologists, and philosophers are becoming aware that experimenters and theoreticians in various mature sciences do not inhabit the same worlds of problems and resources and hence that no single methodological account of a science even in time-slice cross-section can be adequate. Bogen and Woodward's distinction of experimental *data* (the output of instruments, data reduction devices, and the low-level representation and interpretation of this output) from *phenomena* (claims about nature) suggests that theoreticians and experimentalists work up from the experimental data to phenomenal claims, whereas theoreticians are concerned with the relation of theory to phenomena, not data. And within each of these broad groupings

49. What grounds then do certain mathematicians and scientists have for attempting to raise the standards of rigour in their disciplines, to rise above the current community standards? Sometimes such reformist arguments are also couched in terms of economy – a long-term rather than a short-term economy. Indeed, long-term economy was a major motivation for Baconian and Cartesian foundationisms.
50. See the studies in Collins (1981b).

there are specialty commmunities, each in their own microenviron-ments. This is still too simple, for experimentalists do sometimes make provocative theoretical claims, and theorists experimental claims.

The fact that powerful research methods tend to be content- and context-specific provides further evidence for the locality of actual research and, accordingly, evidence of the failure of attempts to provide informative, general methodologies of science. Generatively speaking the widely applicable methods (including simple inductive and H-D methods) tend to be weak and provide insufficient detail about what scientists do or should do. Collectively, human beings can do many things well; individually, they can do only a few; hence my demand for cognitive realisability. The methodological thrust of the social realisa-bility injunction is that although the actual scientific social orders may not be the only possible ones, they are not infinitely malleable. Not just anything goes, socially speaking. But much the same can be said of 'logical' order (methodology). Not just any old 'methodology' or set of local methods will yield the demonstrable achievements attained by modern science. An adequate account of research must combine a viable logical order with a viable, human social order. The social organi-sation and the logical governance attributed to scientific communities must mesh appropriately and must fit our evaluation of the commun-ity's accomplishments.[51]

This makes the theory-of-science problem more difficult than it was as a purely logical–epistemological problem or as a purely psychological or sociological problem: it is now a three-dimensional problem. Yet the problem of understanding and evaluating the self-governance of scien-tific communities now becomes more tractable, because it is more highly constrained. We now have the possibility of cross-checking the different fields of science studies and philosophy against one another. Where this check is successful, we achieve a kind of robustness in our studies of science. What was yesterday fierce competition among these discip-lines is already giving way to co-operation.

51. Ultimately, the methodological order and the social order are inseparable. However, most philosophers have addressed the former in the absence of the latter; and several historians and sociologists have focused on the latter without enough concern for the achievements of modern science (or even have reduced these achievements to a matter of social definition). The two orders are inseparable but irreducible. This section argues the importance of social considerations to methodology, but the claim that methodology reduces to sociology (cf. Latour, 1980) is no more warranted than the parallel claim that physics consists in nothing more than social processes. Sociolog-ical reductionism is no more promising than most other kinds of reductionism.

Whither logic?

The preceding discussion brings out the need for an additional requirement on 'theory of science.' An adequate methodological account must
make psychological and sociological sense; but to be adequate,
psychological and sociological accounts of successful science must also
make logical sense, and not only on grounds of intellectual clarity. For
example, they must not imply untenable methodological strategies. Let
us term this the *methodological tenability requirement.*

Does not this new requirement sneak in through the back door the
a priori methodological assumptions we kicked out the front? It need
not. Consider an account according to which the scientists in a given,
successful specialty regularly but successfully reject any claim that
receives confirmation or any claim shown to be derivable from knowledge already gained. Such an account would be suspect, to say the
least. Justification procedures which are completely arbitrary, logically
speaking, cannot explain why many fields of science have yielded
dramatic results which work in practice. Consider only the limiting
case of those discoverability achievements which result in automation
of experimental or theoretical procedures. We have many such procedures which work very well, by any fair standard. Luck is not an
adequate explanation; nor is success *entirely* a matter of human, conventional attribution. The failure of Lysenkoism surely teaches us this much,
for it certainly enjoyed immense socio-political support.[52]

I have mentioned extreme cases to make a point, but the point itself
holds more generally. For example, it is hard to see how a Popperian
community, adopting the method of conjectures and refutations as its
only methodological strategy could possibly achieve regular success in
meeting its stated goals.[53] The methodological tenability demand can
work against philosophically motivated accounts as well as those which

52. This paragraph expresses the most controversial part of my program, For here I am
    abandoning a purely historicist account of science, an account of what historical
    communities have in fact done, for an attempt to explain why certain bits of history
    were more *successful* than others. Although my account is (or would be, when spelled
    out) 'pragmatic' rather than strongly 'realist', my philosophical move will be criticised
    as an attempt to escape from history and sociology, an attempt that ultimately fails.
    For am I not merely invoking the standards of my historical community in some
    absolute fashion? Yes, I am invoking them, although not uncritically, Surely there is
    more to life than historical contemplation; and if I do not invoke *these* normative
    principles, then which ones am I to invoke? In a life of action, one must be committed
    to something! Besides, we are always operating from within our own community,
    even when we do descriptive history. The two realisability requirements are also
    extracted from results of historical research programmes and then used to judge others.
53. Laudan (1977).

are sociologically motivated. Notice that I do not say that good methodology alone is a sufficient explanation of success. Research is contingent on too many things for that. Nor do I here impose logic *a priori* as an independent constraint like social realisability but, unlike the latter, strangely inaccessible to empirical observation. Even though logic and method are themselves products of human historical investigation, they (or significant components of them) are now sufficiently fixed to provide independent control on inquiry.[54] We have pretty good working ideas about consistency, entailment, fallacious reasoning, instrumentally effective means to goals, reliability, and so on, not to mention mathematical sophistication in dealing with many kinds of problems.

Implications

It would be wrong to say that these logical and methodological research tools add up to a 'positive methodology' of science which is both detailed and universal in application. In my opinion there can be no such methodology. Detailed problem-finding and problem-solving strategies are discipline- and even problem-specific and can only be disclosed by empirical investigation.[55] However, this general methodological heritage does provide an important resource for critical evaluation – a 'negative methodology' if you will. It is a critical tool which helps determine the viability of any account of research.

We need not stop with a negative methodology, however, for the aforementioned empirical studies frequently enable us to go further. They disclose that scientific research produces not only theses about the world but also reliable methods of investigation. The justification of these more or less discipline-specific methods lies not in philosophical *a priorism* but in the pragmatic, scientific determination that they work. Methods as well as individual claims can be, and are, pragmatically justified.[56] The methodological tenability requirement need not impose an *a priori* requirement on an empirical investigation, therefore. As I conceive it, this is a broadly empirical requirement based on our scientific and general problem-solving experience, just as the realisability requirements derive from work in history, psychology, sociology, economics, and political science. How strong the requirement is will

54. After all, the two realisability requirements are also products of human inquiry.
55. The positive side of this last statement is that methodology can be studied empirically; and local, positive methods of problem-solving can be found.
56. Rescher (1977).

depend in part on the methodological resources of the discipline being investigated.

The practical implication of the methodological tenability requirement is that sociologists and psychologists (as well as philosophers) must inquire at some point into the epistemological force of their claims. Philosophers should join their historian, sociological, psychological, and technological colleagues in studying the 'social backing' or cognitive and social realisation of methodological claims. Several essays in this volume (and the literature on which they depend) indicate that progress is being made on both fronts. Questions about the social implications of any methodological claim and the methodological and epistemological implications of any sociological claim have heuristic value. I do not mean to suggest that there does or should exist an exact isomorphism between separately identifiable methodological and social components of an account of science. Clearly, not all methodological features correspond to a social formation, nor do all social formations have logical counterparts. Some broad and elementary examples will convey the flavour of social and logical backing and indicate the relevance of the cognitive and social realisability requirements for a theory of justification, and for experimental research in particular.

1. Popper's methodology permits and encourages wide open *criticism* from all quarters, including criticism of fundamentals. Socially speaking, this excludes no member of the relevant community; potentially, it gives every scientist work to do. Yet this hypercritical methodology makes closure difficult, if not impossible. It precludes the settling out of a body of positive science which fixes the beliefs of the community and serves as an accepted basis for anchoring problems, research techniques, etc. Without positive science, a sufficiently cohesive community seems impossible. Thus Popper's methodology is not socially realisable. It also faces grave problems relating to economy of research.

2. The requirement that scientific claims be empirically *testable* bridges theoretical and empirical interests. It permits a division of labor between theoretical and experimental work without destroying the unity of the community. Even testability divides (roughly) into two levels: the theoretical derivation of phenomenal claims and the experimental determination that these phenomena are detectable (observable).

3. Consilience or *robustness* of results means that two or more communities of investigators are joined in a consensus (or a least share common interests), with the resulting social reinforcement of scientific

commitment (whether doxastic or not).[57] Actually, there are many kinds of consilience and redundancy in scientific work, each with its own type or level of social backing. An investigator may replicate her own results using the same method or different methods; other investigators of the same community may replicate the result; investigators from another community and/or from a *competing* community may replicate it; two different lines of investigation may converge on the same result; a line of investigation may yield unanticipated predictions; and so on. These cases are interestingly different, both logically and socially.

4. In the forward direction, research which unites a current line of work with previously mastered problems and techniques (e.g., by use of analogy) can greatly amplify the social backing for the new work. Although he by no means defends a normative theory of science, Pickering (1981, 1984b) suggestively points out the importance of models and analogies to previous work in high energy physics, from a social-backing point of view.[58]

5. If members of separate communities (often in different fields) decide that their problems are basically identical, the now-common resources can be of immense *heuristic* value. In so far as the separate communities had defined their problems in terms of different constraints, the joint problem is now more highly constrained than before.

CONCLUSION AND PROSPECT

The standard view is that experiments are normally intended to test theories, to confirm or disconfirm theoretical predictions. My central claims have been: (1) Experiment has important uses, even justificatory uses, other than consequential testing. (2) The logic of experimental justification can shift significantly from initial hypothesis testing to 'final' justification as new information becomes available and as old, ill-structured problems and constraints are refined and conceptual and instrumental 'closure' is achieved. (3) The social–dialectical conception

57. A high degree of social activity concerning a body of claims does not imply belief in the literal truth of those claims. But for the same reason we must appreciate that the importance of predictive testability is not purely an epistemological matter of confirming or disconfirming an hypothesis. Especially during the early stages of a programme of research, the heuristic role of experimental testability may be more important than its epistemic role. Standard H-D accounts of science have been overly epistemic, overly 'philosophical', at the expense of understanding 'how science works'
58. Pickering (1981, 1984b).

of justification actually helps answer the main objection to generative justification – that no 'deduction' of theories from phenomena is possible and that derivation of theories from other theoretical assumptions is derivation of the conjectural from the more conjectural and hence justifies nothing. The social conception of justification helps articulate what counts as a body of previously 'established' results and standard procedures and what constitutes an adequate derivation of a claim. Surprisingly, combining 'logic' with 'society' results in a stronger conception of justification than consequential testing alone (which equally depends on appeal to community assent at crucial points). If social judgments sometimes undermine the force of apparent proofs, such judgments can also make such proofs possible and reinforce them. Social relations may be constructive (in an epistemic sense) as well as destructive. (4) More generally, methodology becomes a subject with several constraints beyond the standard logico-epistemic ones. The latter themselves become subject to social and cognitive realisability requirements. The old subject of 'logic of science' yields to a more broadly conceived 'theory of science', although not a global theory in the epistemological sense.

Education and expertise

My discussion has centred on one form of 'discoverability' – derivations of major 'theoretical' results from the phenomena. Such arguments tend to be available only at the 'closure' stage of inquiry, when the problems and solution techniques are well defined. (Discoverability arguments can help to fix the phenomena as well, but sometimes the phenomena in question have long been known to the scientific community.) Both experimentalists and theoreticians (who may be the same people) have by now become fully expert in this previously novel problem area. The Newton example suggests that dialectical–rhetorical moves remain important even here, but they are more evident at the initial stages when new discoveries or constructs are only emerging.

   Much of Gooding's work on Faraday's experimental research is about the path from original discovery to discoverability in experimental contexts. The process takes us from the original, inarticulate gropings in the face of anomalous experience to semi-verbal construals, to personally satisfactory verbal results, to easily communicable results which are routinely replicable and demonstrable to and by other members of the community (and to members of other communities). Both the experimental path and my more theoretical path from crude, conjectural solutions of badly formulated problems to demonstrable theoretical results are cleared by various kinds of socio-cognitive constructions and reconstructions.[59] In both cases the need for explicit dialectical and rhetorical moves, including instrumental instruction, seemingly diminishes as progress is made toward the final, well-structured problems and their solutions. Automation of the problem-solving process is sometimes possible in both cases. In the education process the instrumental, rhetorical and dialectical moves, the social interactions have 'gone underground' or rather 'gone *back*ground'. The process builds the new ways of seeing, representing, and talking into the background of the people working in this area. The more expert they become, the less need for explicit dialectical, rhetorical, and didactical moves and the more 'absolute' the apparent results.[60] From an outsider's viewpoint, specialists wear their expertise lightly. In actuality, behind each piece of expertise is an invisible history of social construction.[61]

If this account, which applies my own terminology to Gooding's work, is correct, it again brings out the locality of research. Stated simply, the point is that the perceived *absoluteness of a scientific result is proportional to the result-specific expertise of the local community*, to the amount of specific knowhow and information 'programmed' into them. This claim is not a tautology, for it is incompatible with the traditional view that by employing correct methods scientists can unproblematically read novel facts off Nature. In fact, it may seem incompatible with another theme of Gooding's work and of this chapter, namely that *reconstruction succeeds in making reconstructed claims unproblematic only in so far as it succeeds in rendering itself invisible*.

Together these two claims express what I shall call the *paradox of expertise*. Applied within the specialist community, the claims say that ease and directness of communication and unanimity of agreement

59. Again, there is no sharp border where construction stops and reconstruction begins. In my terminology, Gooding is mainly concerned with construction: the work-up of results for publication and/or demonstration. Reconstruction concerns the post-publication process of refinement and reinterpretation (discussed by Nersessian, 1984, in the case of Faraday).
60. I am speaking rather loosely here. For example, I should not want to say that experts exist only where there is a complete consensus or that experts may never differ. Again, see Pinch (1986) on 'black-boxed' techniques.
61. An expert rarely will encapsulate the whole history. While students today learn to do paradigmatic experiments, their starting points are quite different, and they tinker with very cleaned-up versions of the original ideas and instruments. They do not have to retrace a Faraday's entire path. One form of the problem I am raising is how the historical acquisition of expertise is somehow transformed into a nonhistorical mode of pedagogical transmission. As Kuhn has noted, scientific education is very unhistorical.

depend essentially on the shared expertise of the group. How can the very scientist who engages in reconstruction or who make use of it be so unaware of what is being done?[62] But both claims (and especially the second) can also be read as holding also for nonexpert outsiders, whence the first reading must fail. The first reading is true if and only if the second reading is false. Thus recounting the scientist's progress from relative novice (when dealing with essential novelty) to expert can be only part of the story. It is another problem to see how the published results of experts can be so successfully transmitted to, and even used by, other individuals who find them so intelligible and so absolute that they remain blissfully unaware of the expertise necessary to produce such claims and still less of the history necessary to produce the expertise. How can it be so difficult for some science-studies scholars, and especially for philosophers, to recognise the expertise and reconstructed history, including the necessary social processes, that are packed into scientific results? Somehow both the meaning and certitude of the claims have been largely unburdened of their technique-ladenness.

Meaning

In this paper I have focused on certitude rather than meaning. A few words about the latter will exhibit the construction–reconstruction problem I have been addressing from another angle. In a nutshell, the problem is: Given that original, creative research is so localised, so socially contextualised, how is the eventual decontextualisation achieved? Roughly, Gooding treats the problem of meaning at the level of experimental meaning and Nersessian[63] at the level of theoretical meaning. The problem is more serious for experimental than theoretical research to the extent that the former requires mastery of more nonverbal skills than the latter. In fact we could restate the problem as one of how experimental meaning gradually gives way to more verbal–theoretical meaning. This will not quite do, for theoretical practice also has its expertise, which is not shared by outsiders. Gooding and Nerse-

---

62. Actually, specialists do have a certain sort of awareness of the constructive nature of their work. (Are they less aware of reconstruction?) In so far as the experts cannot see the results for the constructive contingencies, those constructions are not transparent and the results are less than obvious. Of course, a perceptive expert can always recall contingencies if asked to reflect on the process, but that is largely beside the present point. Pickering (1984a, b) notes how scientists regularly present the results of the preceding generation as given by Nature.
63. Nersessian (1984).

ssian plausibly contend that the meaning of novel scientific concepts is hammered out in the context of local experimental and theoretical practice and is initially inseparable from the latter. How then is the separation eventually accomplished? To see that the problem they address is serious and deep, recall the numerous sociological studies of the 'cultural ethnology' genre which show the large extent to which scientific know-how is acquired by enculturation within the relevant research group. How then is publicly intelligible meaning constructed? How is transgenerational scientific communication of this sort possible? What kinds of knowledge can be communicated effectively by what means?

This problem stands some famous empiricist theories of meaning on their heads. Charles Peirce's pragmatism, P.W. Bridgman's operationism, and the various positivist criteria of meaning can be understood as early attempts to relate scientific theory to experimental practice, for the purpose of determining which 'theoretical' notions have definite, public, observable experimental meaning, and what that meaning is. These empiricists often contrasted their hard-headed conception of scientific meaning, rooted in laboratory practice, with that of the layperson, the businessman, the traditional philosopher, and the theologian. However, at one level these writers could pursue their programme only by assuming that scientists unproblematically read phenomena off Nature, that is by 'putting the phenomena first'.[64] They verbalised the issue in such a way that the tacit dimension of laboratory practice dropped out altogether. Moreover, their criteria of meaning(fulness) became ideological weapons, expressions of a doctrinaire empiricism, according to which experimental meaning is unproblematic and the basis for understanding *all* other meaning (rather than eventually being reconceptualised in terms of it). Observational meaning provided a permanent, neutral framework within which all theoretical meaning could be reconstructed!

The recent studies of the relation of theory and practice have the opposite tendency, to raise rather than to suppress the problem of how commonality of meaning is possible at all, and *especially at the laboratory level*. For to the extent that nonverbalisable (or at least nonverbalised) components of scientific practice are important in determining the meaning of novel ideas; given the differential transmissibility of experimental practice (roughly, know-how) and theory (roughly, knowledge-that); and given the greater locality of experimental work (which usually occurs in small specialty communities), how is it possible that

64. Pickering (1984a).

theoretical claims (and even phenomenal claims, for that matter) are eventually weaned from their matrix of origin? How do they come to be meaningfully broadcast about, taught, criticised, reformulated, and related to other ideas and practices by people who have no knowledge of their original production? Yet such a separation from context of origin is regularly made and *must* be made for scientific work to proceed. How to render the tacit explicit is therefore not the only component of this problem. Another is how meaning can be separated from its origins. If Gooding and others are right that new 'meaning' is wedded to the original context of discovery–construction, by what process is it divorced from this context? A third component of the problem is this: How is a separation of theoretical and experimental work possible, and why is it seemingly more successful in some sciences than in others? In sum: given that all research is local in several respects, but that some of it is eventually transmitted in a more 'cosmopolitan' form far from the source communities, how is this delocalisation, this decontextualisation, of scientific results possible?

ACKNOWLEDGEMENTS

I am indebted to the John Dewey Foundation for research support and to David Gooding and Trevor Pinch for their painstaking criticism of earlier drafts. Nonetheless, I must take 'credit' for all remaining failures.

References

Bloor, D. (1976). *Knowledge and Social Imagery.* London: Routledge and Kegan Paul.
Bloor, D., (1983). *Wittgenstein: A Social Theory of Knowledge.* New York: Columbia University Press.
Bogen, J. & Woodward, J. (1988). Saving the phenomena. *Philosophical Review.* (in press).
Brannigan, A. (1981). *The Social Basis of Scientific Discoveries.* Cambridge: Cambridge University Press.
Brown, J. R. (ed.) *Scientific Rationality: The Sociological Turn.* Dordrecht: Reidel.
Cartwright, N. (1983). *How the Laws of Physics Lie.* Oxford: Clarendon Press.
Collins, H. (1981a). The place of the 'core-set' in modern science: social contingency with methodological propriety in science. *History of Science,* 19, 6–19.
Collins, H., (ed.) (1981b). Knowledge and controversy. *Social Studies of Science,* 11, 158p.
Dorling, J. (1973). Demonstrative induction. *Philosophy of Science,* 40, 360–372.

Earman, J. & Glymour, C. (1980). The gravitational red shift as a test of general relativity: history and analysis. *Studies in History and Philosophy of Science*, 11, 175–214.

Ehrenfest, P. (1911). Welche Züge der lichtquantenhypothese spielen in der theorie der Wärmestrahlung eine wesentliche rolle?. *Annalen der Physik*, 36, 91–118, reprinted in *Paul Ehrenfest: Collected Scientific Papers*, (1959), (ed.) M.J. Klein, pp. 185–212. Amsterdam: North-Holland.

Franklin, A. (1986). *The Neglect of Experiment*. Cambridge: Cambridge University Press.

Fuller, S. (1988). *Social Epistemology*. Bloomington: Indiana University Press.

Galison, P. (1987). *How Experiments End*. Chicago: University of Chicago Press.

Gooding, D. (1982). Epiricism in practice: teleology, economy, and observation in Faraday's physics. *Isis*, 73, 46–67.

Gooding, D. (1986). How do scientists reach agreement about novel observations?. *Studies in History and Philosophy of Science*, 17, 205–230.

Hacking, I. (1983). *Representing and Intervening*. Cambridge: Cambridge University Press.

Herschel, J.W.F. (1830). *A Preliminary Discourse on the Study of Natural Philosophy*. London: Longman, Brown, Green, and Longmans.

Klein, M.J. (1962). Max Planck and the beginnings of quantum theory. *Archive for History of Exact Sciences*, 1, 459–479.

Klein, M.J. (1970). *Paul Ehrenfest: Theoretical Physicist*, vol. 1. Amsterdam: North Holland.

Knorr-Cetina, K. (1981). *The Manufacture of Knowledge*. Oxford: Pergamon Press.

Kuhn, T.S. (1962). *The Structure of Scientific Revolutions*, 2nd edn, enlarged, 1970. Chicago: University of Chicago Press.

Kuhn, T.S. (1970). Logic of discovery or psychology of research?. In *Criticism and the Growth of Knowledge*, ed. I. Lakatos & A. Musgrave, pp. 1–23. Cambridge: Cambridge University Press.

Kuhn, T.,S. (1978). *Black-Body Theory and the Quantum Discontinuity*. Oxford: Oxford University Press.

Larkin, J., McDermott, J., Simon, D.P., & Simon, H.A. (1980). Expert and novice performance in solving physics problems. *Science*, 208, 1335–42.

Latour, B. (1980). Is it possible to reconstruct the research process?: sociology of a brain peptide. In *The Social Process of Scientific Investigation*, ed. K. Knorr, R. Krohn, & R. Whitley, pp. 53–73. Dordrecht: Reidel.

Laudan, L. (1977). *Progress and Its Problems*. Berkeley: University of California Press.

Laudan, L. (1980). Why was the logic of discovery abandoned?. In *Scientific Discovery, Logic, and Rationality*, ed. T. Nickles, pp. 173–83. Dordrecht: Reidel, reprinted in *Science and Hypothesis*. Dordrecht: Reidel.

Medawar, P. (1964). Is the scientific paper fraudulent?. *Saturday Review* (August 1), 43–4.

Musgrave, A. (1974). Logical versus historical theories of confirmation. *British Journal for the Philosophy of Science*, 25, 1–23.

Nersessian, N. (1984). *Faraday to Einstein: Constructing Meaning in Scientific Theories*. The Hague: Martinus Nijhoff.

Newton, I. (1687). *Philosophiae Naturalis Principia Mathematica*. London; translated as *Sir Isaac Newton's Mathematical Principles of Natural Philosophy and his System of the World*, (1934) trans. Mott & Cajori. Berkeley: University of California Press.

Newton, I. (1704). *Opticks*, 4th edn, (1730). London: G. Bell; reprinted New York: Dover Publications.

Newton, I. (1670–72). *The Optical Papers of Isaac Newton, Vol. 1: The Optical Lectures*, (1984), ed. A. Shapiro. Cambridge: Cambridge University Press.

Nickles, T. (1976). Theory generalization, problem reduction, and the unity of science. In *PSA 1974*, ed. R.S. Cohen, A.C. Michalos, & J. Van Evra, pp. 33–75. Dordrecht: Reidel.

Nickles, T. (1985). Beyond divorce: current status of the discovery debate. *Philosophy of Science*, **52**, 177–207.

Nickles, T. (1987a). From natural philosophy to metaphilosophy of science. In *Theoretical Physics in the 100 years since Kelvin's Baltimore Lectures*, ed. P. Achinstein & R. Kargon, pp. 507–41. Cambridge, Mass.: MIT Press.

Nickles, T. (1987b). Twixt method and madness. In *The Process of Science*, ed. N. Nersessian, pp. 41–67. The Hague: Martinus Nijhoff.

Nickles, T. (1987c). The reconstruction of scientific knowledge. *Philosophy and Social Action*, **13**, 91–104.

Nickles, T. (1987d). Lakatosian heuristics and epistemic support. *British Journal for the Philosophy of Science*, **38**, 181–205.

Nickles, T. (1988). Reconstructing science: discovery and experiment. In *Theory and Experiment*, ed. D. Batens & J.P. Van Bendegem, pp. 33–53. Dordrecht: Reidel.

Perelman, C. (1982). *The Realm of Rhetoric*. Notre Dame: University of Notre Dame Press.

Pickering, A. (1980). Exemplars and analogies. *Social Studies of Science*, **10**, 497–502, 507–8.

Pickering, A. (1981). The hunting of the quark. *Isis*, **72**, 216–36.

Pickering, A. (1984a). Against putting the phenomena first: The discovery of the weak neutral current. *Studies in History and Philosophy of Science*, **15**, 85–117.

Pickering, A. (1984b). *Constructing Quarks*. Chicago: University of Chicago Press.

Pinch, T. (1977). What does a proof do if it does not prove?. In *The Social Production of Scientific Knowledge*, ed. E. Mendelsohn, P. Weingart, & R. Whitley, pp 171–215. Dordrecht: Reidel.

Pinch, T. (1981). The sun-set: on the presentation of certainty in scientific life. In ed. H. Collins (1981b), pp 131–58.

Pinch, T. (1985a). Towards an analysis of scientific observation. *Social Studies of Science*, **15**, 3–35.

Pinch, T. (1985b). Theory testing in science—the case of solar neutrinos: do crucial experiments test theories or theorists? *Philosophy of the Social Sciences*, **15**, 167–87.

Pinch, T. (1986). *Confronting Nature: The Sociology of Solar-Neutrino Detection.* Dordecht: Reidel.

Rescher, N. (1977). *Methodological Pragmatism.* Oxford: Blackwell.

Shapere, D. (1982). The concept of observation in science and philosophy. *Philosophy of Science,* **49,** 485–525.

Simon, H.A. (1947). *Administrative Behavior,* 3rd edn (1976). New York: Macmillan.

Spector, M. (1972). *Methodological Foundations of Relativistic Mechanics.* Notre Dame, Ind.: University of Notre Dame Press.

Will, F.L. (1981). The rational governance of practice. *American Philosophical Quarterly,* **18,** 191–201.

Wimsatt, W. (1981). Robustness, reliability and multiple determination in science. In *Knowing and Validating in the Social Sciences,* ed. M. Brewer & B. Collins, pp. 124–63. San Francisco: Jossey-Bass.

Worrall, J. (1978). The ways in which the methodology of scientific research programmes improves upon Popper's methodology. In *Progress and Rationality in Science,* ed. G. Radnitzky & G. Anderson, pp 45–70. Dordrecht: Reidel.

PART IV

THE CONSTITUENCY OF EXPERIMENT

11

EXTRAORDINARY EXPERIMENT: ELECTRICITY AND THE CREATION OF LIFE IN VICTORIAN ENGLAND

JAMES A. SECORD

Whence, I often asked myself, did the principle of life proceed? It was a bold question, and one which has ever been considered as a mystery; yet with how many things are we upon the brink of becoming acquainted, if cowardice or carelessness did not restrain our enquiries.
Mary Shelley, Frankenstein, or, the modern Prometheus[1]
Yet . . . much has been done, and much may be done, by a single individual in whom the necessary qualifications are united. It should be superfluous to notice such great names as those of Newton and Davy, and others who have been an honour to the age in which they flourished. It may be right however to remark that there are few but may contribute their mite to the cause, and then advance a considerable way in it, if they possess sufficient time, ardour, and perseverance.
Andrew Crosse[2]

A few months before Victoria ascended the throne, living mites of the genus *Acarus* unexpectedly crawled out of an electrical experiment in a private laboratory in Somerset. The experiment was conducted by Andrew Crosse, a wealthy English gentleman, political radical and avid natural philosopher. Publication (without Crosse's permission) on the last day of 1836 in a local newspaper under the headline 'extraordinary experiment' led to an international sensation. Crosse was accused of being a Frankenstein, a 'disturber of the peace of families', and 'a reviler of our holy religion'.[3] By others, he was hailed as an enthusiastic genius

1. Shelley (1831), p. 95.   2. C. Crosse (1857), p. 95.   3. C. Crosse (1857), p. 170.

who had broken the ancient boundary between life and matter. Debate about the issue continued for decades.

Much attention has recently been devoted to the ways in which scientists produce facts that are taken for granted as aspects of the natural world. The present episode, viewed as an example of what Collins has called 'extraordinary science', raises a number of issues of wider significance. Extraordinary science has the potential for producing quite fundamental changes in the ways scientists go about doing their work. It is not simply the same as marginal or aberrant science. Rather, it is science conducted without any clear consensus about what can be counted as an experimental failure or success. Collins' extraordinary science can be contrasted with 'normal' science, and parallels Thomas Kuhn's use of the term 'revolutionary'.[4]

As is typical of extraordinary science, the acari experiments were debated in an unusually diverse set of forums. These bring to light equally diverse sets of criteria for establishing the reality of an experimental phenomenon: natural philosophers disagreed, often vehemently, about what counted as a proper experiment. In order to locate the place of experimental practice within these debates, I will devote special attention to those who attempted to replicate Crosse's results and to those reputed to have done so. For some, the discovery represented the foundation of a new scientific discipline; for others, it was a taxonomic puzzle; for still others, it was a stepping stone to atheism, an editorial coup, or a joke. The sensitive nature of research into the origin of life gave the experiments a major role in controversies about miracles and materialism. With so wide a range of possible uses, the creation of life through electricity became the most famous experiment of the first half of the nineteenth century.

Historians, however, have dismissed this episode from serious consideration, taking the criticisms of some contemporary observers as self-evident.[5] In such cases, contamination seems more likely than spontaneous generation; the *Acarus*, technically a member of the Arachnidae or spider family, is too complex to be produced by applying electricity to a chemical solution; no analogous findings are said to have confirmed

---

4. Collins & Pinch (1982); Collins & Shapin (1984); Collins (1985). Thus Collins writes of parapsychology as a normal (albeit marginal) science, while paranormal spoon-bending is extraordinary.

5. See, for example, standard works on the controversies about natural law and evolution in the 'pre-darwinian' period such as Gillispie (1951), p. 155–6; Millhauser (1959), pp. 93–4 (who assigns the experiments to an otherwise unknown 'Edward Crosse'); and Ruse (1979), p. 107. Some recent articles (esp. Richards, 1987) take a more balanced view.

Crosse. As a result, historians have adopted the polemical labelling of Crosse as a crank, and the controversy about his mites is seen as a throwback to an early stage in the long history of debates about the origin of life.

The arguments used by modern historians to laugh off the experiments are palpably inadequate, and would have been perceived in this light by nearly all participants in the debate. For example, it is said that Crosse must have had dirty hands which contaminated the solutions. Yet the mite involved is not parasitic on humans, and his procedures made contamination at so gross a level unlikely. The experimental protocols used in the later phases of the controversy were much more rigorous than is generally recognised, certainly as refined as any being applied in contemporary disputes about the spontaneous generation of micro-organisms.

In these circumstances, it comes as no surprise to learn that the reaction to these experiments was far from simple or straightforward. At least six experimentalists are known to have attempted replications, and one of them, the Kentish surgeon William Henry Weekes, was spectacularly successful. The issue was discussed at the British Association for the Advancement of Science (BAAS), the London Electrical Society, the Royal Institution, the Entomological Society, the Ashmolean Society in Oxford and the Academy of Sciences in Paris. It was widely disseminated in the popular press, from the *Mechanics' Magazine* to the *Athenaeum*, from *The Times* to the *Somerset County Gazette*. Its impact alone would justify taking Crosse's work seriously.

The intervention of the electrical insects in early Victorian science also raises general questions about authority over knowledge and the power of experiment for a diverse range of audiences. I will begin by analysing the conditions, new to the early nineteenth century, that made it possible for a phenomenon to emerge from one private laboratory and became a mass sensation in the popular press.

THE CREATION OF A PUBLIC PROPERTY

The entrance of electrical life on the central stage of Victorian England depended on a very specific set of circumstances. We are used to thinking about experiments as the product of individuals working in conceptually defined traditions. There has been much less attention to the social and material conditions that make experiments possible at all. In the present case, the most important of these was a fundamental change within the technological armoury of electrical science. As John Heilbron has noted, the announcement of the voltaic cell in 1800 opened up a new era in the history of electricity. For the first time experiments could be made on the effects of long exposure to an electrical current. With the voltaic cell, Humphry Davy discovered unexpected new elements like chlorine, iodine and sodium; Jacob Berzelius investigated electrochemical combination; Hans Christian Oersted, Michael Faraday and others revealed connections between the forces of electricity and magnetism. Even before this, in the 1790s, there was a surge of interest in medical electricity.[6]

One individual fascinated by these possibilities was the young English gentleman, Andrew Crosse of Fyne Court in Somerset.[7] Crosse, born in 1785, came from an ancient family with strong radical sympathies. His father had known Priestley and Franklin, and (according to family tradition) had raised the tricolour on the ruins of the Bastille the day it was taken. Not surprisingly, the son found attendance at Oxford 'a perfect hell upon earth'.[8] His chief interest was in the electrical experiments he conducted in his spare time. He became acquainted with several local electricians, particularly George John Singer, and moved in literary circles in the southwest of England noted for their interest in science. Friends included Robert Southey, John Kenyon and Humphry Davy.

Crosse was especially intrigued by the effects of electricity in forming crystalline deposits in nature. After seeing the formations in 1807 at Holwell cavern, a famous romantic beauty spot, he focused on imitating them in the laboratory. The forest around Fyne Court was wired to gather electricity from the atmosphere, and huge arrays of batteries were set up in the converted music hall, where Crosse attempted to create crystals of quartz and other minerals.

The discovery of acari

It was while conducting these researches that Crosse made his unexpected discovery. As he later explained, his aim had been to form crystals from silica.[9] The experimental arrangement is illustrated in Fig. 11.1. A dilute solution (H) of silicate of potassa and hydrochloric acid

6. Heilbron (1979), p. 490–4. For medical electricity, see Walker (1937). The excitement generated by these discoveries, and their connection with the romantic movement in England, are discussed in Levere (1981).
7. Biographical details about Crosse are taken from C. Crosse (1857).
8. C. Crosse (1857), p. 32.
9. My description of these experiments is based on the most detailed published account, A. Crosse (1841).

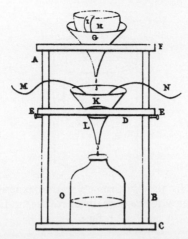

Figure 11.1. The experimental apparatus in which the electrical acari first made their appearance. Wires *M* and *N*, which kept the stone *(K)* electrified, were connected to a series of voltaic cells. These cells are not pictured, suggesting that the once-controversial phenomenon of galvanism had been 'black-boxed' by this time and was considered merely as a source of electrical current. (A. Crosse, 1841, pl.1, fig. 5)

Figure 11.2. From right to left, stages in the emergence of insects from Crosse's first experiment. Similar drawings were sent in letters to leading men of science in March 1837. (A. Crosse, 1841, pl.1)

was set dripping over a porous stone (K) of red oxide of iron from Vesuvius. This stone was kept electrified by platinum wires (M and N) extending from a small voltaic battery. The object was 'to form, if possible, crystals of silica at one of the poles of the battery'.

At first everything went as planned. After the first 14 days 'a few small whitish excrescences or nipples' were visible projecting from the surface of the electrified stone (see Fig. 11.2). Four days later, these had enlarged, and filaments had appeared on each one. By the 22nd day, they had grown still further; four days more, 'each figure assumed the form of a perfect insect'. Crosse claimed that it was only at this point that he suspected that something more than 'an incipient mineral formation' might be involved. When that 'mineral formation' wriggled

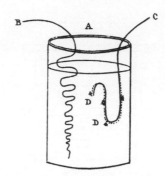

Figure 11.3. Equipment used in Crosse's second experiment. Note that the insects emerge on the wire *below* the surface of the fluid. (A. Crosse, 1841, pl. 1, fig.6)

its legs on the 28th day after the experiment began, Crosse was understandably 'not a little astonished'. Shortly afterwards, the creatures 'separated themselves from the stone, and moved about at pleasure'. A second experiment, shown in Fig. 11.3, was likewise begun without any intention of producing insects. In this case similar animals appeared below the surface of a concentrated solution of silicate of potassa. Current from a voltaic cell had been passed through this solution – a bent iron wire (C) connecting it with the positive pole, a fine silver wire (B) linking it with the negative. After 'some months electrical action', incipient insects (D) began to appear underneath the surface of the solution, initially on the silver wire and then (more abundantly) on the iron. Because the creatures emerged below the surface of a poisonous fluid that killed any that fell back into it, this experiment was more difficult to dismiss than the first.

These were the conditions associated with the appearance of life. During the months and years that followed, many of Crosse's other experiments showed signs of life. But the process of creation had only begun in the isolated laboratory at Fyne Court. Acari had crawled out of Crosse's apparatus, but at the end of 1836 they remained essentially a private phenomenon, known only to Crosse and a few friends. (Apparently the first to hear was the poet Robert Southey, whom Crosse met soon afterwards while walking in the Quantocks.)[10] That the discovery should go further than this was by no means a foregone conclusion.[11]

10. C. Crosse (1857), pp. 170–1.
11. Latour (1983, 1987) provides a helpful perspective on the entire question of networks and allies in science.

Crosse had published almost nothing and, until the preceding summer, his work was known only through the reports of visitors. For nearly three decades he had experimented in the privacy of his spectacular laboratory in Fyne Court, isolated from developments in science and almost unknown to the world at large.

Experimental science and the steam press

The first stage in the creation of electrical life had depended on the voltaic cell, a new experimental technology of the 'second scientific revolution'. The second stage depended on new printing methods, an innovative technology of the Industrial Revolution. The steam press, developed in the first decade of the nineteenth century, transformed the production of newspapers and periodicals, making possible runs in the tens and even hundreds of thousands.[12] The first newspaper to take advantage of these new capabilities was *The Times*, which was to play a key role in the acari controversies. Other, related, technologies revolutionised the printing of books, so that prices fell to the point where they could be bought in large numbers by members of the middle classes and even by educated artisans.[13] The significance of this new genre of 'people's literature' was nowhere more evident than in science. Without the voltaic cell, Crosse's experimental programme would have been impossible; without the steam press, his work could never have entered the public domain as a popular sensation.

There was, however, a tension between the two technologies. The technology of print was only rarely in the control of those who wielded the technology of experiment. The demarcated roles for 'journalists' and 'men of science' which had developed at the end of the eighteenth and beginning of the nineteenth centuries meant that the public at large received its knowledge about science almost entirely through channels distinct from those employed by the specialist practitioners of science. In this sense, 'popular science' – or 'steam intellect' as

12. The rise of industrial publishing and the mass audience is discussed in Altick (1957). In part, my account is an attempt to extend the analysis provided for the seventeenth century by Shapin (1984) and Shapin & Schaffer (1985) into a later period.
13. For example, books like Thomas Paine's *Rights of Man* (1791–92), George Combe's *Constitution of Man* (1828) and the tracts and sermons of the Society for the Promotion of Christian Knowledge all sold in numbers that had previously been reserved for the Bible and the works of John Bunyan. These developments have been discussed in a great many works; see Webb (1955), R. Williams (1961), Thompson (1966), Berg (1980), Cooter (1984) among others. The situation in the natural sciences has had less attention, although an indication of the scope of the literature is available in Sheets-Pyenson (1985).

contemporaries called it – is a category born of the second scientific revolution. Producers of science and their publics stood in a new and more distant relation. In part, the wish to recapture control over the communication of knowledge led the central coterie of scientific men in the British Isles to become involved in the British Association for the Advancement of Science from 1831. Through annual meetings in different provincial towns, the Association's leaders attempted to harness the engines of popular publication to their own ends. These gatherings, reported in newspapers across the country, rapidly became major outlets for defining the image of science among the industrial middle classes.[14]

The discovery of Crosse

Although the gentlemen of science had no wish to become involved in experiments on the origin of life, they unwittingly played a key role in thrusting the acari into the public eye. At the Association's Bristol meeting in 1836, several months *before* his experiments produced insects, Crosse was brought forward by William Buckland, professor of geology at Oxford, and hailed as a scientific 'genius'. Indeed, he seemed just the sort of man the Association wished to bring to notice. 'The wizard of Broomfield' was a local figure, so that any praise for his researches ministered to provincial pride, and he seemed above the petty squabbles that soiled public perception of science. His isolation could be used to demonstrate that science remained an activity of heroic individuals, outside and above the mass culture.[15]

But Crosse was more than 'one of the great show-beasts of the meeting'.[16] The work that he presented to the assembled savants was perceived to have consequences of the highest order for physical geology. As one reporter put it, the effect his work produced upon the geological section of the Association was absolutely 'electrical'.[17] Their enthusiasm was understandable, for geologists were rapidly becoming convinced that electricity was instrumental in crystallisation, stratification, cleavage, mountain building and the formation of mineral veins. They had

14. Morrell & Thackray (1981) provide a superb analysis of the strategy of the gentlemen of science. For more on the image of science they established, see Yeo (1981).
15. For the novelty of the concept of the heroic discoverer in the early nineteenth century, see Schaffer (1986). Morrell & Thackray (1981), pp. 457–8, give a detailed account of the 1836 meeting with reference to the main sources. As shown in a later section of the present paper, however, Crosse could not be 'dropped by the BAAS as rapidly as he had been taken up'.
16. C. Crosse (1857), p. 150.   17. *Athenaeum* (3 Sept. 1836), p. 632.

just heard William Hopkins of Cambridge and Robert Were Fox of Cornwall speak on precisely these subjects. In these circumstances, the researches of the virtually unknown Crosse came with all the power of experimental revelation. It was 'as if the interior recesses of Nature had been of a sudden laid open to them, and her processes, which had been conceived as past all mortal ken, submitted to their inspection'.[18] Many participants hoped that geology could at last take its place as an experimentally-based physical science. A real revolution in the subject was in prospect. No wonder that William Conybeare declined to give his scheduled paper on the South Wales Coalfield (on the grounds that it had been eclipsed by Crosse's discoveries), or that John Phillips believed that the afternoon's proceedings alone justified the work of the Association during its first five years. Not surprisingly the chemists were less ecstatic, pointing out that Antoine-César Becquerel had priority for much of the research. As members of another discipline they had no need to see geology established on an experimental footing. The general feeling, though, was summed up by the professor of geology at Cambridge, Adam Sedgwick, who had long advocated a mathematically-based physical approach to his subject.[19] The results of the Bristol meeting provided 'the most important advances yet made in Geology'. Crosse, hitherto a recluse, must henceforth 'stand before the world as public property'.[20]

Sedgwick, Buckland, Phillips and the other leaders of the Association clearly felt that Crosse was a great discovery. They had captivated their audience with a modest yet striking genius, a man seemingly at one with nature and with experiments that had the potential for revolutionnising the practice of a specialist science. But by giving Crosse such prominence, the geologists unwittingly set the stage for the emergence of his insects as a popular sensation. After such publicity, anything that Crosse said or did became the product of a genius whose authority had been recently validated by the scientific elite.

Acari in the press

Crosse's name, in fact, had become well enough known to sell newspapers, which is how his electrical insects initially escaped into the public domain. At the very end of 1836, the printer and publisher William

18. *Athenaeum* (3 Sept. 1836), p. 632.
19. Sedgwick's interest in relating geology and mathematics (in the context of the Cambridge tripos) is discussed in Smith (1985) and in Secord (1986), esp. pp. 62–8.
20. *Athenaeum* (13 Sept. 1836), p. 652; *Literary Gazette* (17 Sept. 1836), p. 602.

James A. Secord

Bragg of Cheapside in the parish of Taunton St Mary, began issuing a
new title, the *Somerset County Gazette*. Visiting Fyne Court earlier in
December, Bragg had learned that insects were emerging from Crosse's
apparatus. He concluded that a page three feature headlined 'extraordinary experiment' would attract readers. The article briefly described
Crosse's experiments, reflected on their significance and gave a rationale
for bringing them to notice. Although his announcement was unau-
thorised, Bragg stated that he published to protect Crosse's priority.[21]
The newspaper thus claimed one of the roles of a scientific periodical,
that of recording original contributions to knowledge. But Crosse dis-
tanced himself from this new access of fame. He wrote to the *Taunton
Courier* and sent a fuller account to William Stutchbury of the Bristol
Institution, who immediately forwarded it to the *Bristol Advocate*, the
regular outlet for the Institution's proceedings. In turn, the *Gazette*
apologised for any inaccuracies. Bragg now asserted that he had not
used 'scientific and exact terms', and only 'purposed to give a popular
and not a scientific account'.[22] A case like this shows how a publisher
could manipulate distinctions between the 'popular' and the 'scientific'.[23]

Reporting experiments in the press, as Bragg had found, was an
enterprise beset with pitfalls. But there could be rewards, too. As a
strategy for bringing the newborn *Gazette* to public notice, the 'extraordinary experiment' proved tremendously successful. In a season of slow
news, an account of the experimental creation of life by the hero of
the Bristol BAAS was greedily taken up, first in *The Times* and then in
hundreds of other newpapers, magazines and journals. Within a week,
knowledge of the initial report and Crosse's reply had spread throughout
the country. The power of the press was released in a full tide. The
Athenaeum, the *Literary Gazette*, the *Mechanics' Magazine*, the *Gentleman's
Magazine*, the *Mirror of Literature, Amusements and Instruction* and the
daily and weekly press across the British Isles all picked up the story.
A few were sceptical. As the no-nonsense Liverpool *Albion* remarked,
'*Credat Judaeus*, or, "tell it to the marines, for the sailors won't believe
you"'.[24] But the editors of most papers reported the item without com-

21. 'We trust that by thus early publishing and awarding them to their proper owner,
   we shall prevent much unfair appropriation of the honor of his brilliant researches
   by those who have no title to it.' Bragg (1836).
22. A. Crosse (1837a, b); Bragg (1837a, b).
23. Collins & Pinch (1979) provide a study of parapsychology that offers many revealing
   parallels. It is interesting to note that the distinction they draw between 'contingent'
   and 'constitutive' forums for science first becomes relevant in the early nineteenth
   century.
24. 'Extraordinary experiment', *Albion* (9 Jan. 1837), 12, no. 578, p.3.

ment as straightforward (if 'extraordinary') news. For most readers the creation of life was an experimental fact by the end of January 1837.

REPLICATION AND ROMANTIC SCIENCE

The electrical insects were established as a public property, but they had only begun to make their way into any of the diverse groups concerned with the actual practice of science. The 'literary replication' (as I shall call it) of the discovery account took place in a matter of days, sometimes even hours. It depended upon editors, typesetters, pressmen and a rapid distribution network. Experimental replications, refinements and refutations took place much more slowly.

In many respects, the case is similar to that described by Bruno Latour for Pasteur's work on vaccines.[25] Those who had interests in using the phenomenon wanted to have it transferred out of Crosse's country house (or more accurately, out of the newspapers) and into laboratories and other scientific establishments. On one level this transference was literal: the creatures needed to be prepared and sent to expert taxonomists so that their affinities to known organisms could be determined. For the experiments themselves to become available, a written description of the equipment and methods was required. As Steven Shapin has shown, this 'literary technology' makes it possible for others to witness the accomplishment of an experiment by proxy. (It does not, of course, guarantee the possibility of replication.)[26] In response to the demands of his correspondents, Crosse prepared many such accounts – first to satisfy newpaper readers, followed up in March with a circular letter to meet the rather different criteria of those who had encouraged him at the Bristol BAAS, and finally an elaborate report delivered in December 1837.[27] The quantity of written material generated by the experiment was clearly related to its controversial character.

Even with accounts widely available, the experiment did not attract replications or elaborations. In the early months, the scanty details available in the newspapers would scarcely have led others to invest the necessary time and money. More basically, though, the problem arose from Crosse's disinterested public stance. During the crucial years of controversy, he never refined his results, nor did he take certain steps (such as excluding the outside air) that critics pointed to as pre-

25. Latour (1983, 1988).   26. Shapin (1984).
27. A. Crosse (1837b, 1841); copies of the March letter are in a number of scientific archives, e.g. Crosse to Sedgwick, [March 1837], Cambridge University Library Add. ms 7652.I.B.97a. A similar letter was published by Richard Phillips in the *Annals of Electricity*; see A. Crosse (1837c).

requisites to any claim concerning spontaneous generation.[28] Crosse agreed that such a 'necessary precaution' would be obvious 'even to an unscientific view'.[29] He announced at the end of 1837 that he would try the experiments under a glass receiver inverted over mercury. He claimed, though, to be unable to abandon his research on the crystallisation of silica and other minerals, 'in which I am so interested, that none but the ardent can conceive what is not in my power to describe'.[30] Crosse's ambiguous relation to his experiments is evident in the research he undertook during the next few years. With one later exception,[31] he claimed that the generation of life was incidental to experiments on crystallisation. Insects continued to appear in these, but without Crosse having met the objections of critics. Among leading experimenters like Joseph Henry and Faraday, this failure to repeat procedures that had been standardised during earlier spontaneous generation controversies rendered his results worthless.[32] Crosse's naive enthusiasm defused suspicions that he was sensation-mongering; as Faraday told a friend, he 'must not be charged with having pressed himself forward'.[33] But it also made the extension of the acari result into a sustained programme of research exceedingly difficult.

This raises one of the central paradoxes in the debate about the validity of these experiments. In many ways Crosse was perfectly situated to be the author of a controversial phenomenon, but in order to maintain that position he could not engage directly in the debate, nor in the further experimentation that debate would entail. For Crosse's reputation had been built upon his otherworldly isolation, his distance from the squabbles and jealousies of metropolitan science. As Edward W. Cox, a barrister on the western circuit, wrote in 1837 after visiting Crosse at Fyne Court, 'his face is lighted up; his eyes are fixed upon the ceiling; present things seem to have disappeared from him, lost in the greater vividness of the ideas which his full mind throngs before him . . .'. A visit to Crosse calmed the passions, melted the shadows of 'social follies and vices' and left one 'a wiser, better, nobler being'. Even wild animals recognised that the Crosse 'homestead is sacred ground, where hospitality is never violated'.[34]

---

28. For a general survey of disputes about spontaneous generation, see Farley (1977).
29. A. Crosse (1837d), p. 15.   30. A. Crosse (1837d), p. 15.
31. Discussed by Crosse in an appendix to Atkinson & Martineau (1851), pp. 361–7.
32. Henry discusses the issue in his European diary after meeting Crosse on 26 April 1837; see Reingold, ed. (1979), pp. 320–2.
33. Faraday to C.F. Schoenbein, 21 Sept. 1837, in Kahlbaum and Darbishire, ed. (1899), p. 33.
34. Cox (1844), pp. 57–60.

For Crosse himself, experiment served as an escape from the turmoils of family tragedy and local politics.[35] Like his father, he staunchly supported republican causes and worked to see liberal and radical candidates elected to the local parliamentary seat. Many of his political speeches are recorded in local newspapers. As he declaimed at a public meeting in September 1837, 'Shall we submit to see Tory cannon levelled against those who are one people with us? Never.'[36] For Crosse, knowledge was not just for an elite but for all. In this sense, his populist form of scientific practice (with its emphasis on everyday materials and lack of system) and his advocacy of a populist politics were of a piece. Yet in personal terms, the two activities coexisted in contrast. His performances on the hustings were highly public, verging close to demagoguery. His science was a private affair, about which he spoke with enthusiasm and quiet modesty. Fyne Court and its laboratory were open to all comers; as Cox wrote:

Knock fearlessly at that door; the votaries of science are always welcome there. Your name? your station? your calling? your property? Trouble not yourself about any of these things, nor hope thus to commend yourself to the inmates. You are a MAN, you venerate SCIENCE, even if you know little of it; these are your passports into that mansion.[37]

But almost nothing was published, and the emotional satisfaction of experiment for Crosse had little to do with public acclaim.

In the popular press, Crosse became the type of the scientist as original poetic genius, directly in contact with nature. The seeming artlessness of his methods was the key to his success. Certainly no one claimed that he was a brilliant experimentalist. Many of his arrangements were crude, involving chunks of brick and old flowerpots. He melted down the family plate to obtain silver; his laboratory was in the converted hall. Ada Lovelace, Byron's daughter and no paragon of order herself, found the house and its laboratory in chaos.

This is certainly a most extraordinary domicile to visit at. It appears to me to be the most *unorganised* domestic system I ever saw. . . . There is in Crosse the

35. Outram (1984) analyses the career of Georges Cuvier in these terms. The model she develops deserves to be systematically applied to cases in the physical sciences.

36. ' I am convinced', he said on the same occasion, 'that all power ought to emanate from the people – it is well said that the voice of the people is the voice of God; though Providence may at times have maddened them for the purpose of punishing undue oppression, yet I am satisfied that under the sway of increased knowledge, and advancing science, with an enlightened people, revolutions of violence will be no more known'. Report of a 'Reform dinner in East Somerset to Col. Gore Langton, M.P.', *Somerset County Gazette*, (23 Sept. 1837), p. 2. Crosse's political views are briefly characterised in C. Crosse (1857), pp. 106–15, 139–41.

37. Cox (1844), p. 57.

most *utter* lack of *system* even in his Science. At least so it strikes *me*. I may be mistaken. Perhaps I don't see *enough*, as yet, to discover his *system* . . . . I have quite a difficulty to get him to show me what I want. *Nothing* is ever *ready*. All chaos & chance.[38]

Perhaps the most telling point of all was Crosse's failure to keep regular notebooks. Among men of science generally, laboratory or field notebooks were considered an essential part of proper scientific practice. Yet Crosse recorded his experimental memoranda on undated loose slips of paper.[39]

The contrast with the obsessively meticulous Faraday in such matters is especially revealing. David Gooding has shown that Faraday spared no effort in condensing his private research activities into a form that could be demonstrated before the public. The aim, as his famous motto 'work, finish, publish' indicates, was publication and the proof of discovery it provided.[40] An original train of research might involve months of experimental work and highly complex manipulations, but by the time it reached print, the same work might fill only a single paragraph. In this way, Faraday successfully transformed his experiments into observable phenomena. His procedures *became* 'transparent'. Crosse, on the other hand, saw no need for refining his experiments. As his laboratory motto stated, 'it is better to follow nature blindfold than art with both eyes open'. His most famous discoveries appeared before the public almost by accident – in the case of the acari, without even his authorial approval. Exemplifying the romantic idea of the scientist as seer or magus, his procedures were professedly 'transparent' from the start. He was composing what he once called his great '*electric poem*'.[41]

THE USES OF AN EXPERIMENTALIST

Given Crosse's scientific procedures and his diffidence about recruiting allies for the acari, it is not obvious how these could become a subject for further enquiry by others. As it happened, confirmation was announced immediately, and from the highest authority possible. Towards the end of February 1837, Faraday himself was widely reported in the press to have successfully repeated Crosse's experiment with the Vesuvian stone and the silica solution. The source of the report is unknown, but it was soon picked up in many of the papers that had

38. A. Lovelace to W. Lovelace, [24] Nov. [1844], quoted in Stein (1985), pp. 145–6.
39. C. Crosse (1857), pp. vii–iii, 284.
40. Gooding (1985a), see also Chapter 7, this volume.
41. Crosse to John Kenyon, 1817, in C. Crosse (1857), p. 52; 189. For the scientist as sage, see Knight (1967).

carried the extract from the *Somerset County Gazette*. This was 'literary replication' with a vengeance. Judging from the available evidence, Faraday had expressed no firm opinion, but said in conversation after a Royal Institution discourse that 'further trials' and 'clear and confirmatory evidence' would be required before Crosse's work could be seriously considered. This polite ambivalance (amounting, in Faraday's own terms, to a crushing dismissal) became unqualified approval in the popular press.[42] Comments intended for one audience could take on radically different meanings for another.

Faraday had not even tried to replicate Crosse's experiments, but the reports that he had done so successfully had a major impact on the debate. Blanket coverage of the story in periodicals and newspapers meant that this 'result' joined the original report of Crosse's work as an established scientific fact: the leading experimental natural philosopher in England now stood behind the electrical origin of insect life. After the end of the year the account moved out of ephemeral press and into standard works of reference. The *Annual Register*, an authoritative handbook of the year's events, picked up Faraday's alleged replication, as did Harriet Martineau in her widely-used *History of England During the Thirty Years' Peace* (1849). Faraday did his best to counteract these reports. He made a formal disclaimer at a Friday evening discourse, sent letters of disavowal to the *Literary Gazette* and *The Times*, and wrote to Martineau so that a correction could be inserted in her next edition. It is particularly interesting that Faraday had to move outside his usual publications, the *Philosophical Transactions* and the *Philosophical Magazine*, to cancel the effects of the reports. Asked by a continental friend to repudiate the experiment, Faraday replied in blunt terms:

With regard to Mr. Crosse's insects etc. I do not think anybody believes in them here except perhaps himself and the mass of wonder-lovers. – I was said in the English papers to have proved the truth of his statement, but I immediately contradicted the matter publicly and should have thought that nobody who could judge in the matter would have suspected me of giving evidence to the thing for a moment. Contradict it in my name as fully as you please.[43]

But all this was to little avail – in fact, Faraday was invoked in this connection as recently as 1979.[44] The persistence and power of the mass circulation press is revealed again.

42. Stallybrass (1967) provides a full narrative of this incident with many useful references.
43. Faraday to C.F. Schoenbein, 21 Sept. 1837, in Kahlbaum and Darbishire, ed. (1899), p. 33.
44. Roth (1979).

Dislike of popular misapprehension was not the only motive for Faraday's vigorous efforts to disassociate himself from the electrical mites. Within the terms of his own experimental programme it was imperative to keep clear of any meddling with questions relating to the nature of life. He used the occasion of a paper of 1838, 'On the character and direction of the electric force of the Gymnotus' (or electric fish), to argue that science could only deal with the action of the 'nervous power', *not* the 'direct principle of life'. The latter subject, he said, was not open to experimental understanding, and was outside the legitimate boundaries of science. He was evidently fascinated by the problem of the life-force:

Wonderful as are the laws and phenomena of electricity when made evident to us in inorganic or dead matter, their interest can bear scarcely any comparison with that which attaches to the same force when connected with the nervous system and with life; and though the obscurity which for the present surrounds the subject may for the time also veil its importance, every advance in our knowledge of this mighty power in relation to inert things, helps to dissipate that obscurity, and to set forth more prominently the surpassing interest of this very high branch of Physical Philosophy. We are indeed but upon the threshold of what we may, without presumption, believe man is permitted to know of this matter . . . .

But there is a clear sense in the 1838 paper that a self-imposed boundary is being set up, beyond which Faraday was not willing to go.[45]

This personal, private difficulty for Faraday was increased by his public reputation as a heroic discoverer, both at the Royal Institution and in the periodical press. He feared that his own interest in the interconvertibility of the 'powers' of nature – electricity, magnetism, gravity, heat and so forth – would all too easily be seen as an investigation into the 'power' of life itself. The notorious case of the physician, William Lawrence, who had been forced to recant his equation of organisation and life in 1819, during the years of Faraday's scientific apprenticeship, reveals the sensitivity of this question. It involved religious, political and social dangers of the gravest sort.[46] Faraday's public association with the galvanic creations of Fyne Court showed that the issue of the life-force could not be ignored. In a particularly vivid way, the episode demonstrates that even such a distinguished researcher as Faraday had to define the fundamentals of his practice in relationship to issues raised in the popular press.

45. The paper on the gymnotus is reprinted in Faraday (1839–55), 2, pp. 1–17, see esp. pp. 1 and 15.
46. Jacyna (1983). For the social threat from below raised by materialism – a question with which Faraday, with his working-class origins as a bookbinder's apprentice, must have been familiar – see Desmond (1987).

RITUAL REFUTATIONS: THE COUNTERATTACK

Faraday was not alone in this matter. To combat the further spread of Crosse's insects, men of science prepared to wield the technologies of experiment directly against the press. The acari had risen to fame through experiment. They now had to be destroyed by counter-experiments. Otherwise there was a very real danger that men of science might appear to violate their own methodological strictures, dismissing a discovery on political or religious grounds or because of its *a priori* implausibility, rather than appealing to experiment and observation. In this situation successful experiments were those that *failed*. These could then be brought forward as 'ritual refutations' that met the demands of scientific rhetoric. Crosse's findings could not be ignored: they had to be destroyed by experiments. But the results of these experiments could only be effective if targeted at the same audiences and arenas in which the original discovery had taken hold.

The vigilante attack was led by John Edward Gray, conservative curator of the zoological collections at the British Museum. Under his direction, leading men of science, whose work encompassed both the physical and the life sciences, attempted to repeat Crosse's first experiment. John George Children was an evangelical high Tory, a natural philosopher by training and Gray's assistant at the Museum. Golding Bird lectured in natural philosophy at Guy's Hospital in London.[47] As experienced 'electricians' employed by institutions involved with living things, both Children and Bird were well situated to refute the acari experiments. They performed the Crosse experiments in two different ways: once according to the original protocol, the second time in a closed container. None of the trials showed signs of life.

Gray reserved the announcement of these refutations for the Natural History section of the British Association meeting at Liverpool in the summer of 1837. Crosse's reputation, created at the Bristol gathering only a year earlier, was now to be laid to rest. Gray began by reading a paper on a gigantic water lily named in honour of the new queen, *Victoria regina*, and then launched into an unscheduled (but obviously prearranged) attack on Crosse's experiments. He described the failed attempts of Children, Bird and a third experimentalist whose name is not known. The mites, Gray went on to suggest, had hatched in the normal way from eggs which had not been destroyed by Crosse's procedure. As the *Athenaeum* reported, 'a very animated discussion ensued', which focused on the virtually indestructible qualities of insect eggs— even their ability to hatch after being ground to a powder. Here, as

47. Biographical details for all these figures are in the *Dictionary of National Biography*.

was appropriate to the Natural History section, the authority of students of generation and classification was added to the experimental work of the natural philosphers. That evening in the Liverpool Amphitheatre, Cambridge professor of botany John Stevens Henslow summarised the discussion for the entire Association. Experiments had failed to produce acari and entomologists had concluded that they were simply the products of eggs. The electrical insects had suffered a double defeat.[48]

Reports of the debunking session were widely disseminated in the regular newspaper and periodical accounts of the meeting. Some used it as an occasion to attack the British Association itself. As the *Mechanics' Magazine* drily remarked, 'the world of lookers on have been surprised to observe, that it has been one of the objects of the Liverpool Meeting to undermine, if not to pull down, the very reputation which it was the highest boast of the Bristol meeting to have built up!'[49] The destruction of Crosse's reputation was, of course, why the experiments had been conducted in the first place. But Gray and the Association's leaders carefully kept any reference to the subject out of the the bound volume of official proceedings, the permanent record of papers read before the various sections. *Victoria regina*, a genteel compliment to the new queen, was in; but work relating to the once 'immortal' Crosse was now to be relegated to the ephemeral forum of the daily and weekly press. The ritual refutations of Bird and Children thereby gained publicity among the public at large, but without giving electrical life any further legitimacy within the realms of established science.

There is, however, an ironic postscript to the early attempts to refute Crosse through experiment. Because the work of Children and Bird was kept out of authoritative scientific publications, it gradually became inaccessible to those in later phases of the controversy who needed experimental refutations of the electrical origin of life. In consequence, figures like David Brewster, combating the acari in the 1840s and 1850s, turned to the work of the German naturalist Carl Frederic Schultze.[50] But Schultze had dealt with what was actually a very different problem, the appearance of infusoria in flasks reached only by sulphurated air. He employed no electrical action and announced his results well before

48. Accounts of the meeting include the *Literary Gazette* (16 Sept., 1837), p. 590, (21 Oct. 1837), p. 674; and the *Athenaeum* (16 Sept. 1837), p. 671.
49. Anon. (1838), p. 53.
50. [Brewster] (1845), p. 497. This report continues to be passed down in later literature. Stallybrass (1967), p. 614, names Schultze (together with Weekes) as one of only two experimentalists who attempted the Crosse experiments. Schultze's experiments are briefly described in Farley (1977); see also Schultze (1837).

Crosse's results appeared in print. With the work of Children and Bird buried in the back files of newspapers, however, these experiments were the nearest approach to a direct refutation that could meet the demands of experimental rhetoric.

ELECTRICAL LIFE AND SCIENTIFIC DISCIPLINES

Perhaps the chief problem for anyone who wished to eliminate Crosse's mites was the way in which they violated all the disciplinary boundaries of established science. We have already seen that the experimental refutations of Crosse's discovery appeared, not in Section A of the British Association, on Astronomy and General Physics, but in Section E, devoted to Natural History. An experiment that had began as an extension of work on electro-crystallisation was refuted in a discussion following a paper on a water lily.[51] Anomalies like this arose simply because studies of electricity and life had no place on the disciplinary map of Victorian science. Crosse's finding was relevant to at least four groups of specialist practitioners: natural historians (especially entomologists), medical men, geologists and experimental natural philosophers.

For entomologists and other taxonomists, the creatures raised problems of classification: whatever the mechanism of its formation, had a new species been discovered? Practitioners in this community tended to be conservative, for their subject depended upon stability and consensus. They were extremely sceptical of 'bungling' attempts by outsiders, especially experimental natural philosophers or geologists, to make taxonomic statements. As Edward Newman, editor of the *Zoologist*, commented, 'zoological facts are of small value, unless witnessed by Zoologists'.[52]

The first expert to view the acari was Richard Owen of the Royal College of Surgeons. Although he scarcely counted as a specialist on insects and mites, Britain was notoriously lacking in authorities in entomology. Owen received a 'cargo' of the creatures early in 1837 through the agency of William Buckland, together with a long letter

51. An interesting comparison can be drawn with the case of the barometer discussed by J.A. Bennett in Chapter 3 of this volume. During the course of the seventeenth century, this instrument migrated from pneumatics, to the study of tides and finally to meterology. Here as in the acari controversies, specific contexts give experiments meaning.

52. [Newman] (1845), p. 960. Very little has been written on the history of entomological classification during this period; for very helpful background. see Winsor (1976).

and a drawing (Fig. 11.4) from Crosse.[53] Owen concluded that they were common cheese mites. This early judgment was replaced within a few months by the comments of Pierre Turpin before the Academy of Sciences in Paris. Turpin was a botanist – a specialist in fruit-trees – but had published extensively on problems of organisation and vitality. The paper he presented was a remarkable double-edged document.[54] On the one hand, its claim that a new species of *Acarus* ('*Acarus horridus*') had been found was widely accepted. Needless to say, this was a godsend to those who argued that Crosse's results could not be explained by mere contamination. On the other hand, Turpin disassociated himself from such conclusions, urging that the newness of the species reflected the rudimentary state of the taxonomy of the Arachnidae, rather than a genuinely novel 'creation'.

Much more fundamentally, though, it was not at all clear to English readers that the paper was really to be taken as anything but a joke. For example, Turpin captioned the plate (Fig. 11.5) of the type specimen: 'female individual containing an egg, created artificially by M. Crosse!!'. This form of words could hardly instil confidence, especially in the austere context of the *Comptes Rendus* of the Academy. Partly because of the ambiguities of Turpin's paper, the classificatory problem surrounding the new mite became extraordinarily complex as the controversy continued into the 1840s. There were *two* main translations of the paper into English, one in the electrical press, the other for natural historians; each omitted different passages and led to radically different views of Turpin's meaning.[55] By 1845 communication between the different communities had become so garbled that Adam Sedgwick, a geologist and opponent of spontaneous generation, clinched his case against the Crosse experiments by claiming that they had produced nothing more exciting than the *Acarus horridus*.[56] He (and many others since) thought that this was the established name for the common domestic mite, not a novel name given to Crosse's own productions.

53. A. Crosse to R. Owen, 10 Feb. 1837, British Museum (Natural History), Owen correspondence, 9, 108–9, 112–13; Anon. (1837), p. 148. For Owen's ambivalence on the issue of spontaneous generation, although not the acari, see the important article by Richards (1987), pp. 155–6. Andouin (1836) provides a helpful contemporary discussion of the Arachnidae.
54. Turpin (1837). The situation was further complicated by the fact that the Academy initially stated that the subject did not deserve any report at all; see *Comptes Rendus*, 5, p. 640. Turpin's career is described in Hocquette (1976).
55. Turpin (1838a, b).
56. [Sedgwick] (1845), p. 71. Sedgwick's mistake, ironically, is derived from one made by the self-conscious 'zoologist' Edward Newman – see [Newman] (1845), p. 960. Newman is basing his account on Bladon (1842) rather than the original paper by Turpin.

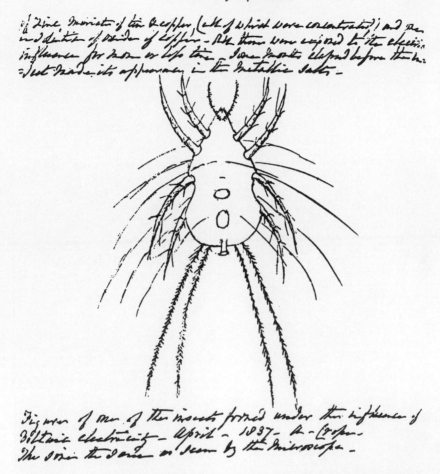

Figure 11.4. Crosse's own drawing of 'one of the insects formed under the influence of voltaic electricity', as seen under the microscope and sent to the comparative anatomist Richard Owen in April, 1837. (Stallybrass, 1967, p. 609)

Even with all this confusion, few if any taxonomists in Britain were willing to believe that a new creation had taken place. As the naturalist Edward Forbes said, the facts were 'rubbishy'.[57] The medical community had a much less united view. Opinion here was bitterly divided on the entire question of spontaneous generation, as witnessed by the fact that the London physician Golding Bird experimented to refute Crosse, while the important provincial surgeon William Henry Weekes carried

57. From an address at the Royal Institution, quoted in Wilson & Geikie (1861), p. 450.

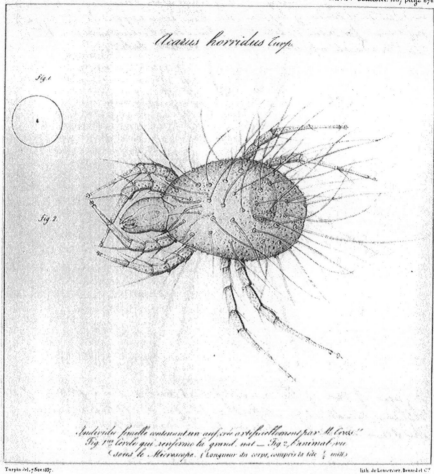

Figure 11.5. Pierre Turpin's illustration of the *Acarus horridus*. Note the difference between this drawing, prepared by an expert naturalist, and the one in the preceding figure prepared by Crosse. For the English audience, the combination of accuracy and facetiousness in Turpin's paper made it all the more difficult to tell if it was a joke. (Turpin, 1837, p. 676)

out years of work to support him. It deserves to be noted, though, that many British medical men remained much more sympathetic than most gentlemen to the spontaneous generation of complex organisms, largely because of their familiarity with the problem of the origin of gut parasites.[58] The medical author, Dr James G. Davey, writing in the *Lancet* in 1850, claimed that the 'experiments of Messrs. Crosse and Weekes are conclusive'.[59] Not least, the leading Unitarian physiologist William B. Carpenter was willing to reserve judgment on the experiments until further trials had been made. Carpenter even planned his own physiological research on the acari, although this does not appear to have been published. As he wrote in 1845:

Not entertaining the opinion, which a certain clergyman is reported to have expressed. . . – that 'it was a very dark business, and such as no christian man ought to engage in,' – we please ourselves with the hope of being able to throw some light, ere long, upon the history of this interesting creature, by the careful study of the physiological peculiarities of its development; leaving it to those eminent in physical science to investigate the physical conditions under which it is produced. Should the author's views be realized, the *acarus Crossii* will indeed be worthy to take rank as one of the most – if not *the* most – physiologically important species of the whole animal creation; affording proof of the *possible* origin of living beings of complex structure and varied actions, from the combination of particles of inorganic matter under peculiar conditions; and thus outshining, in the eyes of the physiologist, the transendent merits of frogs, rabbits, dogs, asses, &c as subjects of experiment.[60]

Carpenter's understanding of comparative anatomy did not rule out the generation of insect life from inorganic matter. He undertook no experiments in this area, because he acknowledged the authority of other disciplines over that subject: he was not 'eminent in physical science'. But Carpenter believed the question was within the bounds of legitimate inquiry. Like many Unitarian physicians, Carpenter wished to rid science of miracles and to see God acting through general natural laws.[61]

A third group involved with the acari was the geologists. Their enthusiasm at Bristol for the work on crystallisation put them in a dangerous position with regard to the new discovery. To avoid this, the geologists attempted to find a respectable genealogy for the mites. The initial report in the *Somerset County Gazette* noted that Crosse 'has contented himself with stating the fact, but without attempting to account

58. For this point, and a full analysis of attitudes in medical circles to questions of materialism and spontaneous generation, see Desmond (1988) and his forthcoming *Politics of Evolution.*
59. Davey (1850), p. 415. I owe this reference to Professor Dennis Dean. See also Hytch (1837) for further support in the *Lancet.*
60. Carpenter (1845), esp. pp. 170–2.
61. The changing views of Carpenter on this question are apparent in Carpenter (1838, 1850). For the Unitarian physicians, see Raymond & Pickstone (1986) and Desmond (1988) and his forthcoming *Politics of Evolution.*

for it'. But it went on to say that Oxford geologist William Buckland had suggested (in correspondence with Crosse) a possible mode of formation. The German naturalist Christian Ehrenberg had recently shown that silicaeous rocks were composed of the remains of 'insects', tiny infusoria. Buckland wondered if electricity might have acted in Crosse's experiments to revivify remains of this kind. As the editor wrote, 'May not the germs of some of them released from their prison-house, and placed in a position favourable to the development of vitality, have sprung to life after a sleep of thousands of years?'[62] Because of its inclusion in the *Gazette*, Buckland's view received wide publicity. Fellow geologist Henry De la Beche, carrying the idea to its logical extreme, drew a cartoon that brought to life not just a few insects, but an entire museum of fossil bones. It was, as the Quaker diarist Caroline Fox wrote, 'grotesquely horrible'.[63]

For anyone concerned about the religious and ideological implications of the creation of life from inert matter, Buckland's interpretation had much to recommend it. If he was right, the principle of life had simply lain dormant in the siliceous solution, so that direct spontaneous generation was avoided. The hypothesis attracted much attention. In *The Newtonian Philosophy*, a standard children's science book of the period, the young lecturer Tom Telescope commended Buckland's interpretation, as lessening 'the extraordinary discovery of insects by Mr. Crosse, amid his experiments on crystallisation.'[64] Revivification was certainly preferable to creation, especially as it fitted neatly with other work being done at this time on the growth of ancient seeds and grains of wheat found in Egyptian tombs. Buckland, Conybeare, Darwin and other geologists were very interested in this work,[65] and at least some of them seem to have been as enthusiastic about Crosse's new result as they had been about his work on crystals.

The revivification hypothesis of the geologists persisted for some time, particularly in the popular press, but serious naturalists soon became convinced of its deficiencies. In particular, after competent authorities like Owen and Turpin examined Crosse's specimens they could no longer be seen as simple infusoria of the type described by Ehrenberg. Moreover, Buckland's friend Conybeare was rebuked by Faraday at an informal get-together of natural philosophers in London, for giving Crosse's work too much publicity.[66] There can be little doubt

62. Bragg (1836). 63. Diary entry for 7 Feb. 1837, in Pym, ed. (1882).
64. Telescope (1838), p. 286. For more on the book, see Secord (1985).
65. As evidenced in Rupke (1983), pp. 166–7, and in Burkhardt & Smith, ed. (1986), pp. 355–6, 389, 450–3.
66. The meeting is described in the entry for 26 Apr. 1837 in Joseph Henry's European diary; see Reingold ed. (1979), pp. 320–2.

that the willingness of some geologists to consider the relation between electricity and life had drawn fire both from natural philosophers and from natural historians. Geology had a reputation as a speculative science, and its leading representatives had trespassed into other domains of expertise. Buckland, unhappily unable to suppose that the little animals had been brought to life after several millenia, withdrew his support from Crosse's whole experimental programme at a meeting of the Ashmolean Society in Oxford in March 1837.[67]

The final group whose practice impinged upon the acari was concerned with experimental natural philosophy. No disciplinary boundary created during the early nineteenth century was stronger or more significant than the line between matter and life, between the physical and the biological sciences. The separation of experimental studies of living beings from natural philosophy had been a key element in redefining the latter subject at the turn of the century.[68] Here Faraday, whose unwillingness to consider the 'life force' has already been discussed, was typical of many of his contemporaries. Even those like Bird and Children, whose career patterns would seem to lead to an interest in the study of the physical basis of life, in fact used their expertise to counter suggestions that experiments could explain (or even explore) the power of life. Faraday, Bird, Children and others recognised that the autonomy of the new specialist disciplines needed to be fought for, and they maintained a tight boundary around the problem of life and its origins.

The issue was thus a contentious one, although it would be a mistake to think that any definitive consensus existed on this point – or on any other relating to the science of electricity. Compared with other areas of experimental science in the early nineteenth century, the study of electricity remained remarkably diverse and fragmented. As William Sturgeon, one of the leading figures in the field, lamented in 1837, electricians remained 'disunited and insulated from each other'.[69] The subject was pursued in England almost entirely by individuals, either wealthy gentlemen able to afford the expensive equipment, or those with access to the facilities of large public institutions. Others, like Sturgeon himself, were public lecturers, inventors, or practical men.[70]

Co-operation, in the form of an organised society, was held up by

67. Reports of the meeting can be found in the *Literary Gazette*, (25 Mar. 1837), p. 194, and in the *Magazine of Popular Science*, (1837), 3, p. 234.
68. Heilbron (1979), pp. 9–19.   69. Sturgeon (1836), p. 68.
70. Gooding (1985b; Chapter 7 this volume) characterises the London 'electrical network' in the 1820s, although he arguably de-emphasises its diversity and lack of agreement over fundamentals.

Sturgeon as the best way of transforming this fragmented community into a coherent discipline. The result was the formation of the London Electrical Society in 1837. This group soon became the central forum for positive discussion and debate in England about the relation between electricity and life. The Society met in the Adelaide Gallery on Regent Street, a centre for middle-class scientific entertainment. The bulk of its membership were 'amateur experimentalists', as founder and first president Sturgeon admitted. Typical of the group were the instrument-maker Edward M. Clarke, the young Manchester electrician James Prescott Joule, and William Leithead, employed by the Gallery as a demonstrator. Others, like Walter Hawkins, William G. Lettsom, John Peter Gassiot and Crosse were either successful men of business or independently wealthy. Conspicuously absent were the real controlling interests in the London scientific world, men like John Herschel, William Grove and Faraday.[71]

With its relatively populist middle-class basis, the Electrical Society encouraged a more inclusive notion of its subject than that followed by the gentlemen of science. Much attention was devoted to electricity in nature, to topics such as lightning, galvanism and the effects of electricity on crystals, geological stratification and the growth of plants. The members had no time for electrical theories of mesmerism or phrenology – man was outside their subject matter – but generally encouraged study of the borderline between electricity and life.

This made the group an ideal scientific forum for discussing Crosse's electrical insects. Crosse himself became a member in June 1837, a little over a month after the Society had been founded. (This was probably through his contacts with the first secretary William Leithead, with whom he had been in correspondence.) Crosse attended meetings whenever he was in London and published the fullest account of his famous experiments in the first (and only) volume of the Society's *Transactions and Proceedings*. As almost the only scientific forum open to consideration of electricity and life, the Electrical Society welcomed the acari onto its pages. By providing an outlet of this kind, it ensured that Crosse's work was not limited to the ephemeral context of popular

71. There is no history of the London Electrical Society. My account is based on the minute book, Institute of Electrical Engineers, Special Collections ms 42, and the published proceedings in the *Annals of Electricity, Magnetism and Chemistry* (edited by Sturgeon), and the *Transactions and Proceedings of the London Electrical Society, from 1837 to 1840* (London: 1841). As David Gooding has pointed out to me, Faraday was not totally aloof from the Adelaide Gallery, which loaned him its gymnotus for his experiments.

magazines. It is equally telling, however, that none of the members – at least initially – was willing to extend this work further with actual experiments. For example, Leithead, in a book based on articles for the weekly *Atlas* newspaper, detailed Crosse's inorganic crystallisation experiments and referred somewhat vaguely to the effect of electricity in 'imparting regular form to bodies'. He then dealt with the vegetable kingdom, but denied the possibility of an electrical creation.[72]

This seems to have been about as far as most members of the Society were willing to go in considering the origin of life. Their encouragement of these unorthodox experiments indicates the continuing fluidity of the boundary between popular interest in the origin of life and the actual practice of electrical science. But the fact remains that no one interested in a positive result had yet found it worth their while to reproduce Crosse's original experiments.

At the end of 1837 the *Annual Register* recorded the emergence of the acari as one of the year's events and mentioned that Faraday had successfully produced them. The insects were firmly in the public realm. They were also way into the realm of scientific practice, as a result of Electrical Society meetings and Crosse's growing contacts with the metropolitan world of learning. But in general, by failing to interpose a barrier between his experimental set-up and the outside air, Crosse had left not only his solutions but his science open to contamination. Like Robert Boyle's early air pumps, Crosse's apparatus leaked.[73] It violated some of the most fundamental boundaries of the new specialist disciplines; and the geologists, who seemed most willing to consider the issue, were soon brought into line. In a world that drew a sharp limit between the sciences of life and matter, Crosse talked about crystallisation and organic materials without distinction. And in a world that demanded specific protocols for any experiment dealing with spontaneous generation, he spoke of all-pervasive nature of electricity and refused to intervene more directly in his experiments. For this reason, together with fears about the consequences for morals, religion and the social order, the electrical insects failed to find any part in scientific practice. Debate about them in scientific periodicals and books rather abruptly ground to a halt about a year after it began. Although occasionally mentioned after 1838 (largely as a nightmare to be forgotten), the controversy seemed destined for oblivion.

72. Leithead (1837), pp. 230–41. Leithead's scepticism is also apparent from a letter to J. Henry, 8 Oct. 1839, in which he says that 'the magician is still busy with his electrical incantations': see Reingold, ed. (1981), 268–9.
73. Shapin & Schaffer (1985).

EXTRAORDINARY EXPERIMENTS AND THE STRUGGLE FOR ELECTROBIOLOGY

That the experiment was not forgotten is due to a surgeon and lecturer in natural philosophy, William Henry Weekes of Sandwich in Kent. Weekes, born in 1789, was (as one contemporary put it) 'the greatest dealer in thunder and lightning . . . in the three Kingdoms' save Crosse himself.[74] Like Crosse, Weekes was interested in atmospheric electricity, electrical instrumentation, and the relation between electricity and life. Although his residence in the provinces precluded attendance at Electrical Society meetings, he became an enthusiastic supporter and regular correspondent. Working in a private laboratory behind his house, Weekes experimented on the medical uses of electricity, the effect of currents on plant growth, and a variety of related issues, including the electrical insects – the *Acarus Crossii*, as he termed them.

Weekes commenced his experiments in 1840. His aim, announced in his first paper on the subject, was to vindicate Crosse from charges of atheism, to 'render nugatory all objections urged against the original experiments of Mr. Crosse'.

Painful as it must ever be to the genuine lover of science to witness the diligence and research of its votaries assailed by culumny and misrepresentation, I feel assured that no remarks of mine are needed to disengage the Broomfield experiments from the abuse of those who have sought to crush and misrepresent facts which, if carefully investigated, will ultimately conduct us to a more intimate acquaintance with the sublime laws of Creative Wisdom. The experiments, however, which it is the object of this paper to detail, will, I trust, constitute a more powerful appeal to conviction than the most persuasive eloquence could command for the occasion.[75]

To meet the objections, Weekes wished to 'submit a simple detail of every fact and circumstance' of the arrangements and precautions he had taken. As shown in Fig. 11.6, the most important step was placing his silica solution in a bell jar over mercury, with two mercury-filled cups on top. Gases produced by electrochemical decomposition could thereby escape, while incoming air was purged of contaminants by the mercury bath. The fundamental objections against Crosse's arrangements had at last been countered.

The experiment took a long time to show any results. Over a year passed without anything happening; Children and Bird would have abandoned it as a 'failure' months before. Then, on 25 November 1841, Weekes noticed that 'five perfect insects' had appeared inside the jar

---

74. Anon. (1844), p. 61. For a notice of Weekes (1789–1850), see the 1847 issue of *The Provincial Medical Directory* (London: John Churchill), p. 295; for his death, *Gentleman's Magazine*, (1850), n.s., **34**, p. 339.
75. Weekes (1842), pp. 241–2.

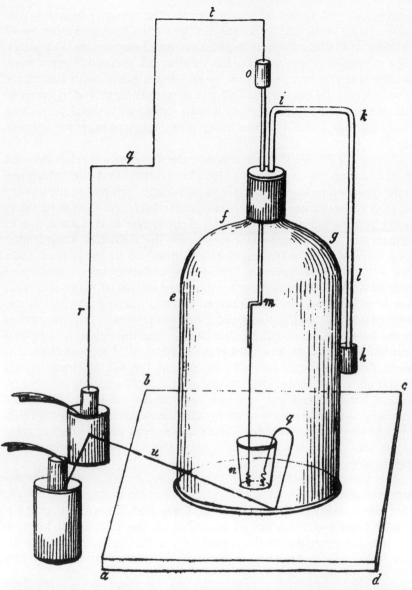

Figure 11.6. Weekes' first apparatus, designed to meet the criticisms levelled at the experiments of Crosse. The electrolytic solution was placed inside a bell-jar about eight inches tall; this was set in a circular mercury-filled groove on a sealed block of beech-wood. Mercury cups at *h* and *o* provided further seals for incoming wire and escaping gases. All parts of the experiment were carefully cleaned and baked, and there was no obvious way that contaminants could reach the inside of the jar. (Weekes, 1842, pl. v, fig. 24)

amid a cluster of pyramidal quartz crystals. The presence of insects was subsequently witnessed by others. Unlike the half-hearted attempts of Children and Bird, the new replication had been begun with every hope of positive results. And like almost all successful replicators, Weekes was the first person to try the experiments who had direct experience of the equipment and arrangements employed in the original. He had been in close contact with Crosse for several years, had visited his 'hospitable mansion' many times, and was eager to vindicate him.[76]

Closing off the outside air was not the only way Weekes avoided contaminating his experiments. He also ensured that the announcement appeared before a relatively expert forum before it made its way into the popular press. Weekes first wrote to his friend George Newport at the Entomological Society, who had the letters read at a meeting in January 1842. The result was a victory for the vigilantes. Their leader, John Edward Gray, was present as a visitor, and he mentioned again that Children had repeated Crosse's experiments over a period of months without success. (The fact that Weekes had let his experiments run for at least twice as long was conveniently ignored.) An announcement of Weekes' finding, 'balanced' by Gray's comments, duly appeared in the February issue of the *Entomologist*.[77] But this report in a limited circulation journal created almost no effect at all.[78] Weekes then sent his findings to another specialist society, the London Electrical, which by this point was fighting for its very existence. The Society clearly saw a fresh batch of mites as a route to renewed public attention and (perhaps) favour. The new secretary, Charles Walker, made certain that a notice was sent to *The Times*.[79] The paper eagerly ran this new chapter in the sensational story and once again the acari made their way to the farthest reaches of the realm.

The Electrical Society was a marginal institution and Weekes was something of a marginal figure within it. But for all that, the challenge that critics like Faraday and Joseph Henry had thrown at Crosse had finally been met. Announced according to the established forms of science, presented within the confines of a scientific body and following the form of an 'acceptable' experiment, Weekes' efforts literally gave

76. Collins (1985) discusses the necessity of this kind of tacit knowledge in any successful replication.
77. The meeting is described in Westwood (1842).
78. See, however, Bladon (1842).
79. *Times* (14 Mar. 1842), p. 14. Walker's efforts to exploit the popular press are detailed in his 'Report of the Honorary Secretary of the London Electrical Society for the session 1841–2', London Electrical Society minute book, Institution of Electrical Engineers, Special Collections, ms 42.

the acari a new lease of life. A number of active experimentalists now found it worth their while to begin their own trials. At least two were established figures on the metropolitan electrical scene, Henry Noad and Alfred Smee.[80] Noad's efforts were especially significant, for he had a substantial reputation and had authored a widely praised textbook, *Lectures on Electricity*, first published in 1839.[81] The second edition, in 1844, was dedicated to Crosse, whom Noad had first visited shortly after the announcement of Weekes' success. Noad not only included a full account of Weekes' and Crosse's work in his *Lectures*, but also announced that three experiments of his own had been in progress for 16 months. At the time of writing, these had not yet produced acari, but Noad remained hopeful. Despite Noad's enthusiasm, however, these experiments occupied an anomalous position in the book, for they were tacked onto the section about crystallisation through electricity, not integrated with any other results.[82] Thus despite their appearance in a textbook, they remained in the realm of what Collins has called 'extraordinary science'.

The acari had a much more immediate revolutionary potential for the research of Alfred Smee, a major figure in Victorian electricity now remembered chiefly as the inventor of a new kind of battery.[83] Smee was extremely interested in the relation between life and electricity. His major book on the subject, *Elements of Electro-biology, or the Voltaic Mechanism of Man; of Electro-pathology, Especially of the Nervous System: and of Electro-theraputics* (1849), summed up a decade of research.[84] The book viewed the human body as a complex series of batteries, with the brain as the 'central battery' and the nerves as 'bio-telegraphs'. Smee gave elaborate names to the various sub-divisions of his subject: 'Aistenics', 'Phreno-aisthenics', 'Syndramics' and so forth. 'Pneuma-noemics', for example, was the 'Totality of Combinations of Totalities of each Sense and both Sides'.[85] In coining these hard words, Smee had in mind nothing less than the birth of a new scientific discipline. His ambition was similar to that of his German contemporary Emil Du Bois-Reymond, who (as Timothy Lenoir has shown) hoped to become the Faraday of a new discipline of 'electrophysiology'.[86] Smee called

80. A third individual, who corresponded with Crosse about the experiments in 1842 but does not seem to have followed up her plan to carry them out, was Ada Lovelace, scientific enthusiast, Charles Babbage's protogée, and frequent visitor to Fyne Court. See Stein (1985), p. 143.
81. Noad (1839), which did not mention Crosse at all.
82. Noad (1844), esp. pp. 215–19.
83. For biographical information on Smee, see O[dling] (1878).
84. Smee (1849).   85. Smee (1849), p. xiv.
86. Lenoir (1986); see also E. Du Bois-Reymond to Faraday, 15 Nov. 1849, in Williams, ed. (1971), pp. 564–5.

for others to join him in forging the new science, pursued at present in one tiny laboratory in the heart of the Bank of England:

In submitting this volume to the Public, I regard it as a beginning of a subject, not as an end; as a commencement of the study of Physiology upon the laws of physics, which will require the united efforts of physiologists to render it more complete. It is a worthy subject for the talent of the country to be engaged upon, and nothing but the associated labours of many can render it of ultimate value.[87]

The foundations of electrobiology lay in cell theory, and specifically the power of electricity over the form of individual cells. The work of Crosse and Weekes pointed an important way forward here, by showing that life might have a physical basis. Smee thus made serious efforts in the mid-1840s to replicate Weekes' successes. The experiments failed, and Smee never reported any details about them. But he viewed these as a temporary setback, due more to his cramped facilities than to any inherent implausibility in the electrical generation of insect life.[88] Smee *did* succeed in a related enterprise in the study of electricity and organic form: in the summer of 1848 he effected a transmutation of one variety of insect into another through the use of current action. Smee believed that an insect (which he called the *Aphis vastator*) had been responsible for the Irish potato famine. As part of an experimental programme related to the famine, he subjected it to continuous electrical action. After some time he 'found at one pole the Vastator; at the opposite, the Bean Aphis'. This remarkable discovery occupied a prominent place in the Smee's chapter on the 'Electro-biology of cells'.[89] Again, he emphasised the need for repeated trials before this result could be used as part of the experimental scaffolding for electrobiological science.

That practical experimenters of the calibre of Smee and Noad attempted serious replications after Weekes' triumph in 1842 is a crucial point. Had they found insects too, the electrical life would have made its way into standard scientific works other than Noad's *Lectures*, and become more closely tied to a body of related results. A subject like Smee's electrobiology would have lost its 'extraordinary' status and become

87. Smee (1849), p. xi.
88. 'As far as appertains to this subject, I am of opinion, that the facts should be neither believed nor disbelieved, but kept in abeyance for fresh experiments. The subject has been taken up with much ill-judged acrimony; for the question really resolves itself into this proposition – Has, or has not, the Creator endowed inorganic matter with the power of assuming, under the influence of certain forces, an organic form?' Smee (1850), p.15. See also Smee (1849), pp. 74–7.
89. Smee (1849), p. 76.

part of mainstream science. But this did not happen.[90] Only one person successfully repeated the experiments in a closed container, and that was Andrew Crosse. At some time during the 1840s Crosse abandoned at least some aspects of his disinterested stance, perhaps at the urging of Ada Lovelace. He now sought to eliminate potential contamination by placing his solution in a closed retort with mercury seals, not in a bell jar as Weekes had done. Once again the experiment took a long time, but after several months he found 'one fat acarus' on the inside of the retort. A letter to Harriet Martineau, written in 1849 and published in 1851, detailed his procedures. As before, Crosse refused to interpret his observations, preferring to let the experiments speak for themselves.[91]

Crosse's call for others to follow up these trials met with silence from the experimental community. The reasons are not difficult to find. An experiment by Crosse, whose name was intimately attached to the insects, merely turned the cycle of replication back upon itself. In addition, Crosse (like Weekes in some of his later publications) extended the potential importance of the experiments to an extent which contemporaries found implausible. For example, fungus growing in his apparatus was announced as exciting evidence for 'a close connection between animal and vegetable life'. For a sceptical audience, such a widening of significance only cast doubt on the entire enterprise. Moreover, the letter was first published in one of the most notorious free-thinking tracts of the nineteenth century, scarcely a propitious environment for convincing the gentlemen of science.

No one else repeated Weekes' work successfully. Insects did not appear in other laboratories, and electrobiology proved stillborn. This was not because the experiments were necessarily flawed,[92] but because they failed to generate viable scientific practice. There was nothing further one could *do* with them. In the twentieth century the only experimenters hoping to try the work of Crosse and Weekes have been American school-children, led by books like Frank Edwards' *Stranger than Science* (1959) to believe that Faraday had repeated the experiments successfully. The Royal Institution was deluged with their enquiries during

90. Smee disavows the experiments outright in Smee (1875), p. 51, and they were gradually removed from later editions of Noad's textbook.
91. Atkinson & Martineau (1851), pp. 361–7.
92. As the work of Farley & Geison (1974) on the Pasteur–Pouchet debate shows, spontaneous generation claims were too protean to be disproved in any straightforward way.

the early 1960's.[93] After all, the creation of insect life through electricity would make a superb project for a student science fair.[94]

NATURAL LAW AND DIVINE EXPERIMENT

We have seen that the electrical acari proved useful to a specialist society devoted to electricity. They were even more appealing to those hoping to found a borderline discipline like electrobiology. But the acari really came into their own in the great early Victorian campaign for natural law, development and the 'science of progress'. Traditionally debates about these issues have been viewed almost entirely in terms of geology and natural history, as part of the 'background to Darwin'. But contemporary discussions demonstrate that experiment – especially in electricity – had a major role to play. The results of Weekes and Crosse not only threatened disciplinary barriers between the physical sciences and those dealing with the organic world; they dissolved old distinctions between matter and life, one of the chief barriers blocking a full-blown naturalistic cosmology. The result, inevitably, was an extended, bitter and highly politicised controversy about the wider meanings of experiment.

As mentioned earlier, Buckland and other geologists had pointed out that the creatures might be 'living fossils' revivified by the power of electricity. In the early months of the debate, this hypothesis had the advantage of sidestepping the entire problem of the creation of life through experiment. However, this compromise was no longer available after March 1837, and lines of debate soon became more sharply drawn. From the perspective of conservative clerics, evangelicals and a great many ordinary English men and women, the experiments were precisely equivalent to atheism. As the itinerant popular science lecturer John Murray put it, 'this was *atheism* with a witness, and it is well to call things by their real and right names'. Even granting Crosse's own innocence in the affair, Murray had no doubts about the impact of the experiments: they led down the same slippery track as the nebular hypothesis, Geothean plant morphology and anti-Mosaic geology.[95]

93. Stallybrass (1967), p. 597; Edwards (1959), pp. 123–30.
94. For another example of American high school science laboratories as the final repository for troubled experiments, see Holton (1978), esp. p. 28, on the 'Millikan oil drop experiment'. Travis (1981) provides another case involving school-children as replicators.
95. Etheldred Benett, an expert student of fossil sponges and member of an ancient Wiltshire family, told Murray that she shared his view of this 'wild theory'. E. Benett to J. Murray, 16 Apr. 1838, University of Edinburgh Library, ms Gen 1971/2/5; Murray (1837), esp. p. v.

Most prominently, the Cambridge professor of geology Adam Sedgwick made an attack on the acari a central element in his half-century campaign against naturalistic explanations of species origins. 'But is there so much as one good physiologist or chemist who now adopts the first interpretation of these galvanic experiments?' he asked. 'I believe not so much as one. Ridicule is the only weapon we can condescend to use against the outrage on common sense and universal experience implied in this mockery of a creative power.'[96] He advised Crosse 'not to meddle again with animal creations; and without delay to take a crow-bar and break to atoms his obstetrico-galvanic apparatus'.[97] Sedgwick passionately believed that studies of the origin of life were beyond the boundaries of legitimate science. When the issue involved actual experiments, this meant that laboratories needed to be destroyed and equipment smashed. For Sedgwick, extraordinary science was dangerous science.

As a consequence of their discoveries, Crosse and Weekes received anonymous letters and threats of violence. The 'persecution' of Crosse became notorious. The case was compared with classic stories of the oppression of science, such as William Lawrence's humiliation by Lord Justice Eldon and Galileo's treatment by the Inquisition. 'And think you those times are passed, and men now pure and tolerant?' asked one author.[98] Crosse's enemies among local high Tory agriculturalists even blamed his insects for a blight on the wheat crop in the southwest of England[99] – an early example of the kind of fears that have re-emerged with the rise of experiments in genetic engineering. An exorcism – possibly the only one in the entire history of experimental science – may even have been carried out by a local priest on the hills above Fyne Court.[100] From perspectives like these, Crosse was nothing less than a Frankenstein, a lost soul wandering beyond the proper limits of human knowledge.

On the other side, a small group of middle-class intellectuals saw no threat in the acari, but rather the promise of an experimental basis for a new cosmology based on progress, evolution and natural law. The central work here was the anonymous best-seller *Vestiges of the Natural History of Creation* (1844).[101] The book was written by Robert Chambers,

96. Sedgwick (1851), pp. xxiv–v.   97. [Sedgwick] (1845), p. 72.
98. Atkinson & Martineau (1851). For the persecution of Lawrence, see Jacyna (1983).
99. C. Crosse (1857), p. 170.
100. See the highly coloured account of the incident in Haining (1979), pp. 125–7, although I have not yet been able to confirm this from original sources.
101. [Chambers] (1844), esp. 185–90. For *Vestiges*, the best places to start are articles by Hodge (1974), Yeo (1984) and Secord (1988); see also Millhauser (1959).

a popular author and publisher in Edinburgh, who followed precisely the new kind of journalistic career opened up by the technologies of mass publication. The *Vestiges* ranged from the nebular hypothesis of the formation of the universe, through the geological history of the earth, and finally to the origins and destiny of the human race. The acari had a critical role in this scheme, in showing that life (under suitable conditions) might have originated directly from matter. Miracles were not needed to explain the introduction of life or new species. In the new science of the *Vestiges*, the experimental study of life and electricity (Smee's electrobiology?) would occupy an important place. To advocates of this new science, the general drift of the experiments of Crosse and Weekes mattered more than their details, which is why Chambers kept them in editions of the book published long after their plausibility began to fade.

For any science in mid-nineteenth century England to be truly a 'people's science', it could not directly violate the scruples of religion. No one was better aware of this than Chambers, who had achieved consumate success in meeting the expectations of his Victorian middle-class audience. The *Vestiges* skillfully defused fears that experiments might lead to a reduction of God's role in the world:

The experimentalist could never be considered as the author of these creatures, except by the most unreasoning ignorance. The utmost that can be claimed for, or imputed to him is that he arranged the natural conditions under which the true creative energy – that of the Divine Author of all things – was pleased to work in that instance. On the hypothesis here brought forward, the *acarus Crossii* was a type of being ordained from the beginning, and destined to be realized under certain physical conditions. When a human hand brought these conditions into the proper arrangement, it did an act akin to hundreds of similar ones which we execute every day, and which are followed by natural results; but it did nothing more. The production of an insect, if it did take place as assumed, was as clearly an act of the Almighty himself, as if he had fashioned it with hands.[102]

In a later book Chambers discounted Sedgwick's objections and argued that Weekes' careful repetitions (further detailed in a long appendix to the volume) deserved to be met by more than ridicule. Against *a priori* conclusions from the Cambridge cloisters, Chambers wrote, 'I adduce careful experiment'. As Richard Yeo has noted, the rhetoric of experiment could be turned against men of science by someone who was essentially a journalist.[103]

102. [Chambers] (1844), pp. 187–8.
103. Yeo (1984), p. 17; [Chambers] (1845), pp. 119–21, 189–98. In his appendix Weekes also tentatively claimed the electrical production of a new species of fungus.

Chambers' willingness to keep the acari in the *Vestiges* has often been used as *prima facie* evidence of his inability to appreciate the high experimental standards of Victorian science. In fact the consensus on this issue, even after practising electricians had abandoned their efforts to replicate Weekes' experiments, was by no means clear. The support of medical men for spontaneous generation meant that expert opinion remained divided. William Carpenter, whose ambivalence to the acari has already been discussed, actually helped behind the scenes in revising the *Vestiges*, so it is hardly surprising that Chambers felt justified in keeping them in his book. He was much more concerned that his work might be perceived as a frontal assault on divine omnipotence. As in the case of the nebular hypothesis, the troubled history of the acari was a definite liability.[104]

No such scruples, though, restrained another major work which used the experiments to further the science of progress. This was Henry George Atkinson and Harriet Martineau's *Letters on the Laws of Man's Nature and Development* of 1851. This remarkable book printed a series of letters between Atkinson, who assumed the position of teacher, and Martineau, who asked questions and approved the replies in rapturous terms. Atkinson was convinced that religion was a fable, forced upon men by a 'puritanical priestcraft'. Enquiries into the nature of the divinity and subjects of religion were a waste of time; the cause of causes was an 'unfathomable mystery'. The authors hailed the discoveries of Crosse and Weekes as 'those noblest experiments of the age', precisely because they collapsed the distinction between life and matter, removing the finger of God from the origin of organic beings.[105] Electrical science had provided what were in effect 'crucial experiments' in the debate over natural law. The material tools of science would forge a new system of belief to replace the old.

What was the position of Crosse and the other active experimenters in these disputes? Crosse did not share Atkinson and Martineau's exhilerated embrace of theological scepticism, although his confirmation of Weekes was appended to their *Letters*. On the other hand, there can be no doubt that Crosse countenanced the use of his work in the campaign for natural law. (His position is analogous to that of the reclusive Charles Darwin in the debates about scientific naturalism in the wake of the *Origin of Species*.) Crosse announced his findings publicly long after their potential as weapons in the debates about natural law became clear;

104. The difficulties faced by the nebular hypothesis are analysed in Yeo (1984), pp. 17–19 and Schaffer (1988).
105. Atkinson & Martineau (1851), esp. pp. 180–1, 210, 240.

he responded at length to the request of known free-thinkers for experimental details; and he penned lengthy poems on the persecution of Galileo and other scientific martyrs. A late, unpublished poem entitled 'Science' spoke from bitter experience:

> Strange? that the boundless works of God
> Such gratitude may win,
> As that to trace their hidden laws
> Should be denounced a sin!

Crosse wrote in scathing terms of 'the canting bigot's art', 'the noisy war of vulgar throats', and 'the sneer of those who knowledge hate'.[106] His public statements present a similar picture.

Thus despite his reputation for naive simplicity, Crosse favoured the extension of natural law into the organic realm and wished the acari used towards that end. Although he had told Martineau that he did not claim to 'form' insects in the same manner that he had 'formed' crystals, the overwhelming force of his experimental descriptions made that possibility almost irresistable. As he put it, a six-sided prism might turn into a quartz crystal – or it might develop filaments and legs and move about 'at pleasure'. Because of this, every crystallisation experiment that he undertook was potentially an inquiry into the origin of life. Thus, by continuing his inorganic crystallisation experiments Crosse *was* also following up his acari work: both were ultimately part of the same programme of research. As he once told a visitor to Fyne Court, to bottle electricity was no more impious than bottling rainwater;[107] by an extension of the same reasoning, to form living insects was no more sacrilegious than forming crystals. But Crosse and the other replicators saw no danger of materialism, for in their view God was immanent in the universe, proceeding always according to law.

It is here that some of the ambiguities surrounding the Victorian notion of experiment begin to arise. Certainly Crosse, Weekes, Noad and Smee had no wish to play God, and they claimed to be observing divine laws under particular conditions – *not* playing God by intervening, distorting or creating. As Crosse told Martineau in 1849:

As to the appearance of the acari, under long-continued electrical action, I have never, in thought, word, or deed, given any one a right to suppose that I considered them a creation, or even as a formation, from inorganic matter. To create, is to form a something out of a nothing. To annihilate, is to reduce that something to a nothing. Both of these, of course, can only be the attributes of the Almighty.[108]

106. C. Crosse (1857), pp. 254–65.   107. C. Crosse (1857), p. 161.
108. Crosse to Martineau, 12 Aug. 1849, in C. Crosse (1857), p. 174; also printed in Atkinson & Martineau (1851), p. 362.

From this point of view, God himself was the only true experimentalist. Smee agreed, arguing that some future Paley might employ experiments of this kind 'as the most powerful proof deducible from nature of the infinite power and wisdom of God'.[109] Rather than taking creation out of God's hands, as Frankenstein had claimed to do, men of science could use experiments to show that the material and organic realms were equally within the realm of his law. As Crosse put it, 'I am neither an "Atheist," nor a "Materialist," nor a "self imagined creator," but a humble and lowly reverencer of that Great Being, whose laws my accusers seem wholly to have lost sight of'.[110] The natural philosopher was like a priest, someone who displayed God's laws, not someone who created them.

From the perspective of Crosse and those who used his results, the opponents of the acari were inconsistent. Typically the successful experimentalist was granted a status akin to that of an observer, a humble inquirer into nature's hidden recesses, or as Faraday put it, a pupil 'in nature's school'. But when the origin of life was at issue, the same individual was denounced for attempting to rival God rather than patiently revealing his secrets. The public applauded the ingenious manipulations required for artificial crystallisation, but drew back in horror when life appeared instead.[111] The meaning and limits of legitimate experiment had emerged as a major battleground in the campaign for scientific naturalism.

THE POWER OF EXPERIMENT

What is the power of experiment? The issues involved are especially clear in the early nineteenth century, a period that saw many of the characteristics of science established in a form that persists to the present day. Compared with a modern project like CERN, discussed by John Krige in Chapter 12 of this volume, the experiments dealt with here exemplify what Derek J. de Solla Price would have termed 'little science'.[112] Yet for men like Weekes, Children or Smee, keeping an experiment going for many months was no small matter. Experiments, both in modern post-war 'big science' and in cases like the present one, take place not in relation to a disembodied world of theoretical traditions, but because individuals and groups are forced to choose from a variety of options. No one has to invest in any experiment; certainly no one *had to* invest in Crosse's acari. To do this required a positive decision

109. Smee (1849), p.77.   110. A. Crosse (1841), p. 10.
111. For an excellent discussion of creation and manipulation in the Romantic era, particularly in relation to the Frankenstein myth, see Cantor (1984).
112. Price (1963).

to apply resources of time, status, money, philosophy and faith. Such commitments give experiments their power.

As we have seen, the resources brought to bear on a single experiment can be very diverse. Even on the relatively small scale of Victorian science, one is struck by the lack of a unified common culture accessible to all those interested in science. The general impression is of a set of distinctive practices with relatively little communication between them. When disciplinary communities did talk to one another, they often failed to agree. Taxonomic zoologists ignored the concerns of medical men, geologists spoke out on taxonomic issues before consulting specialist zoologists, experimental electricians chided the geologists for their supposed credulity. The deepest division of all was between sciences of life and sciences of matter: hence the problematic nature of experiments that juxtaposed electricity and organic creation.

Although Victorian science was specialised, its ideas, practices and concerns still tended to be constructed from elements of the general cultural environment. Thus participants in the acari controversies could formulate their views only in part as geologists, medical men, electricians or naturalists. More fundamentally, they spoke from a diverse range of orientations towards the established religious and political order of Victorian society. For the individual this meant that everything could be at stake, a point demonstrated vividly by responses to the use of electrical life in the *Vestiges* and the Atkinson–Martineau *Letters*. What kind of natural knowledge, and what kind of scientific practice, was appropriate to the new industrial age?

The persistence of disagreements at this fundamental level suggests that the electrical insects, like the more famous problem of the origin of species, cannot be taken as representative pieces of Victorian science. (Although one would never suspect it from the historical literature, evolutionary theorising in the first half of the nineteenth century is a classic case of extraordinary science.) More typical examples of 'normal' practice would be the formation of crystals in electrolytic solutions or studies of the taxonomy of the Arachnidae. Yet even the ways that mundane subjects like these were studied had reference to the same political and religious controversies involved in the acari debates. The disciplinary divisions within science which made taxonomy and chemistry possible had been drawn precisely to put dangerous questions safely on the margins.[113] No wonder that the most important boundary was the line, drawn by Faraday and many others, between life and matter.

113. Porter (1977); Morrell & Thackray (1981); Outram (1984).

With active practitioners engaged in these wider disputes, it is not surprising that their right to legislate for popular conceptions of the natural world was constantly challenged. We have seen this illustrated many times, in cases ranging from Faraday's difficulties in disclaiming his alleged replication, to Chambers' attempts to base the alternative cosmology of his *Vestiges* on the foundations of the acari. Chambers and other popular authors maintained that the authority of experiment was questioned partly because of the esoteric practices and specialisation of leading scientific men.[114] Even more important, though, was the phenonenon that Chambers himself represented: the emergence of a mass culture during the early phases of the Industrial Revolution, based on new techniques of printing, publishing and distribution. As a result, the scientific elite confronted middle and working class audiences of unparalleled diversity and size. 'Steam intellect' could sometimes be controlled, through Mechanics' Institutes, the British Association or reviews in the major quarterlies. But for the most part contact was mediated by journalists, editors and the whole network of industrial publishing.

In the seventeenth century, experiment before a competent audience had been established as an effective means for creating consensus within science. Its authority depended on a small group of like-minded gentlemen of polite manners and similar status agreeing about what they had seen.[115] The social and economic circumstances of the early nineteenth century were radically different. Experiment had yet to prove its ascendancy over the new printing technologies and heterogeneous reading public of the industrial age. During the present century channels of communication and the size of the audience have multiplied further, so that tensions between the power of experiment and the community at large have become still more acute. The period that witnessed Crosse's 'extraordinary experiment' holds the roots of one of the most fundamental problems in the relations between popular culture and the culture of science. The struggle for authority continues.

ACKNOWLEDGEMENTS

I am indebted to Geoffrey Cantor, Dennis Dean, Adrian Desmond, David Gooding, Sarah Jane Lefort, Trevor Pinch, Simon Schaffer and Anne Secord for help of various kinds in the preparation of this essay.

114. Yeo (1984).   115. Dear (1985); Shapin & Schaffer (1985).

Archivists at the British Museum (Natural History), Cambridge University Library, Edinburgh University Library, and the Institution of Electrical Engineers gave access to manuscripts in their possession; I am particularly grateful to Lenore Symons of the Institution of Electrical Engineers for making it possible for me to consult the record book of the London Electrical Society. All figures are reproduced by permission of the Syndics of the Cambridge University Library.

References

Altick, R.D., (1957). *The English Common Reader: A Social History of the Mass Reading Public, 1800–1900*. Chicago: University of Chicago Press.

Anon. (1837). Accidental production of animal life. – Mr. Crosse. *Magazine of Popular Science*, 3, 145–8.

Anon. (1838). Remarks on the sixth report of the British Association. *Mechanics' Magazine*, 26, 51–6.

Anon. (1844). Noad's lectures on electricity. *Mechanics' Magazine*, 40, 57–61.

Andouin, V. (1836). Arachnida. In *The Cyclopaeia of Anatomy and Physiology*, ed. R.M. Todd. London: Sherwood, Gilbert, and Piper.

Atkinson, H.G. and Martineau, H. (1851). *Letters on the Laws of Man's Nature and Development*. London: John Chapman.

Berg, M. (1980). *The Machinery Question and the Making of Political Economy, 1815–1848*, Cambridge: Cambridge University Press.

Bladon, J. (1842). Note on *Acarus horridus*. *The Entomologist*, 1, 307–8.

Bragg, W. (1836). Extraordinary experiment. *Somerset County Gazette* (31 Dec. 1836), 1(1), 3.

Bragg, W. (1837a). To correspondents. *Somerset County Gazette* (7 Jan. 1837), 1(2), 2.

Bragg, W. (1837b). Mr. Crosse's experiment. *Somerset County Gazette*, (25 Feb. 1837), 1(9), 4.

Brewster, D. (1845). 'Review of Vestiges of the Natural History of Creation'. *North British Review*, 3, 470–515.

Burkhardt, F. & Smith, S. ed. (1986). *The Correspondence of Charles Darwin*, vol. 2., 1837–43. Cambridge: Cambridge University Press.

Cantor, P.A. (1984). *Creature and Creator: Myth-making and English Romanticism*. Cambridge: Cambridge University Press.

Carpenter, W.B. (1838). On the differences of the laws regulating vital and physical phenomena. *Edinburgh New Philosophical Journal*, 24, 327–53.

[Carpenter, W.B.] (1845). Natural history of creation. *British and Foreign Medical Review*, 19, 155–81.

Carpenter, W.B. (1850). On the mutual relations of the vital and physical forces. *Philosophical Transactions of the Royal Society of London*, 1, 727–57.

[Chambers, R.] (1844). *Vestiges of the Natural History of Creation*. London: John Churchill.

[Chambers, R.] (1845). *Explanations: A Sequel to 'Vestiges of the Natural History of Creation'*. London: John Churchill.

Collins, H.M. (1985). *Changing Order: Replication and Induction in Scientific Practice*. London: Sage.

Collins, H.M. & Pinch, T.J. (1979). The construction of the paranormal: nothing unscientific is happening. In *On the Margins of Science: the Social Construction of Rejected Knowledge*, Sociological Review Monograph 27, ed. R. Wallis, pp. 237–70. Keele: University of Keele.

Collins, H.M. & Pinch, T.J. (1982). *Frames of Meaning: The Social Construction of Extraordinary Science*. London: Routledge and Kegan Paul.

Collins, H.M. & Shapin, S. (1984). Uncovering the nature of science. *The Times Higher Education Supplement*, **612**, 13.

Cooter, R. (1984). *The Cultural Meaning of Popular Science: Phrenology and the Organization of Consent in Nineteenth-Century Britain*. Cambridge: Cambridge University Press.

Cox, E.W. (1844). Andrew Crosse, the electrician. *The Critic of Literature, Art, Science and the Drama*, **1**, 57–60.

Crosse, A. (1837a). To the editor of the *Taunton Courier. Taunton Courier*, 4 Jan. 1837, **1481**, 7.

Crosse, A. (1837b). Mr. Crosse's experiments. *Bristol Advocate*, 4 Feb. 1837, 1(21), 165.

Crosse, A. (1837c). On the production of insects by voltaic electricity. *Annals of Electricity, Magnetism, and Chemistry*, **1**, 242–4.

Crosse, A. (1841). Description of some experiments made with the Voltaic Battery ... for the purpose of producing crystals; in the process of which experiments certain insects constantly appeared. *The Transactions, and the Proceedings of the London Electrical Society, from 1837 to 1840*, 10–16.

Crosse, C. (1857). *Memorials, Scientific and Literary, of Andrew Crosse, the Electrician*. London: Longman.

Davey, J.G. (1850). On the physiology and pathology of the ganglionic nervous system. *Lancet*, **2**, 413–15.

Dear, P. (1985). *Totius in verba*: rhetoric and authority in the early Royal Society. *Isis*, **76**, 145–61.

Desmond, A. (1987). Artisan resistance and evolution in Britain, 1819–1848. *Osiris*, **3**, 77–110.

Desmond, A. (1988). Lamarckism and democracy: corporations, corruption and comparative anatomy in the 1830s. In *History, Humanity and Evolution*, ed. J.R. Moore, Cambridge: Cambridge University Press. (in press).

Edwards, F. (1959). *Stranger than Science*. New York: L. Stuart.

Faraday, M. (1839–55). *Experimental Researches in Electricity*, 3 vols. London: R. and J.E. Taylor.

Farley, J. (1977). *The Spontaneous Generation Controversy from Descartes to Oparin*. Baltimore: Johns Hopkins University Press.

Farley, J. & Geison, G.L. (1974). Science, politics and spontaneous generation in nineteenth-century France: the Pasteur–Pouchet debate. *Bulletin of the History of Medicine*, **48**, 161–98.

Gillispie, C.C. (1951). *Genesis and Geology: A Study in the Relations of Scientific Thought, Natural Theology, and Social Opinion in Great Britain, 1790-1850.* Cambridge, Mass.: Harvard University Press.

Gooding, D. (1985a). In nature's school: Faraday as an experimentalist. In *Faraday Rediscovered: Essays on the Life and Work of Michael Faraday, 1791-1867,* ed. D. Gooding & F.A.J.L. James, pp. 105–36. Basingstoke: Macmillan.

Gooding, D. (1986b). Experiment and concept formation in electromagnetic science and technology in England in the 1820s. *History and Technology,* 2, 151–76.

Haining, P. (1979). *The Man who was Frankenstein.* London: Frederick Muller.

Heilbron, J.L. (1979). *Electricity in the 17th and 18th Centuries: A Study of Early Modern Physics.* Berkeley: University of California Press.

Hocquette, M., (1976). Pierre Jean Francois Turpin. *Dictionary of Scientific Biography,* ed. C.C. Gillispie, New York: Charles Scribner's Sons.

Hodge, M.J.S. (1974). The universal gestation of nature: Chambers' *Vestiges* and *Explanations. Journal of the History of Biology,* 5, 127–51.

Holton, G. (1978). Subelectrons, presuppositions, and the Millikan–Ehrenhaft dispute. In *The Scientific Imagination: Case Studies,* ed. G. Holton, pp. 25–83. Cambridge: Cambridge University Press.

Hytch, E.J. (1837). Mr. Crosse's revivification of insects contained in flint. *Lancet,* 1, 710–11.

Jacyna, L.S. (1983). Immanence or transcendence: theories of life and organization in Britain, 1790-1835. *Isis,* 74, 311–29.

Kahlbaum, G.W.A. & Darbishire, F.V. (ed.) (1899). *The Letters of Faraday and Schoenbein, 1836–1862, with Notes and References to Contemporary Letters.* Bale: Benno Schwaber; London: Williams and Norgate.

Knight, D. (1967). The scientist as sage. *Studies in Romanticism,* 6, 65–88.

Latour, B. (1983). Give me a laboratory and I will raise the world. In *Science Observed: Perspectives on the Social Studies of Science,* ed. K.D. Knorr-Cetina & M. Mulkay, pp. 141–70. London: Sage.

Latour, B. (1987). *Science in Action: How to Follow Scientists and Engineers through Society.* Milton Keynes: Open University Press.

Latour, B. (1988). *The Pasteurisation of French Society.* Cambridge, Mass. Harvard University Press. (in press).

Leithead, W. (1837). *Electricity: Its Nature, Operation and Importance in the Phenomena of the Universe.* London: Longman.

Lenoir, T. (1986). Models and instruments in the development of electrophysiology. *Historical Studies in the Physical and Biological Sciences,* 17, 1–54.

Levere, T.H. (1981). *Poetry Realized in Nature: Samuel Taylor Coleridge and Early Nineteenth-century Science.* Cambridge: Cambridge University Press.

Millhauser, M.(1959). *Just Before Darwin: Robert Chambers and Vestiges.* Middleton, Conn.: Wesleyan University Press.

Morrell, J. & Thackray, A. (1981). *Gentlemen of Science: Early Years of the British Association for the Advancement of Science.* Oxford: Oxford University Press.

Murray, J. (1837). *Considerations on the Vital Principle; with a Description of Mr. Crosse's Experiments.* London: E. Wilson.

[Newman, E.] (1845). Notice of the natural history of creation. *The Zoologist: A Popular Miscellany of Natural History,* **3**, 954–63.

Noad, H.M. (1839). *A Course of Eight Lectures on Electricity, Galvanism, Magnetism, and Electro-Magnetism.* London: Scott, Webster, and Geary.

Noad, H.M. (1844). *Lectures on Electricity, Comprising Galvanism, Magnetism, Electro-magnetism, Magneto- and Thermo-Electricity.* London: G. Knight.

O[dling], E.M. (1878). *Memoir of the Late Alfred Smee, F.R.S.* London: George Bell.

Outram, D. (1984). *Georges Cuvier: Vocation, Science and Authority in Post-Revolutionary France.* Manchester: Manchester University Press.

Porter, R. (1977). *The Making of Geology: Earth Science in Britain, 1660–1815.* Cambridge: Cambridge University Press.

Price, D.J. de Solla (1963). *Little Science, Big Science.* New York: Columbia University Press.

Pym, H. N. (ed.) (1882). *Memories of Old Friends: Being Extracts from the Journals and Letters of Caroline Fox of Penjerrick, Cornwall, from 1835 to 1871,* 2nd edn, 2 vols. London: Smith, Elder.

Raymond, J. & Pickstone, J.V. (1986). The natural sciences and the learning of the English Unitarians: an exploration of the roles of the Manchester College. In *Liberty, Truth, Religion: Essays Celebrating Two Hundred Years of Manchester College,* ed. B. Smith, Oxford: Manchester College.

Reingold, N. (ed.) (1979). *The Papers of Joseph Henry,* Vol. 3, Washington, DC: Smithsonian Institution Press.

Reingold, N. (ed.) (1981). *The Papers of Joseph Henry,* Vol. 4, Washington, DC: Smithsonian Institution Press.

Richards, E. (1987). A question of property rights: Richard Owen's evolutionism reassessed. *British Journal for the History of Science,* **20**, 129–71.

Roth, N. (1979). Bugs and blasphemy: Andrew Crosse and the Acarus electricus. *Medical Instrumentation,* **13**, 357.

Rupke, N. (1983). *The Great Chain of History: William Buckland and the English School of Geology, 1814–1849.* Oxford: Clarendon Press.

Ruse, M. (1979). *The Darwinian Revolution.* Chicago: University of Chicago Press.

Schaffer, S. (1986). Scientific discoveries and the end of natural philosophy. *Social Studies of Science,* **16**, 387–420.

Schaffer, S. (1988). The nebular hypothesis and the science of progress. In *History, Humanity and Evolution,* ed. J.R. Moore. Cambridge: Cambridge University Press. (in press).

Schultze, F. (1837). Notice of the result of an experimental observation made regarding equivocal generation. *Edinburgh New Philosophical Journal,* **23**, 165–6.

Secord, J.A. (1985). Newton in the nursery: Tom Telescope and the philosophy of tops and balls, 1761–1838. *History of Science,* **23**, 127–51.

Secord, J.A. (1986). *Controversy in Victorian Geology: The Cambrian–Silurian Dispute.* Princeton: Princeton Univesity Press.

Secord, J.A. (1988). Behind the veil: Robert Chambers and *Vestiges*. In *History, Humanity and Evolution*, ed. J.R. Moore, Cambridge: Cambridge University Press. (in press).

[Sedgwick, A.] (1845). Natural history of creation. *Edinburgh Review*, 82, 1–85.

Sedgwick, A. (1851). *A Discourse on the Studies of the University of Cambridge*. London: J.W. Parker.

Shapin, S. (1984). Pump and circumstance: Robert Boyle's literary technology. *Social Studies of Science*, 14, 481–520.

Shapin, S. & Schaffer, S. (1985). *Leviathan and the Air-pump: Hobbes, Boyle, and the Experimental Life*. Princeton: Princeton University Press.

Sheets-Pyenson, S. (1985). Popular scientific periodicals in Paris and London: the emergence of a low scientific culture, 1820–1875. *Annals of Science*, 42, 549–72.

Shelley, M. (1831). *Frankenstein, or, the Modern Prometheus*, (1985), ed. M. Hindle. Harmondsworth: Penguin Books.

Smee, A. (1849). *Elements of Electro-biology, or the Voltaic Mechanism of Man: of Electro-pathology, Especially of the Nervous System: and of Electro-theraputics*. London: Longman.

Smee, A. (1850). *Instinct and Reason: Deduced from Electro-Biology*. London: Reeve.

Smee, A. (1875). *The Mind of Man: Being a Natural System of Mental Philosophy*. London: George Bell.

Smith, C. (1985). Geologists and mathematicians: the rise of physical geology. In *Wranglers and Physicists: Studies on Cambridge Mathematical Physics in the Nineteenth Century*, ed. P.M. Harman. Manchester: University of Manchester Press.

Stallybrass, O. (1967). How Faraday produced living animalculae: Andrew Crosse and the story of a myth. *Proceedings of the Royal Institution of Great Britain*, 41, 597–619.

Stein, D. (1985). *Ada: A Life and a Legacy*. Cambridge, Mass.: MIT Press.

Sturgeon, W. (1838). Address. *Annals of Electricity, Magnetism, and Chemistry*, 2 64–72.

Telescope, T. (1838). *The Newtonian Philosophy, and Natural Philosophy in General, Explained and Illustrated by Familiar Objects, in a Series of Entertaining Lectures*. London: T. Tegg.

Thompson, E.P. (1966). *The Making of the English Working Class*. New York: Vintage Books.

Travis, G.D.L. (1981). Replicating replication? Aspects of the social construction of learning in planarian worms, *Social Studies of Science*, 11, 11–32.

Turpin, P. (1837). Note sur une espèce d'Acarus présentée à l'Académie, dans sa séance du 30 octobre. *Comptes Rendus Hebdomadaires des Séances de L'Académie des Sciences*, 5, 668–78.

Turpin, P. (1838a). Note on a kind of Acarus presented to the Academy. *Annals of Electricity, Magnetism, and Chemistry*, 2, 355–60.

Turpin, P. (1838b). Note on a kind of Acarus presented to the Academy. *Annals and Magazine of Natural History*, 2, 55–62.

Walker, W.C. (1937). Animal electricity before Galvani. *Annals of Science*, 2, 84–113.

Webb, R.K. (1955). *The British Working Class Reader, 1790–1848: Literacy and Social Tensions*. London: Allen and Unwin.

Weekes, W.H. (1842). Details of an experiment in which certain insects, known as the Acarus Crossi, appeared incident to the long-continued operation of a voltaic current upon silicate of potass, within a close atmosphere over mercury. *Proceedings of the London Electrical Society*, 1, 240–56.

W[estwood], J.O. (1842). Entomological Society of London; January 3rd, 1842. *Entomologist*, 16, 264.

Williams, L.P· (1971). *The Selected Correspondence of Michael Faraday*. Cambridge: Cambridge University Press.

Williams, R. (1961). *Culture and Society, 1780–1950*. Harmondsworth: Penguin.

Wilson, G. & Geikie, A. (1861). *Memoir of Edward Forbes, F.R.S.* London: Macmillan.

Winsor, M.P. (1976). The development of Linnaean insect classification. *Taxon*, 25, 57–67.

Yeo, R. (1981). Scientific method and the image of science, 1831–1890. In *The Parliament of Science: The British Association for the Advancement of Science, 1831–1981*, ed. R. MacLeod & P. Collins, pp. 65–88. Northwood, Middlesex: Science Reviews.

Yeo, R. (1984). Science and intellectual authority in mid-nineteenth century Britain: Robert Chambers and *Vestiges of the Natural History of Creation*. *Victorian Studies*, 28, 5–31.

12

WHY DID BRITAIN JOIN CERN?[1]

JOHN KRIGE

This paper presents some key elements of a detailed, historical study
of the process whereby a group of British scientists, science administrators, and government officials came to believe that their country
should join CERN – the European Organization for Nuclear Research
as it was then called. It aims specifically, though not always explicitly,
to undermine the conception of such a decision as the rational outcome
of a series of successive steps in a logically interconnected chain of
events. Instead it depicts a process unfolding through several phases
which are loosely interconnected with one another, phases in which
different groups of actors addressed themselves to the question of Britain's
relationship to the European laboratory project, had *different reasons*
for being for, against, or indifferent to it, and adopted their attitudes
in the light of a situation as they defined it *at a particular point in time.*
In short, where the rational model abstracts an ordered sequence I find

---

1. The main archival sources used for this article were the files deposited at the Public
   Record Office, Kew, London (PRO), Sir James Chadwick's personal papers at Churchill
   College, Cambridge (CC), and the Science and Engineering Research Council's boxes
   on CERN stored at Hayes, Middlesex (SERC). Of particular interest at the PRO were
   – Series AB6/1074 and 1076 of the United Kingdom Atomic Energy Authority, which
   was our basic source for Cockcroft's correspondence;
   – Series DSIR 17/559 and 560, covering the Department of Scientific and Industrial
   Research's dealing with the European laboratory project during 1951 and 1952;
   – Series ED157/302, the Department of Education and Science file on the 'Proposed
   European Nuclear Physics Laboratory 1951–2';
   – Series FO371/101514–101517, for Foreign Office attitudes in 1952;
   – Series CAB134/943 for the minutes of the meeting of the Cabinet Steering Commit-
   tee on International Organizations in 1952.
   The article is based on a far more comprehensive and fully documented study of why
   Britain joined CERN which has recently appeared (see note 3). To avoid unnecessary
   repetition and to contain its length, we generally only give references for quotations
   in what follows.

a messier, more haphazard process, and where it suggests that events were suffused with necessity, I insist on their aspect of contingency.[2]

The story focuses on the attitudes and motivations of the scientists themselves: they were the main (though by no means the only) actors, their perception of Britain's scientific needs carried the greatest weight, and it was they who shaped policy. In line with this choice the paper divides the period of interest into three phases, each characterised by a different attitude towards the scheme by the national physics community. Put somewhat crudely, this evolved from relative indifference (up to mid-1951), through hostility by some influential members of the establishment (for roughly the next year), to determined support by a group at Harwell (from mid-1952 onwards). As the scientists were barely involved in the negotiations around the text of the Convention establishing CERN, its signature, and its ratification by Parliament in 1953, these will not be dealt with here. The interested reader is referred to my more extensive study of how and why Britain joined CERN, on which this paper is based.[3]

EARLY RETICENCE, JANUARY – JULY 1951

The background

Towards the end of 1949 the first of a number of important steps was taken to foster European collaboration in nuclear science. In December of that year Raoul Dautry, Administrator-General of the French Commissariat à l'Energie Atomique, had a resolution passed at the European Cultural Conference in Lausanne proposing that the possibility of setting up a European institute for nuclear science 'in its applications to daily life' be explored.[4] In June 1950 a similar proposal was made independently by Isidor I. Rabi, Nobel prizewinner and founder of the Brookhaven National Laboratory NY, USA, at the 5th General Conference of UNESCO in Florence. Rabi's resolution asked the Director-General 'to assist and encourage the formation and organization of regional

2. For general criticisms of the rationalistic model see, classically, Schilling (1961); Allison (1971); Allison & Morris (1975). For its application to the history of CERN see Krige & Pestre (1987); Pestre (1987a,b). For a different illustration of what it means to study a historical process, see Rudwick (1986).
3. Krige (1987).
4. Mouvement Européen de la Culture, Lausanne, 8–12 décembre 1949, Résolutions et déclaration finale, edited by Bureau d'Etudes pour un Centre Européen de la Culture, (1950?), in the Centre's archives, Geneva.

research centres and laboratories . . .'.[5] Elaborating, he identified western Europe as a suitable region, high-energy physics as a possible field, and Pierre Auger, French cosmic-ray physicist and the Director of UNESCO's Department of Exact and Natural Sciences, as the man to push the project ahead. This Auger did with relish. He discussed and refined his ideas in several circles, notably with a number of those present at the Harwell Physics Conference in Oxford in September 1950. And along with Dautry he convened a meeting of influential European science administrators and scientists on 12 December 1950 at the newly formed European Cultural Centre in Geneva. This meeting recommended 'the creation . . . of a European laboratory for nuclear physics centred on the construction of a large machine for accelerating elementary particles'.[6] Its power was to be greater than the design energy of any similar device then under construction, the biggest of which were the 3 GeV Cosmotron at Brookhaven and the 6 GeV Bevatron at Berkeley, California.

No British representatives attended the meeting on 12 December. However, its proceedings were funnelled via the Council of Europe to the Overseas Liaison Division of the DSIR (Department of Scientific and Industrial Research). The latter, in turn, asked the Nuclear Physics Committee of the Ministry of Supply 'to examine the proposals and to express its view on them'.[7] This occasioned the first *formal, collective*, response to the developments taking place on the continent by some of Britain's leading physicists.

The response of the Nuclear Physics Committee

On 7 February, 1951 the Secretary of the Nuclear Physics Committee (NPC), J.H. Awbery, circulated a number of documents pertaining to the 'formation of a European Laboratory for Nuclear Physics Research', including the resolution passed at the European Cultural Centre some two months before. The members of the NPC were asked to comment on the proposal itself, and on 'whether this country should support the project and join it, by contributions of money and men'.[8] By the end of March 13 physicists had responded to Awbery's circular. Their views were summarised by the chairman, Sir James Chadwick, Nobel

5. Records of the General Conference of UNESCO, 5th session, Florence 1950; Resolutions, Section B, resolution 2.21, 38. Copy available in United Nations Library, Geneva.
6. Resolution adopted on 12 December, 1950, at the European Cultural Centre in Geneva, a copy of which was attached to document O.S.R.(51)1, 3/1/51 (PRO–AB6/912).
7. Annex I to document N.P.C.54, undated but probably 7/2/51 (PRO–AB6/912).
8. Documents N.P.C.54 and N.P.C.55, 7/2/51 (PRO–AB6/912).

prizewinner, Master of Gonville and Caius, Cambridge, and Britain's leading elder statesman of science, in his official reply to Alexander King in the DSIR on 23 April, 1951. Though the views expressed by the members of the NPC varied widely, wrote Chadwick, there was

a very definite balance of opinion that this country should not join directly in the establishment of such a Laboratory and should not promise support either with men or money; but there [was] at the same time a feeling that we should not dissociate ourselves entirely from the scheme (if it [was] adopted) but give informal support by means of advice and help, if requested, more particularly perhaps in the preliminary studies on the scope of the laboratory and on the design of the equipment.[9]

At this stage, therefore, Britain's physicists were not prepared to commit themselves to the European laboratory project, either by joining it or by supporting it financially or with personnel. On the other hand, as their replies to Awbery made clear, they were not *opposed* to the scheme, 'at any rate as far as Continental scientists [and countries were] concerned'.[10] Hence their willingnesss to co-operate with the venture by offering technical advice and know-how. This latter attitude is easily understood: it was consistent with the 'internationalism' of science which had evolved after World War I, an attitude which expressed itself *inter alia* in the informal (cf. Chadwick above) circulation of expertise within the community. But why did British physicists not want to go further? Why this reticence when, by contrast, Auger was pressing ahead, and collecting together a group of experts to implement the project, and when, by April/May 1951, science administrators in France, Italy and Belgium, had respectively secured 2 MFF (million French francs), 1 MFF and 0.4 MFF for preliminary studies?

The reasons for British reticence

'The chief reason', as Chadwick saw it, for the position adopted by the NPC, was related to the imminent maturation of the domestic accelerator programme launched immediately after the war. These new machines, he wrote to King, were 'adequate both in number and in performance to provide full scope for all our own research workers in nuclear physics'. Their exploitation would 'absorb all our available manpower', and Britain could only play a 'prominent part in a European

---

9. Letter Chadwick to King, 23/4/51 (PRO–DSIR17/559).
10. For the detailed replies of the scientists see document N.P.C.57, 31/3/51 (PRO–AB6/912).

Laboratory' at the expense of using them to the full.[11] In short, it was not in Britain's *scientific interest* to participate in Auger's project, nor, he believed, did she have the resources to do so.

One may—and later some did—dispute Chadwick's claim that his country's physics community was *adequately* served by the existing national accelerators. Yet whether or not she needed to expand her resources, the fact remains that in the early 1950s Britain led (western) Europe in this respect. A wide variety of machines was available or nearing completion. Harwell was equipped with a 170 MeV synchro-cyclotron, an electron synchrotron and an unusual type of linear accelerator. In addition, five British universities were building a variety of machines, the more important being a 300 MeV electron synchrotron at Glasgow (under P.I. Dee), a 400 MeV synchro-cyclotron at Liverpool, initiated by Chadwick and continued by H.W.B. Skinner when he succeeded Sir James in 1949 and the giant 1000 MeV (1 GeV) proton synchrotron being constructed by Mark Oliphant in Birmingham. The only comparable device on the continent was the 200 MeV synchro-cyclotron in Uppsala, Sweden. By virtue of this lead it was only to be expected that Britain would be relatively diffident about collaborative ventures with continental countries. Correlatively, the alacrity and enthusiasm with which the European laboratory project was initially supported in, say, France, owes much to the relative backwardness of this country's science in the decade after the war, and to the recognition that she could not afford to build big accelerators on her own.

It was not only the scientific strength of British nuclear physics which accounts for her physicists' reticence *vis-à-vis* the European laboratory project; there were also the perceived scientific *weaknesses* of the scheme itself. Several members of the NPC, including its chairman, thought the idea of building a 6 GeV bevatron on the continent was 'too ambitious' and 'technically impracticable'.[12] Patrick Blackett, Nobel prizewinner and one of Britain's leading cosmic-ray physicists went further; his 'first reaction' was that the scheme was 'quite crazy', a sentiment echoed by Skinner.[13] This scepticism was not peculiar to the British, and was not without foundation. Those who had met in Geneva in December 1950 had formulated their proposal less in the light of an appraisal of Europe's scientific capabilities than in the hope of making the old continent once more competitive with the United States, of reversing the brain-drain, of restoring her erstwhile glory in nuclear physics, and so on.

11. Letter Chadwick to King, 23/4/51 (PRO–DSIR17/559).
12. Document N.P.C.57, 31/3/51 (PRO–AB6/912).
13. Letter Blackett to Cockcroft, 13/2/51 (PRO–AB6/912).

These scientific considerations were the single most important factor accounting for the initial lack of enthusiasm among British physicists for the European laboratory project. Three other concerns also played an important role in distancing them (and Britain's administrators) from the scheme at this time, though of course the weight to be attached to each, and the degree of awareness in the mind of the actor, differed considerably between individuals. These were:

a. *Psychological attitudes.* These reflected British arrogance ('the reason why we didn't want to join with the Europeans', said Sir Ben Lockspeiser, the then Secretary of the DSIR afterwards, 'was because we thought that those foreigners were inferior and we had always been top dogs and we were going to be top dogs'),[14] and xenophobia, here in the form of a distrust of French motives ('if the French want to have a nuclear physics research laboratory why don't they go ahead with the co-operation of any other nation interested', wrote Skinner, implying that they were simply using UNESCO as a front to further their own sectional interests);[15]

b. *Policy considerations.* In the immediate post-war years both scientific and governmental circles repeatedly stated that they were not in favour of the 'foundation of new centralized international laboratories in subjects where active research [was] being carried out on a large scale'–they preferred the traditional form of scientific collaboration between *existing, national,* institutes.[16] We have seen this factor at work in Chadwick's reply to King;

c. *The general political climate.* This was not conducive to close ties with the Continent. The Labour Party, which was unexpectedly swept to power over Churchill's Conservatives in 1945, and remained in office until 1951, was singularly unenthusiastic about European ventures, opposing for example the Schuman Plan for a European Coal and Steel Community. This was consistent with the general

---

14. Lockspeiser made this remark in an interview with Margaret Gowing at CERN on 5/6/73. There is a copy in the CERN archives.

15. Quoted by Gowing (1974), II, P. 227.

16. Document IOC(50)95, 26/6/50 (PRO–CAB134/405). British reluctance to join in a collaborative venture is consistent with the attitude Salomon (1971) attributes to the superpowers:

Apart from those fields which, by their extranational nature, require joint research efforts (meteorology, oceanography), the two 'big ones' expect from scientific cooperation only what it has always provided – the exchange of ideas and new information. On the other hand, all the other countries are obliged to see in cooperation the indispensable path to the more economic or more rapid achievement of their national objectives.

lines of British Foreign policy at this time, which was dominated by the concern to maintain the 'special relationship' with the United States established during the war, and to preserve the 'traditional' links with the Commonwealth. Each of these associations were likely to be strained if Britain moved too close to Continental Europe, particularly in nuclear affairs.

DEEPENING INVOLVEMENT, AUGUST 1951–JUNE 1952

The background on the Continent

Throughout 1951 Auger pushed ahead steadily with the European laboratory project under the auspices of UNESCO. In May he gathered together a number of scientific consultants who discussed in more detail the type of apparatus the laboratory should have (an accelerator 'of at least 3 BeV and possibly up to 6 BeV' – a tacit admission that the Geneva proposal for more than 6 GeV was somewhat overambitious), and the chronological development of the project.[17] In July he discussed the scheme further with the cream of Europe's physicists in Copenhagen, and apparently managed to quell the doubts which Niels Bohr, the doyen of them all, and Hendrik Kramers, the eminent Dutch physicist, were having about the financial implications of the project. And in August he took the first steps towards convening an intergovernmental conference of European member states of UNESCO with a view to raising funds for detailed design studies of the laboratory and its equipment.

Towards the middle of August, too, Kramers returned to Copenhagen, where his and Bohr's doubts about cost surfaced again. On 23 August he wrote to Auger proposing that Bohr's Institute for Theoretical Physics 'could really act as a most effective nucleus for an international European laboratory', a scheme which, he said, had many advantages including the financial benefits of having 'an international cooperation . . . grow naturally out of an existing institute (instead of having to build up at a place where as yet there is nothing and nobody . . .)'.[18] At this time Kramers apparently accepted the UNESCO consultants' idea of equipping the laboratory with accelerators; two months later that too was being questioned. On 26 October, 1951 Bohr wrote to Auger suggesting that if the institute in Copenhagen was used as a

17. UNESCO General Conference, 6th Session, Programme Commission, Report 6C/PRG/ 25, 19/6/51, Annex 1, 5, copy in the United Nations Library, Geneva.
18. Letter Kramers to Auger, 23/8/51 (UNESCO archives, Paris).

'pilot' centre for the European laboratory project he would propose setting up a board of directors which would nominate 'a group of experts in experimental and theoretical physics' 'to consider thoroughly how the new institution would best stimulate the intereuropean cooperation in the field of atomic physics . . .'. Effectively then, Bohr was proposing to put the UNESCO project on ice, and to reconsider from scratch the scientific aims and appropriate equipment of the European laboratory. 'We expect', Bohr concluded somewhat ominously, 'that Kramers' proposal will be discussed at the coming Paris meeting . . .'. i.e., the intergovernmental conference now scheduled to begin on 17 December.[19]

Although originally unruffled by these initiatives, the UNESCO camp became increasingly concerned by Bohr's activities in particular, and the divisions were real, and the atmosphere tense when the Paris conference opened. After several days of difficult negotiations it was decided to reconvene the conference in Geneva in February 1952, and a number of working groups were set up to prepare the ground for the second session. This met from 12 to 15 February, and produced an 'Agreement Constituting a Council of Representatives of European States for Planning an International Laboratory [essentially the scheme favoured by the Auger/UNESCO camp] and organising other forms of co-operation in Nuclear Research [the Bohr project]'. By thus *juxtaposing* their proposals the conference successfully papered over the rift between the two schools of thought. Nine delegations (from Denmark, the Federal Republic of Germany, France, Greece, Italy, the Netherlands, Sweden, Switzerland and Yugoslavia) signed the Agreement forthwith; some $211,000 were promised for the next stage of the project. The Agreement entered into force soon thereafter, and on the 5 May the Council of Representatives met for the first time.

This brief and select chronology highlights the developments around the European laboratory project which occurred *on the Continent* between about May 1951 and May 1952. Thus the absence of any explicit reference to British physicists so far should not be taken to imply that they played no role in the events just described. For example, consistent with the recommendations of the NPC, Sir John Cockcroft, the Director of Harwell, and a man relatively close to Auger and in sympathy with the aspirations of Continental scientists, released one of his young accelerator builders, Frank Goward to serve as a member of the UNESCO group of expert consultants. This gesture of technical support was, however, overshadowed by the far more substantial offer of access to

19. Letter Bohr to Auger, 26/10/51 (UNESCO).

the Liverpool cyclotron proposed by Chadwick and Thomson—an offer whose motivations were quite different and whose effects on British policy were far more profound.

Chadwick, Thomson, and the Liverpool cyclotron

In August 1951 Sir James Chadwick happened to arrive in Copenhagen while Kramers and Bohr were sharing their doubts about Auger's accelerator-based project. Sir James' interest in Kramers' ideas was soon apparent, and on 24 August the Dutch physicist sent him a copy of the letter he had written to Auger the day before, asking him to show it to Darwin (recently retired director of the National Physics Laboratory). Chadwick replied at once. Thanking Kramers for the copy of his letter to Auger suggesting the creation of 'an international centre at Copenhagen', Chadwick added 'I shall support your proposal strongly', going on to say that he would soon be in touch with Darwin, and could 'take some other steps, in confidence, to prepare people in this country to receive your suggestion favourably'.[20]

Acting on his offer, Chadwick spent the next few weeks discussing Kramers' idea with several influential people. Darwin (like Chadwick himself, in fact) was not particularly keen on establishing a new European research centre, but felt that if there was to be one it should be in Copenhagen. On the other hand Sir George Thomson, then Professor of Physics at Imperial College, and an enthusiastic proponent of nuclear fusion, was 'definitely attracted' by the scheme.[21] On 16 November, 1951, both he and Chadwick were invited to a meeting of the British Committee for Co-operation with UNESCO in the Natural Sciences to discuss it further. Expressing confidence in Kramers' idea, they claimed that if 'the proposed [European] laboratory be located in Denmark and associated with Professsor Nils Bohr [it] would command the support of British nuclear physicists'.[22] The meeting thought that the 'use' of the Liverpool cyclotron could be offered to the envisaged centre, and asked Chadwick, who still retained strong links with his old university, to negotiate the terms of the arrangement. He discussed the matter with Skinner and several senior university officials, and it was formally agreed that

if a central laboratory for nuclear physics [were] established in Copenhagen, and if it [were] under the direction of Professor Niels Bohr, the University of

20. Letter Chadwick to Kramers, 30/8/51 (CC–CHADI, 1/13).
21. As reported in letter Chadwick to Bohr, around 20/9/51 (CC–CHADI, 1/3).
22. The minutes of the meeting are document C/113(51) or NS(51)20, undated (PRO-ED157/303).

Liverpool would be prepared in principle to consider the possibility of collaborating [. . .] with the central nuclear physics laboratory so established.[23]

During the autumn of 1951, then, there was a marked shift in the attitudes of several leading members of the British physics community towards establishing a European laboratory on the Continent. Their earlier scepticism about the merits of the European bevatron project remained. But whereas before they had been prepared to stay detached, to leave the scheme to run its course, now they allied themselves unambiguously with the *opponents* of that project and agreed to support them materially if the European centre was set up in Copenhagen and directed by Bohr. Why did they take such clear sides in the ripening conflict across the Channel? To begin with there was the confidence which Chadwick had in Bohr, a confidence apparently shared by those with whom he discussed the matter. The scheme put forward by Auger was, as he put it to King, 'so ambitious and so divorced from reality as to be quite impracticable'. By contrast, wrote Chadwick, 'I have no doubts that if Kramers and Bohr produce a scheme it will be a practicable one, which will not impose a heavy burden in men and money on the participating countries'.[24] Clearly the as yet (September 1951) unformulated *details* of the Kramers/Bohr scheme were irrelevant. It was enough to know that it was *their* scheme; it was *they* who were being supported.

This confidence in Bohr and Kramers, here transformed into an alliance against the Auger/UNESCO project, reflected the deep-rooted bonds of respect and friendship first built up between these men between the wars. Longstanding scientific colleagues, all members of the scientific establishment, with considerable 'political' influence in their own countries and indeed beyond, they were not used to spending huge sums of money on a single item of equipment for basic research, they were not convinced that the biggest was necessarily the best, and they drew back from the prospect of draining national resources 'just . . . to build a big accelerator of a type now under construction in the U.S.A.', as Kramers put it. Out of their shared traditions of scientific practice, out of their shared values, grew a solidarity between these physicists which cemented them together in opposition to the European bevatron project.[25] Reinforcing this bond, and even more fundamental than it, was the need, already referred to, not to appear to be unwilling to co-operate. As we have just said, neither Chadwick nor Darwin were

23. Letter Mountford to Chadwick, 3/12/51 (CC–CHADI, 1/8).
24. Letter Chadwick to King, 17/9/51 (PRO–ED157/302).
25. The quotation is from letter Kramers to the Director-General of UNESCO, 24/9/51, reproduced in document UNESCO/NS/NUC/1, Annex 7 (UNESCO). Galison (1987) has some interesting comments on the weight of a scientist's habits of work.

'strongly attracted to the idea of an international centre'. Yet both felt that if it was 'generally desired on the Continent we should help in its formation'.[26] This felt imperative to assist their scientific colleagues, despite their misgivings, and against the chauvinistic instincts of some, had led the members of the NPC to offer technical support to Auger. Now it clinched the alliance between those members of the European scientific establishment who were against the UNESCO scheme.

The Paris conference and the meeting of Chadwick's advisory committee

Two official representatives from 13 European member states of UNESCO – one a scientist and the other an administrative and financial expert – attended the 'Conference on the Organization of Studies Concerning the Establishment of a European Nuclear Physics Laboratory' in Paris from 17 to 20 December 1951. The United Kingdom's delegates were Sir George Thomson and A.H. Waterfield, the British Scientific Attaché in Paris. Sir George Thomson made the offer of access to the Liverpool cyclotron in his official statement to the conference. The machine, he said, should be ready by the late summer or autumn of 1952, and could 'be used for a period of years to give the Centre a machine of substantial size, over which it would have special rights, both as regards sending men to work with it and as regards the direction of researches to be undertaken'. The one 'necessary condition' of the offer, Sir George went on, was that the new centre 'be associated with an existing research organization'. Specifying, he suggested that Bohr's laboratory 'would be an ideal base for such collaboration'.[27]

The British offer was warmly received by both factions at the conference and, in Thomson's view, 'went far to make an agreement possible' between them. It was embodied in a lengthy resolution hammered out by the end of the meeting – though no mention of the 'necessary condition' was made. Indeed any reference to the location of the new laboratory was eschewed. Several study groups were planned, two for designing the accelerators, and one for theory, the latter to be based at Bohr's institute in Copenhagen to 'provide theoretical guidance for experimental work, to be carried out with the machines', including the machine at Liverpool. Thomson returned home satisfied that the resolution represented 'a compromise between the two views, that work at Copenhagen and Liverpool might start without much delay, but that the plans for one or two large machines will go forward and

---

26. Letter Chadwick to Thomson, 14/9/51 (CC–CHADI, 1/9).
27. For Sir George Thomson's statement see Annex 1 to his Notes on the conference, 20/12/51 (PRO–ED 157/302).

that the question of their construction will have to be faced in a year's time'.[28]

The terms of the Copenhagen – Liverpool link were discussed with Niels Bohr and Paul Scherrer, the eminent Swiss physicist, during a visit they paid to England on 9–10 January 1952. They were formally spelt out by a new advisory committee especially set up to handle developments around the European laboratory project. It was chaired by Chadwick; its other members were Cockcroft, Lockspeiser, Skinner and Thomson, and it first met on 23 January, 1952. In line with the resolution passed in Paris, the committee agreed that the experimental programme of the Liverpool cyclotron would be formulated in consultation with Bohr's group. The university also accepted to provide places for a number of physicists and engineers recommended by the Council of Representatives. There is little to surprise us here. More significant is the framework within which the committee discussed the Bohr – Skinner co-operation: according to the minutes, the committee considered 'the terms on which this country could *become a member*' of the Council, and thought the 'provision of facilities on an international basis' for work around the Liverpool machine would be an 'appropriate contribution in kind' (author's emphasis).[29]

With this the senior members of the British physics community took a decisive step towards their country *joining* the European laboratory project. As this account has stressed, their main motive for doing so was to consolidate their alliance with Bohr and the alternative scheme of which he was now the leading Continental exponent (Kramers died in April 1952). This behaviour had two aspects. On the one hand there was the desire to create a framework for collaboration with the Continent, and with Bohr in particular, with a view to exploiting fully the scientific potential of the Liverpool machine. On the other there was the wish to work alongside Bohr and the Danes in the Council with a view to shaping the development of the accelerator schemes along lines which they thought fit.[30] The committee members were 'not convinced on the present evidence that a machine as large as the largest U.S. Bevatron [was] necessary or desirable'.[31] It was only by joining the Council, they believed, that they could hope to have their argument prevail.

28. See Thomson's Notes (note 27), for the first and last quotations in this paragraph. The resolution is in Annex I to document NS/269.950 (UNESCO).
29. The committee's minutes are document NS(52)1, 23/1/52 (PRO–ED157/302).
30. As Chadwick put it in a letter to Bohr – 9/2/52 (CC–CHADI, 1/3) – 'I think I can assure you that our delegate will support you on every important point and will act in concert with you as far as possible'.
31. See the minutes (note 29).

When Sir George Thomson arrived in Paris in mid-December his idea had been to use the offer of the Liverpool cyclotron as a way of locating a European research centre in Copenhagen. Now, at the end of January, the physicists found themselves thinking of the offer as a contribution in kind entitling them to membership of the Council of Representatives. They might have preferred it otherwise. But confronted with the compromise reached between the rival groups in Paris, and unwilling to appear unco-operative, the British had little option but to accept the context of the debate as it evolved on the Continent—even though that context was not of their choosing, and its terms carried implications that they did not necessarily agree with.

Reactions in governmental circles

The recommendations of Chadwick's advisory committee were quickly passed to the Cabinet Steering Committee on International Organizations – a top-level interdepartmental body whose chairman and secretariat were provided by the Foreign Office. The committee considered the question of British membership of the envisaged European laboratory on 8 February, 1952 – thus in time to prepare a brief for the British delegates to the reconvened intergovernmental conference scheduled to start in Geneva a few days later. After lengthy discussion, the committee 'were unable to approve the association of Her Majesty's Government with the present proposals for an agreement constituting a Council of representatives of States for the establishment of a European Nuclear Research Centre . . .'.[32] As a result the United Kingdom's delegates to the Geneva conference found themselves without powers to sign the Agreement setting up the Council, and Britain was represented by an unofficial observer at the first session in May. Why had the Cabinet committee gone against the recommendation of senior British physicists, and placed the country's delegates to the Geneva meeting in such an embarrassing position? The main reason for this was that the committee recognised that the proposed research laboratory was likely to be equipped with big machines. Since signature of the agreement 'seemed to involve at least the moral obligation to support such a project', 'it was probable that the United Kingdom would be drawn into more expenditure on the Centre' than that involved in the offer of the Liverpool cyclotron. However 'our scientific advisers were opposed to the building of large new machines', and 'if the encourage-

32. The minutes of the Cabinet Committee are document I.O.C.(52)2nd Meeting, 8/2/52 (PRO–CAB134/943).

ment of European scientists to take part in work on the synchro-cyclotron at Liverpool University was the main consideration, this could no doubt be achieved by other means' – the proposed machinery, which involved 'the setting up of a Council . . . by means of a document in the nature of a Treaty, imposing obligations on participating states and requiring signatures and ratification', 'was unnecessarily cumbersome for the task in hand'.[33]

The logic of the Cabinet Committee's position was strong: what was the point of joining a body which would almost certainly recommend the construction of expensive scientific equipment when Britain's senior physicists were not convinced that such equipment was necessary or desirable? Undaunted, officials first in the Ministry of Education and then in the DSIR, continued to press the case. The scientists, too, were active. Chadwick, Cockcroft and Thomson pleaded it in person with Sir Roger Makins, a very senior official in the Foreign Office, and with Lord Cherwell, Churchill's highly influential science advisor. Their efforts paid off. On 11 June the Cabinet Committee met again. Chadwick, invited to attend, reiterated that 'at present United Kingdom scientific opinion was against the building of large cyclotrons . . . '.[34] The chairman, for his part, rehearsed the now standard objections of the Foreign Office. But his tone had changed. And the committee in fact approved the recommendation that the government adhere to the Agreement, provided that the Chancellor of the Exchequer and the Lord President (the 'science minister') gave their approval. They did not. The Chancellor turned down the Cabinet Steering Committee's recommendations on 23 June, and the Secretary of the DSIR, who advised the Lord President, refused to fight the case because, he said, 'there would be little hope of its being successfully contested'.[35] Plainly there was no chance of Britain joining CERN until her senior physicists were convinced of the need for a big machine.

TAKING THE PLUNGE, JUNE – DECEMBER 1952

The emergence of the Harwell group

A marked feature of the period we have just discussed is that the articulation of Britain's policy *vis-à-vis* the European laboratory project

33. Ibid. The committee in fact felt that it was unnecessary to establish a Council for *any* of the envisaged tasks, including planning the accelerators.
34. The minutes of the meeting are document I.O.C.(52)10th Meeting, 11/6/52 (PRO–CAB134/943).
35. (Draft) letter Lockspeiser to Chadwick, 20/6/52 (PRO–DSIR17/559).

was in the hands of a few highly influential members of the scientific establishment – Chadwick, Cockcroft, Thomson and to a lesser extent, Skinner. This monopoly was to be broken during the latter part of 1952 by the involvement of a new group of actors in the decision-making process. The change was triggered by an informal conference held on 7 June at Buckland House near Harwell. The topic was 'High energy accelerators for nuclear research', and it was attended by over 20 people, many of them experimental physicists and engineers from Harwell and the universities equipped with accelerators. The aims of the meeting, as spelt out in a letter from Cockcroft to Chadwick (who apparently did not attend it) were to discuss future accelerator policy in Britain and her relationship to the European laboratory project. A number of papers were read, and although no formal resolutions were passed, at the end of the day most of those present agreed that Harwell should press on with its plans for building a 450 MeV high intensity proton linear accelerator, and that British physicists *should have access* to a bevatron of at least 10 GeV. This could be achieved either by building one at home–and 'Lord Cherwell thought that it was not impossible to get a million pounds a year for a few years' for this purpose–or by contributing to the European effort.[36] It was not clear to them at this stage which course was preferable. Four days later, Sir James Chadwick, although (probably) aware of what had transpired at Buckland House, was to tell the Cabinet Steering Committee that Britain's scientists were still against building powerful accelerators.

Chadwick's attitude was indicative of the rift that now separated him (and Thomson) from the majority opinion at Buckland House, and from an active and organised group at Harwell in particular – people like Donald Fry, the Head of General Physics, Gerry Pickavance, the Head of the cyclotron group, John Adams, Jim Cassels, Frank Goward and Mervyn Hine. These men were a generation younger than Chadwick (b. 1891) and Thomson (b. 1892): the oldest was Fry (b. 1910), the youngest Cassels (b. 1924), and their average age was about 35. Many of them (Adams, Fry, Goward, Hine . . .) had worked on the development of radar during the war, and had transferred the skills acquired to accelerator construction thereafter, coming to constitute the most experienced nucleus of machine-builders in Europe. They were used to spending a lot of money on one item of scientific equipment, they were accustomed to working in teams and they took it for granted that engineers and physicists worked side by side in a basic science project. They were the children of the age of big science. By

36. For the proceedings of the Buckland House Conference, see the report written by M. Snowden, 7/6/52 (CC–CHADI, 26/2).

contrast the 'old guard' of the 20s and 30s, men like Chadwick and Thomson, were schooled in an entirely different laboratory world, a world in which one worked alone, or perhaps with a technician, in a university environment, ekeing the best possible results one could out of the limited resources one had.[37] For them this passion for a new accelerator was misplaced, this willingness to spend a huge amount of money was irresponsible.

The rift we have described was to sharpen in the weeks to come. Immediately after Buckland House, Cassels and Goward participated in an international physics conference held at Bohr's institute in Copenhagen. It was followed by the second session of the provisional Council attended by Fry at which it was officially decided to equip the European laboratory with a proton synchrotron of 10–15 GeV and a synchro-cyclotron of 600 MeV. Fry and Cassels came home even more convinced that British physicists should have access to a big bevatron. Both prepared papers arguing the case for a specially convened meeting of Chadwick's advisory committee. This was held on 16 July, 1952, but to no avail.

The committee 'agreed that it was too early to make any recommendation' on whether a powerful new accelerator was necessary. As Cockcroft explained afterwards, its permanent scientific membership (himself, Skinner, Chadwick and Thomson) was 'evenly divided on the question of the large machine. Chadwick and Thomson [were] strongly opposed to this project, but [were] favourable to a cooperation with the Bohr group' – so ignoring the opinion of 'the majority of the nuclear physicists in this country . . .'.[38]

The decision is taken

This situation obviously could not persist much longer. It was clear to all that the European bevatron project was going ahead, and that it was technically viable: the successful operation in the spring of the first of the new generation of giant accelerators, the Brookhaven Cosmotron, removed any lingering doubts about that. The need for Britain's physicists to resolve their differences became even more imperative with the discovery of the 'strong focusing' principle at Brookhaven in the summer of 1952. This innovation promised to reduce drastically the cost/GeV of an accelerator. It was based on the idea that, by using

37. See again Galison (1987), esp. chapter 5 on Theoretical and Experimental Cultures.
38. The minutes of the meeting are document NS(52)3, 16/7/52 (PRO–ED157/302). For Cockcroft see his letter to Peierls, 18/8/52 (PRO–AB6/1074).

magnets with high alternating field gradients, one could focus the accelerated proton beam more tightly. This meant that one could reduce the dimensions of the vacuum chamber in which the beam circulated, which in turn meant that one could reduce the size of the magnets (i.e., the amount of iron) used, and so the cost. The idea was immediately communicated by Brookhaven machine-builders to a group of CERN visitors who promptly decided to increase the energy of their 'bevatron' to 30 GeV. There was no doubt then: CERN was going to try to build the biggest machine in the world and the Americans were prepared to help them.[39]

The CERN bevatron group were given official permission to embark on what was in fact a research and development project at the third Council session early in October. Blackett and Fry attended the meeting on behalf of the Royal Society, stressing afterwards that if the government did 'not intend to join CERN fully', it was 'highly desirable that the decision not to do so [was] made quickly and definitely . . .'. 'If nothing is done', they warned, 'we stand the chance of neither having our own machine nor full access to the European one'. This 'would be a bad blow to British physics, and deeply resented by many British physicists . . .'.[40] A few weeks later, on 1 November, 1952, Chadwick's advisory committee, enlarged by the presence of 11 more physicists including Blackett, Cassels, Fry, Goward and Pickavance, met for the last time. It was unanimously resolved 'that British nuclear physicists should have access to a proton synchrotron of the size and character' projected by CERN, and that this could best be achieved by joining the organisation.[41] Their recommendations were endorsed by the Chancellor of the Exchequer on 25 November, 'in view of the clearly expressed view of British scientists that our physicists should have access to the more powerful machines now being planned . . .', as he put it.[42] With a number of remaining legal and financial issues clarified, he went on to authorise British membership of CERN on 29 December, 1952.

39. For more detail on the strong-focusing principle see my section 8.5 in Hermann *et al.* (1987). Why were Brookhaven engineers so willing to share their new findings with the CERN visitors? We have little documentary evidence to explain their behaviour, but an answer would include at least an engineers' desire to see his or her new idea put to the test as soon as possible, American scientists' wish to co-operate unselfishly with their European colleagues, and the belief that if Europe got funds for a strong-focusing accelerator it would be that much easier to persuade the US government to provide money for a similar machine on American soil.
40. For Blackett and Fry's report to the Royal Society, see the copy received by Cockcroft on 16/10/52 (PRO–AB6/1074).
41. The minutes of the meeting are document C/105(52), 1/11/52 (PRO–AB6/1074).
42. Letter Butler to Brunt, 25/11/52 (PRO–DSIR17/551).

How general was this consensus? In particular where did Chadwick stand? Unfortunately our documents do not allow us to answer these questions with any precision. Whatever he may have felt privately, he certainly gave his formal endorsement to the decision reached on the 1 November, and indeed went out of his way to ensure that subsequent negotiations between British and CERN administrators went smoothly.

Why did British physicists agree to join CERN in the autumn of 1952?

Now that we are aware of the general evolution of the situation in Britain in the second half of 1952, let us look more closely at some of the more important considerations which persuaded British physicists, firstly, that they needed access to a big proton synchrotron, and secondly, that rather than build one at home, the United Kingdom should join CERN. Why the need for access to a machine of at least 10 GeV? First and foremost there was the interest of the *scientific work* that could be done with such a device. From the late 1940s onwards it became increasingly clear that accelerators had an important role to play in the study of 'elementary particles', dominated until then by cosmic-ray physicists. It was recognised that the extremely high energies accessible in cosmic rays made them valuable for *qualitative* observations, e.g., the discovery of new particles. However the more controlled conditions obtainable in an accelerator were essential for *quantitative* measurements of particle properties. A new and exciting field of fundamental research was blossoming around these machines, the focus being on the study of a group of particles known as the mesons, the kind and number of which increased dramatically in the early 1950s. Cosmic-ray work suggested that an energy of at least 10 Gev was needed to study their properties carefully; at 15 GeV meson yields would be considerably enhanced; beyond that one did not know what to expect, though it was 'reasonable to hope that several lines of research which [were] qualitatively new [might] be opened up . . .' (Fry).[43]

Secondly, there was the *need to remain competitive*. As was only to be expected, the Americans were planning to build a number of new machines in and around the energy ranges being contemplated on the Continent. At the same time, by mid-1952, it became clear to at least Fry and Cassels that the European laboratory would be built with or without British help. As a result, noted Cassels, if British physicists had no access to a powerful bevatron, by 1960 they would probably 'have

43. In Fry's memo on British accelerator policy received by Cockcroft on 31/10/52 (PRO–AB6/1074).

to be content to take no part in a field of research accessible not only to the U.S.A. but also to . . .', and Cassels listed the 10 member states of the provisional CERN.[44] It was one thing to be second to America. It was another to be eleventh after Yugoslavia.

The discovery of strong focusing only increased the fear that Britain would be left out in the cold. Its rapid dissemination and immediate adoption by the CERN group created, wrote Pickavance, 'the interesting situation that the Europeans now start level with the Americans'.[45] Thus in a stroke it removed the lead which Britain had had in accelerator design and construction over her Continental neighbours. One other reason for building a strong-focusing machine should also be mentioned – the wish to construct the biggest device which was *technically feasible*, to take on the challenge posed by a research and development project, to do something more exciting than 'simply' scaling-up the Cosmotron or Bevatron to 10–15 GeV. Other research I have done in collaboration with Dominique Pestre suggests that this argument is widely used by physicists and engineers to set the upper limit to the 'size' of an experimental tool.[46]

Granted the interest in having access to a big proton synchrotron, why did British physicists recommend CERN membership – with all the political and bureaucratic complexities that that entailed – rather than going it alone? Here too we can identify three main concerns. Firstly, there was the *question of cost*. Skinner estimated that the big European machine would cost some £4 m; the consensus seems to have been that Britain could not build her own for much less. This was to be compared with the estimated £2 m for Britain's contribution to CERN. Secondly, there was *the need to preserve the domestic programme*. It was clear that if Britain were to design and construct her own big strong-focusing machine, Harwell would have to provide the bulk of the staff and expertise for it. However, calculations made by Fry indicated that this was not possible without Harwell abandoning the planned 450 MeV linear accelerator – which, said Pickavance, it should build 'no matter what happens'.[47] By contrast, if the United Kingdom joined CERN, only about 15 members of staff were needed, and Harwell could participate in the building of a powerful synchrotron while pressing ahead with its own linac (linear accelerator). Finally, there was the need for the physicists *to hedge their bets*. The strong-focusing prin-

44. See Cassels' paper on Great Britain and the European laboratory, 27/6/52 (PRO–AB6/1074).
45. Letter Pickavance to Chadwick, 15/9/52 (CC–CHADI, 26/2).
46. Krige & Pestre (1986).
47. In his memo headed Accelerator Programme, 10/6/52 (PRO–AB6/1014).

ciple, it must be remembered, was untried in practice. In Fry's view it was 'still only about 70% certain' that it would actually work, and in fact in December 1952 Harwell engineer John Lawson raised serious doubts about its feasibility.[48] Thus, on the one hand it seemed advisable to proceed with caution, while on the other the need to remain competitive meant that one had to keep in touch with the evolution of the strong-focusing concept. Joining CERN would resolve the dilemma. British physicists and engineers participated in research on the new idea, all the while retaining sufficient resources to build their own scaled-up bevatron of 10–15 GeV if that idea turned out to be impracticable.

To conclude, it is worth noting one argument that was hardly, if ever, used for British membership of CERN, either among physicists or among government officials. It was the argument that, by joining, the United Kingdom would be participating in a *European* venture. On the contrary, bearing in mind Britain's ties with the Commonwealth and the United States, the Foreign Office felt that CERN's Convention should be so drafted as 'to show to the world that the Organization had in fact *no political significance as a European body*' (author's emphasis).[49]

CONCLUDING REMARKS

Reflecting on the above account, the reader may be tempted to summarise what I have said by trying to draw up a 'list of reasons' or 'factors' purportedly explaining why Britain joined CERN. The urge to do this – to which I have fallen prey on more than one occasion myself – must be resisted. If one's aim is to understand the process which culminated in this decision, a general list of this kind has no meaning. As my account makes clear, during each of the phases through which that process passed, different groups of actors shaped British policy, and their attitudes differed depending on *who* they were and the state of play *when* they entered the scene. To detach reasons for their behaviour from the context in which they emerged, to locate them in the disembodied realm of a 'list', is to violate the *heterogeneity* of such an historically-evolving process, and to fail to grasp the *specificity* of its various moments. One example will suffice to make the point. Consider the stock argument that British reluctance to join CERN compared to,

48. For Fry see note 42. For J.D. Lawson's 'The effect of magnet inhomogeneities on the performance of the strong focusing synchrotron', 2/12/52 and (revised) 10/12/52, see (CC–CHADI, 1/2).
49. Memo Verry to Lockspeiser, 16/4/53 (SERC–Box NP24).

say, that of France, was yet another example of typical British caution towards all things European. Now as we have seen this certainly helps to explain attitudes in the country early in 1951. On the other hand, it was far less important as a factor shaping British physicists' attitudes than the relative position of domestic accelerators *vis-à-vis* those on the Continent. There is no better proof of this than the strong pressure for British membership of CERN which built up as soon as a group of Harwell scientists and engineers realised that they stood a real chance of dropping seriously behind the Continent; their erstwhile caution, such as it was, evaporated. No abstract list of reasons, no matter how sophisticated, can discriminate in this way, can grasp the weights and meanings which actors in a particular situation give to any of its elements. At best it can provide historians with a very general set of ideas, ideas to be refined as they work to understand why people behaved just as they did when they did.

References

Allison, G.T. (1971). *Essence of Decision: Explaining the Cuban Missile Crisis*. Boston: Little, Brown & Co.

Allison, G.T. & Morris, F.A. (1975). Armaments and arms control: exploring the determinants of military weapons. *Daedelus*, 104, 99–130.

Galison, P. (1987). *How Experiments End*. Chicago: University of Chicago Press.

Gowing, M. (1974). *Independence and Deterrence, Britain and Atomic Energy, 1945–52*, Vol. I & II. London: MacMillan.

Hermann, A., Krige, J., Mersits, U. & Pestre, D. (1987). *History of CERN, Vol. I, Launching the European Organization for Nuclear Research*. Amsterdam: North-Holland.

Krige, J. (1987). Britain and the European laboratory project. In Hermann *et al.*, chapter 12, 13.

Krige, J. & Pestre, D. (1975). The how and the why of the birth of CERN. In Hermann *et al.*, chapter 14.

Krige, J. & Pestre, D. (1986). The choice of CERN's first large bubble chambers for the proton synchrotron (1957–1958). *Historical studies in the Physical and Biological Sciences*, 16, 255–78.

Pestre, D. (1987a). *La Seconde Génération d'Accélérateurs pour le CERN, 1956–1965, Etude Historique d'un Processus de Décision de Gros équipement en Science Fondamentale*. Geneva: CERN Report CHS-19.

Pestre, D. (1988). Comment se prennent les décisions de trés gros équipements dans les laboratoires de 'science lourde' contemporains: Un récit suivi de commentaire. *Revue de synthèse*, **4th series**. 1, 97–130.

Rudwick, M.J.S. (1985). *The Great Devonian Controversy.* Chicago: University of Chicago Press.
Salomon, J.-J. (1971). The *Internationale* of science. *Science Studies,* 1, 23–42.

PART V

HALLMARKS OF EXPERIMENT

13

FROM KWAJALEIN TO ARMAGEDDON? TESTING AND THE SOCIAL CONSTRUCTION OF MISSILE ACCURACY

DONALD MACKENZIE

INTRODUCTION

Kwajalein Atoll in the Marshall Islands is the western end of the main American nuclear missile test range. One of the few journalists ever to have visited it describes it thus:

[Kwajalein] is the strangest looking South Sea island in the world, bristling with enormous radar dishes and radio antennae among the coconut palms . . . it houses a town of 3,000 expatriates . . . totally modelled on the American dream . . . [The Pentagon] cleared the inhabitants off nearly every island in the atoll and they are now crowded on to a tiny speck about two miles north of Kwajalein island. Ebeye island is just 67 acres in size and is home for the 8,000 people who were originally on the other islands or in other parts of the Marshalls chain . . . without a reliable water supply, without proper medical care, and even without sufficient room to bury their dead . . . No Marshallese are allowed to stay on the main island . . . workers are allowed to take back no more than five gallons [of water] to their homes on Ebeye . . .[1]

The Marshall Islanders cleared from Kwajalein suffer in order that facts be constructed. Some 30 times a year, a US Air Force Inter-Continental Ballistic Missile (ICBM) is fired from Vandenberg Air Force Base, on the Californian Coast north of Santa Barbara, down the 4500 mile trajectory to Kwajalein. Results from these minutely monitored tests form a key part in the determination of the characteristics of US missiles, particularly their accuracy. The outcome of this process is facts – facts that some say would hold for these missiles even if the circumstances of their use approximate to the biblical image of Armageddon:

And he gathered them together in a place called in the Hebrew tongue Armageddon . . . And there were voices, and thunders, and lightnings; and there was

1. Jackson (1984), p. 17.

a great earthquake, such as was not since men were upon the earth, so mighty an earthquake, *and* so great. And the great city was divided into three parts, and the cities of the nations fell ... And every island fled away, and the mountains were not found. And there fell upon men a great hail out of heaven ...[2]

Others vigorously contest this inference from Kwajalein to Armageddon, and deny the facticity of missile accuracy claims in particular. Similar debate has taken place over inferences from US intelligence observations of Soviet strategic missile tests, which are mainly over a range from Tyuratam near the Aral Sea east to the Kamchatka Peninsula or on into the Pacific Ocean.

Missile accuracy matters because it is understood to be a key factor in the capacity of nuclear missiles to destroy 'hard targets' such as opposing nuclear missile silos and underground command posts. The mathematics of the destruction of such targets is complex, but two statements based upon it would gain wide assent. The first is that the probability of destroying such a target is *very* dependent on missile accuracy, 'accuracy' appearing in the key conventional equation in squared form. If missile *A* is twice as accurate as missile *B*, then its 'counter-military potential' is four times as large.[3] The second consensually agreed statement is that the probability of destroying such a target is less sensitive to variations in the total explosive force, or yield, of the nuclear warhead used – yield appearing in the conventional equation only in its two-thirds power. Thus the yield of the warhead of missile *A* would have to be eight times as large as that of missile *B* to achieve the same increase in 'counter-military potential' as arises from a doubled accuracy. Beyond certain limits, accuracy is of little significance in nuclear attacks on people or industry, given the very great destructive power of nuclear weapons. But for the above reason, accuracy is central to questions of 'nuclear war fighting' – whether the US need fear a Soviet attack on its Minuteman missiles, or whether the US could *in extremis* hope to conduct a disabling first strike against the Soviet nuclear force.

As American nuclear strategy has over the past decade-and-a-half shifted decisively away from retaliatory, city-destroying notions of 'Assured Destruction' towards various versions of nuclear war-fighting, and as the Soviet missile force has become more formidable, the facticity of missile accuracy claims has moved to the centre of American defence politics. Previously an 'insider's' preoccupation, in the early 1980s the issue received wide coverage in the American press. One major aim of

2. *Revelation*, xvi, 16–21.   3. IISS (1985), p. 179.

this article is to outline the various positions adopted in debate, and to describe a previous, more 'fundamental' debate of the early 1960s.

My second aim is more theoretical. It is to show the relevance of recent sociology of science to the understanding of issues such as this. Sometimes this sociology of science has appeared rather introverted in its detailed preoccupation with scientific papers, scientists' talk and what goes on within laboratory walls. But this apparent introversion is, I hope, only a temporary phenomenon. For the key reason for focus on the minutiae of science is to grasp the process of the construction of scientific facts. And it is not only within those settings that we recognise as 'scientific' that facts are constructed.

A major locus of the construction of facts, I would argue, is the *testing* of technologies. Here is one important site where knowledge is produced, knowledge not of the natural world but of the world of human artifacts. Here it is decided whether artifacts 'work', how well they work, and what their characteristics are. The credibility of the knowledge generated is reasonably often of crucial significance, as the case of missile accuracy demonstrates. And all the issues that recent sociology of science has raised about *experiment* in science can be raised about *testing* in technology.

TESTING – PRODUCING FACTS ABOUT ARTIFACTS

Given the significance of testing, there are surprisingly few studies of technological testing at anything like the level of detail reached by studies of scientific experiments. Testing, perhaps has seemed a rather dull and peripheral topic by comparison with such themes as the effects of technology on society, the relationship between science and technology, and so on. But far from being peripheral, it is central. The part played by technical artifacts in social life is intimately connected to our knowledge of them. I use the word 'knowledge' here not to imply 'true belief', but simply belief that is taken as true within a social group; a 'fact' is a part of such knowledge that is taken to be an unproblematic description of 'the way things are'. What we know about artifacts (how 'safe' they are, what 'uses' they have, what their 'technical characteristics' are) will have a major influence on how widely we adopt them, what we do with them, how we design them.

Formal testing is, of course, only one aspect, and often a very minor one, of the process in which knowledge of artifacts is constructed. But that does not dislodge consideration of testing from analytic centrality. All previous use of an artifact could be considered as a process of testing it, and thus what follows is true where knowledge of artifacts is based upon use as well as upon testing. Furthermore, testing grows in significance the closer we approach to the generation of new technology. Indeed, that is where most formal testing is found, either as an aid to the design of a new artifact or as a way of determining its characteristics in advance of widespread use. In addition, testing becomes more important the more closely we approach the present, with the growth of legal or normative requirements for formal testing of efficacy, safety and of a wide variety of particular characteristics (for example, the petrol consumption of cars).

Experiment and Testing

The relevant analogy between scientific experiment and technological testing is as follows. Recent sociology of science, following sympathetic tendencies in the history and philosophy of science, has shown that no experiment, or set of experiments however large, can on its own compel resolution of a point of controversy, or acceptance of a particular fact. A sufficiently determined critic can always find reasons to dispute any alleged 'result'. If the point at issue is, say, the validity of a particular theoretical claim, those who wish to contest an experimental proof or disproof of the claim can always point to the multitude of auxiliary hypotheses (for example about the operation of instruments) involved in drawing deductions from a given theoretical statement to a particular experimental situation or situations. One of these auxiliary hypotheses may be faulty, critics can argue, rather than the theoretical claim apparently being tested.

Further, the validity of the experimental procedure itself can also be attacked in many ways. Successful experiment rests on a variety of procedures and competences. Each of these procedures can be challenged – was it correctly carried out? Is it in fact the correct procedure? And the experimenter's competences are equally open to challenge. Because scientific activity cannot be reduced to an algorithmic set of instructions that can be followed automatically, and always involves tacit skills, the only way of knowing whether an experiment has been competently performed is to know whether it produces the right result. But when the right result is controversial, then the competence of experimentation becomes inextricably bound up with beliefs as regards results. Those who believe in a controversial phenomenon will judge those experiments that fail to demonstrate it incompetently performed; those who disbelieve will attribute incompetence to those who claim to have found it. Collins summarises this consequence of the dependence of experiment on tacit skills as 'the experimenters' regress'.[4] Schaffer's paper in Chapter 2 of this volume shows a closely related phenomenon in disputes in seventeenth- and eighteenth-century optics. For Isaac Newton and his followers a properly working prism was one that demonstrated his claims. An experiment that failed to replicate his results was, *ipso facto*, inadequate.

All this, I would argue, is true of technological testing also. Any test, or set of tests, is always open to challenge, just as any experiment or set of experiments is. To simplify the discussion, let us concentrate on testing aimed at deciding the characteristics of an artifact – deciding the efficacy and safety of a drug, the petrol consumption of a car, the accuracy of a missile, the efficiency of a propeller,[5] the power of a turbine.[6]

How might critics challenge such a test or tests? In the most sophisticated testing, it will often be found that test procedures rest on the assumption of the validity of certain scientific theories. The artifact in question may, for example, be too big to test, and thus a scale model will be built and tested. This procedure is common with waterwheels, windmills, ship hulls, propellers and the like.[7] But 'scale effects' then obviously need to be taken into account, and that is a theoretical process, dependent on, say, understanding of fluid dynamics. Similarly much testing of mechanical devices is dependent on the assumption of the validity of Newtonian mechanics, or statistical theory will be employed (as is common in drug tests) in interpreting test results. So it is open to a critic to challenge these theoretical assumptions, either by asserting that the relevant theories are false or that they have been incorrectly applied to the situation in question.

Then there are the questions of instrumentation, procedures and competences. If a test uses an instrument or instruments – say a Prony brake to measure the motor output of a turbine[8] – then it is possible to question whether the instrument measures correctly what it is believed to measure (either in general, or for this particular instrument in this case), or whether the very presence of the measuring instrument alters the phenomenon being measured. Amongst the questions that can be raised about procedures, a particularly interesting set concern the preparation of the artifact for testing. Does it receive special attention? Does what is done to it invalidate test results? There will also be at least a category of tests which cannot be reduced to algorithmic rote. where competent testing depends on unverbalised knowledge. And so here too, questions of testers' competences become entangled with

4. See Collins (1985).   5. See Vincenti (1979).   6. See Constant (1983).
7. See Vincenti (1979).   8. Constant (1983).

questions of test results. If the skills involved are tacit, and the procedures not algorithmic, then the only way to evaluate the testers' competences is to see whether their tests are 'successful' and produce the 'correct results'. But if there is no agreement as to what constitutes successful testing, and what the correct results are, then it becomes impossible independently to evaluate competence. Here is a 'testers' regress' to parallel Collins's 'experimenters' regress'.

Further, there is the problem of inference from test performance to actual use – or, if we are not considering formal testing, the problem of inference from past use to future use. At one level, this is a version of the problem of induction. We can only test an artifact a finite number of times, ultimately because we have limited life-spans, but normally also because tests cost money and money is finite; similarly, any artifact, however common, has been used only a finite number of times. Can we be certain it will perform the next time as it has performed in the past?[9] Of course, there are probabilistic techniques here for expressing degrees of certainty based on amounts of past experience, but their straightforward application can be vitiated by more basic questions of the validity of inference from testing to use, or from past use to future use. Is the test situation, say, sufficiently like use to allow inferences to flow? Typically, there is a strain here between two contradictory impulses. The obvious answer to this potential line of criticism is to make testing as like use as possible. But testers typically also want clearcut tests, careful monitoring, as much information as possible gathered, tight control over potential disturbing factors, procedures as near algorithmic as possible – precisely those things that might make testing *unlike* actual use.

The removal of Modalities and similarity Judgments

Two general points about testing emphasise parallels between the situations in science and in technology. The first concerns what (socially) successful testing consists of. Its product, analogously to that of successful scientific work, is a credible statement of a certain form. In both, 'facts are constructed through operations designed to effect the dropping of modalities which qualify a given statement'.[10] In technological testing, typical modalities to be removed would be reference to the particular person or people doing the test, the particular geographical location

---

9. The original 1985 draft of this paper here contained the sentence, 'Can we be confident that no nuclear power station will experience a melt-down because none yet has?'
10. Latour & Woolgar (1979), p. 200.

of the test, the particular time of the test, the particular artifact or artifacts being tested. If one or more of these modalities cannot credibly be dropped, then what is produced will be less than a full fact. In other words, socially successful testing consists in moving from statements like 'on the 7th of March Ford Escort number 2345 was driven by Anne Smith along the B749 road for what its odometer registered as 35 miles on what Joe Bloggs said was a gallon of four-star petrol' to statements like 'the petrol consumption of the Ford Escort is 35 miles per gallon'. As we shall see, missile testing involves precisely this kind of move from one type of statement to another. The argument of the critics of missile accuracy claims is that one or more of the modalities cannot justifiably be dropped. Just as a critic of the Ford Escort test might insist that reference to the B749 road cannot be dropped (if, for example, it was a markedly downhill road!), critics of missile accuracy claims argue against the stage of fact construction that involves dropping reference to the Vandenberg-Kwajalein trajectory.

Second, judgments about *similarity* are as crucial in technological testing as they are in science.[11] One typical issue involved in judging the validity of test results is whether the particular artifact(s) tested are sufficiently *like* others of their class to allow inferences to flow (are tested artifacts 'gold-plated'?). Another is whether test situations are sufficiently *like* use to allow inference. There will *always* be ways in which test artifacts can be seen as differing from 'the real thing', or test situations from actual use, given sufficient ingenuity in seeing them. No amount of modification of test procedures can wholly remedy this. Hence debates about testing are *potentially* endless.

Interests, expectations, convention and credibility

Of course, debates about testing do (often) end. Indeed, they generally do not even begin. Just as the vast majority of reports of scientific experiments go wholly unchallenged, so the result of most testing is routinely accepted as fact. If the preceding argument is correct, neither 'reality' nor 'logic' are sufficient to explain this. Other elements must be involved, and in them 'the technical/scientific' is inextricably interwoven with 'the social'.

One basic question is whether there is a social interest that may be served by the challenge. In many cases there will simply be no significant group with an interest in defeating the knowledge claims made as a

11. See Barnes (1982a).

result of an experiment or test. As we shall see below, a challenge to missile test results more fundamental than that raised by the critics of accuracy claims is possible – but it is hard to identify a contemporary constituency for that challenge. Another issue is expectations. Most of the time, the relevant scientific and technological communities have consensual expectations as to the general nature of the results to be expected from a given experiment or test. A knowledge claim grossly at variance with these expectations can expect the most careful scrutiny and challenge, while one in line with expectations will not normally have its foundations probed.

A further aspect is convention. Here the discussion of testing by Constant is of particular relevance.[12] He notes the relationship between what he calls 'traditions of technological testability' and the technological communities that sustain them. That something is testable, and what it is to test it adequately, cannot simply be an individual matter, but will be defined by the consensual practices of groups of accredited practitioners. Often this will sharply curtail the possibility of successful challenge to test results. That a test has been conducted according to 'accepted practice' will under most circumstances be a weighty reply to any criticism. To go on to challenge 'accepted practice' will be to challenge the community of practitioners itself. Note, however, that accepted practice may be quite local. The 'community structure of technological practice' is 'differentiated', notes Constant, and 'particular techniques, technologies, and practices relating to testing ordinarily will be commensurate with, and to some degree will define, a specific community's normal technology'. Just as experiment can prove inadequate to decide between different scientific paradigms, so 'traditions of testability, and the techniques that express them, thus frequently become enmeshed in, rather than resolve, competition among systems'.[13]

More generally, the challenges that are practically (rather than abstractly) possible will be constrained by the beliefs of the relevant social groups. Take a process of quality control testing where someone chooses a selection of artifacts coming off an assembly line and subjects them to a test. An abstractly possible challenge to such a process would be that the mental act of selecting an artifact influences by psycho-kinetic means the characteristics of the artifact. Only in certain limited social circles would that challenge not be dismissed out of hand. Consider by comparison the challenge that those making the selection may

12. See Constant (1980), especially pp. 20–24, and Constant (1983).
13. Constant (1980), p. 22.

not be doing it randomly, but may be subconsciously influenced by some aspect of the artifact that is correlated with the characteristic being tested. That challenge might well possess much greater credibility.

The question of *credibility* is central, and the issues discussed so far (interests, expectations and convention) could all be considered simply as aspects of it. For it is above all credibility that separates the challenges that are practically feasible from those that are abstractly possible. Science, and to a large extent technology too, is about creating *hard* facts – precisely, facts that are in practice resistant to challenge and change. Scientific and technological facts are not free-floating knowledge claims; they are tied to older, even harder facts. Imagine what it would be like in practice to dispute a technological test by disputing Newton's laws. Only if one could tie the challenge to an equally established area of physics ('relativity, or quantum theory, is what is relevant here, and it really makes a difference') would one have a chance of success in scientific circles. And the links are not just to knowledge, which is in one sense cheap, but to knowledge reified in instruments, which is very expensive. As the author of the best study of this puts it:

The layman is awed by the laboratory set-up, and rightly so. There are not many places under the sun where so many and such hard resources are gathered in so great numbers, sedimented in so many layers, capitalised on such a large scale . . . confronted by laboratories we are simply and literally impressed. We are left without power, that is, without resource to contest, to reopen the black boxes, to generate new objects, to dispute the spokesman's authority.[14]

A great virtue of the sociology of scientific knowledge is thus precisely that it reminds us that this hardness of fact is a creation, not Nature's direct effect. But this literature misleads if it implies that in practice belief is a matter of *will*. Sometimes the sociological literature about scientific knowledge uses voluntaristic language like 'a collective *decision* to stop arguing' or 'the parties to the debate *chose* to confine their discourse to a culturally-available conceptual space'.[15] This language does indeed accurately capture the crucial sense that things could have turned out otherwise. But it is far from clear that the phenomenology, as it were, of such events is of the exercise of free will. Belief is not a voluntary matter: I cannot by an act of will cause myself to believe a psycho-kinetic hypothesis like that suggested above, and I suspect that many readers, if they attempt the thought-experiment of trying to

14. Latour (1987), p. 93.
15. The quotations are from Pickering (1981, pp. 64–5), with emphasis added in the first. Note, though, that elsewhere Pickering stresses the constraints under which scientists operate. See Pickering (1985) and Chapter 9 of this volume.

make themselves believe it, will likewise fail. And the sanctions risked by those who do *not* 'confine their discourse to a culturally-available conceptual space' are truly awesome – ultimately, the diagnosis of madness. Convention, it is worth recalling, is not '*mere* convention'.[16] It includes our deepest convictions and most strongly held principles. To those involved, 'the conventional' will under most circumstances seem like a description of the ways things are and the ways things have to be – matters to be challenged only for very good reason and often at extreme peril.

TESTING STRATEGIC MISSILES

To understand the dispute that is my central focus here, it is first necessary to discuss the process of missile testing in more detail. What I shall be describing is the process of operational testing of US inter-continental ballistic missiles. Though the distinction is not hard-and-fast, operational tests are those where the purpose is to help determine the characteristics of a missile system, once it has been designed, as distinct from research and development tests, where the goal is information to help in the process of modifying the original design. The test procedure for submarine-launched ballistic missiles is different from that for inter-continental ballistic missiles, but since the accuracy of the latter, not of the former, is the key issue at stake here, I shall concentrate on the testing of the latter.

Although in the past American inter-continental ballistic missiles have been fired from Cape Canaveral in Florida, current testing is done from Vandenberg Air Force Base in California. Nearly all firings are towards Kwajalein Atoll in the Marshall Islands. The Vandenberg – Kwajalein test range is heavily instrumented, and therefore much more information can be gained from tests on it, though on occasion missiles are fired towards other targets. The tenth MX test, on 13 November, 1985, for example, was targeted beyond Kwajalein to a broad ocean area 400 miles northeast of Guam,[17] presumably to verify MX's performance at the greater range.

Though early test flights are normally above-ground 'pad launches', Vandenberg is equipped for silo-firings. In the operational testing of missiles that are already deployed, a missile is selected at random from those in the missile fields. If it can successfully be brought to alert status in its operational silo, its re-entry vehicles are removed and transported to Texas, where the live nuclear warhead is taken out and replaced by

16. See Bloor (1976), pp. 37–9.   17. Anon. (1985).

telemetry equipment. The purpose of the latter is to monitor and transmit information about the workings of the missile during the test flight. The missile itself is transported from the missile field to the test silo at Vandenberg, which is operated by 'randomly selected crews from the operational missile fields'. The re-entry vehicles are replaced, 'the missile is "wired" so that it can be destroyed should it go awry', and the guidance system is aligned and calibrated. The missile is brought to alert, and if that can be done successfully it is fired at targets, which are normally points within the large enclosed lagoon at Kwajalein. Ground-based radars, telescopes and instruments within the re-entry vehicles track the test, and a 'splash-net' of hydrophones is used to determine where the re-entry vehicle impacts on the surface of the lagoon.[18]

Statistical analysis of these impacts provides a 'circular error probable', officially defined as the radius of the circle round the target within which 50% of warheads will fall in repeated firing. Since flight testing will involve trajectories with different geometries and to some extent different ranges, in fact a set of circular error probable figures will be produced. And since the number of tests is finite, figures for the circular error probable will have a statistical uncertainty – they will have a 'confidence interval' associated with them.[19]

A further process is needed, however, in order to construct circular error probable figures for wartime firings. This process has never, to my knowledge, been fully described in published sources.[20] A *theoretical model* of error processes has to be constructed out of the results of test firings, laboratory tests of components, estimates of the errors in mapping and in models of the gravitational field, etc. Such a model is necessary because errors with different causes are understood to extrapolate differently. Some, for example, will vary more or less linearly with range to the target. Others will not. A degree of 'boot-strapping' is also involved, because increasingly the theoretical model of error processes is also used by the missile's on-board computer to correct for errors that are believed to be predictable. Because of this centrality of the theoretical understanding of errors, much more is expected from the test range than just a final, total miss distance. Attempts are made to develop and verify the theoretical model by, for example, comparing externally-derived estimates of missile position and velocity with what the guidance system reports, through telemetry, it 'believes' these to be.

18. The description in this paragraph is drawn from Bunn & Tsipis (1983), quotes on pp. 132–3.
19. See Moore (1960).
20. Though see Moore (1960), Drucker (1962) and Slay (1978).

American knowledge of Soviet missile tests

The centrality of the theoretical understanding of errors is even greater in the construction of American knowledge of Soviet missile accuracies, for here a crucial datum is obviously under normal circumstances missing: certain knowledge of the exact location of the intended test target. 'They don't paint a bullseye on Kamchatka', said more than one interviewee, referring to the eastern end of the main Soviet strategic missile test range. Nevertheless, a considerable amount of other information is available to American intelligence analysts.

Soviet missile tests are closely observed by a variety of means.[21] Firings are monitored by land-, sea-, air- and space-based sensors, radars and telemetry receivers. Before the Iranian revolution, Iran was a key tracking site, but also important are the US base at Diyarbakir in Turkey and, interestingly, a covert telemetry interception facility in China.[22] Geosynchronous 'Rhyolite' satellites may have had the capacity to intercept telemetry, while current generation 'Chalet' satellites almost certainly do; low-altitude 'ferret' satellites are also, apparently, used to do this. Radars at Shemya Island in the Aleutians, at Kwajalein and also on a naval vessel, the USS Observation Island, track the latter stages of Soviet missile tests. For those tests impacting on land, 'Key Hole' intelligence satellites have the capacity to observe the impact holes, although Kamchatka's persistent cloud cover limits the feasibility of doing this immediately.

Impact patterns (whether deduced directly from satellite observation, or indirectly inferred from tracking trajectories) are of *some* use in estimating Soviet missile accuracies. Though the methods used are not publicly revealed,[23] it is, for example, clear that if impact points in repeated tests of the same system form a cluster, then the size of the cluster could be taken as providing some indication of an upper bound on circular error. Again, though, observation of testing provides intelligence analysts with much richer data than this alone. Although its encryption has in recent years been an increasing problem, telemetry is crucial. Intercepted telemetry from the missile's on-board computer can reveal a great deal about the workings of the guidance system. Telemetry from components in the system is used in order to attempt to deduce the nature of these components (and sometimes even their

21. There is a useful, though now dated, description in Moncrief (1979).
22. Arkin & Fieldhouse (1985), p. 73.
23. See, for example, the security deletions in testimony to the House of Representatives Armed Services Committee (1979), p. 99.

performance). Tracking can indicate how rapidly re-entry vehicles travel, an important aspect of accuracy since fast, streamlined re-entry vehicles are understood to be more accurate than slower ones. The deduction that the Soviets had begun to use such re-entry vehicles was an important aspect of the growing American concern in the 1970s about Soviet missile accuracies.[24]

There is no straightforward and simple way of turning all this data into an actual accuracy figure. It is again a matter of constructing, and drawing conclusions from, a model of error processes. It is therefore not surprising that there has been considerable disagreement about accuracy estimates. In 1985, for example, the Defense Intelligence Agency was estimating the circular error probable of the Soviet SS-19 missile to be 325 yards, while the Central Intelligence Agency's estimate was 435 yards[25] – a difference of some significance, given that in calculating the threat posed by such a missile to an American hard target the accuracy estimate appears in squared form.

Challenging testing

The most basic challenge to the results of testing nuclear missiles had, however, nothing directly to do with the finer points of defining accuracies. It arose from the normal procedure of testing missiles without nuclear warheads on board. Warheads were themselves tested, but separately – either in fixed locations above or below ground, or by being dropped from aircraft or towers. It could be argued that the circumstances of ballistic missile flight were not the same as that form of testing. Rocket boost into space and re-entry to the atmosphere involve much vibration and large variations in termperature and pressure. Could one be sure that the missile's warhead would explode following this? At its most extreme, might all nuclear missiles turn out to be 'duds' in operational use?

In the early 1960s this challenge was raised in the United States. *No* American missile test up to that point had carried a live nuclear warhead. The pressure was significant enough to lead to one live test of a Polaris missile on 6 May, 1962. Its warhead did, reportedly, explode, but that did not still all doubts – it was argued that the missile had been modified for the test. And by 1963 the Partial Test-Ban Treaty, prohibiting nuclear explosions in the atmosphere, ruled out an attempt

24. See Smith (1982).   25. Keller (1985).

to replicate the test; there has been only the one live firing of an American missile.[26]

Interestingly, though, the abstract possibilities for challenge that this creates do not seem to have been taken up since around 1964. The issue is occasionally alluded to, for example, in a letter in the *Bulletin of the Atomic Scientists*:

Does a test conducted dead still at the bottom of a hole provide reliable assurance that the same weapon will work after travelling several thousand miles an hour in a reentry vehicle through extremes of temperature?[27]

Nevertheless it has not in recent years been raised as a major challenge to test-based knowledge of nuclear missiles.

The nature and fate of this challenge exemplifies many of the issues about testing discussed above. A 'similarity judgment' was central to it; was the difference between stationary testing of a nuclear weapon and the circumstances of ballistic missile flight significant enough to cast doubt over whether nuclear missiles would 'work'? A 'crucial experiment' designed to resolve the issue did not do so: the adequacy of the experiment was challenged. A social interest existed that provided a constituency for the challenge. Those opposed to the decisive shift then underway from reliance on the manned bomber to reliance on the missile were able to make use of the argument that the missile was in a profound sense an untested (and after 1963 an untestable) weapon system.[28] And it is difficult to attribute the decline of the challenge after 1964 to 'logic' or changes in the 'evidence'. There is a defensible-case to be made that, abstractly considered, matters remain unchanged since 1964.

Beliefs *have* shifted, of course, since 1964. The ballistic missile is no longer novel but an accustomed feature of the strategic world. The credibility of the fundamental challenge to it has declined. From being definitely seen as 'worth testing' in the early 1960s, the challenge if reactivated now would probably seem implausible, certainly in the strongest form that maybe *no* nuclear ballistic missile might work. Paradoxically, it may be that the political impossibility of replicating the one live firing test – because of the entrenchment of the Partial Test-Ban Treaty – has contributed to the decline of the challenge's credibility, even while it has maintained its 'abstract' status unaltered.

26. See MacKenzie (1987) for more details. According to Hussain (1981) the Soviet Union also conducted some live tests before the signing of the Partial Test-Ban Treaty; he does not report whether these were 'successes'. Fieldhouse (1986, p. 104) reports two live missile tests by China in 1966 and 1976.
27. Meyers (1986), p. 67.
28. See MacKenzie (1987).

Harvey noted that in a case he studied the 'plausibility' of a hypothesis was raised by the very existence of moves to test it. Here, perhaps, the plausibility of the hypothesis has declined because testing it has seemed impossible.[29]

What has replaced it is the less fundamental, but nevertheless highly significant form of doubt concerning the facticity of missile accuracies: are the accuracies deduced from performance over test-ranges such as Vandenberg – Kwajalein reliable guides to what would be achieved in operational use of nuclear missiles?

This more complex episode further illustrates the analysis of testing outlined above.[30] The form of the dispute is straightforwardly one in which the dropping of modalities is contested. A host of specifics about firing over particular test ranges have been suggested as arguments against the move from statements of the form 'this set of MX warheads fired from Vandenberg to Kwajalein landed on average within 100 yards of their targets' to statements of the form 'the circular error probable of MX is 100 yards'. Also, similarity judgments are central to those specifics. Is firing west towards Kwajalein sufficiently like firing north over the Pole towards the Soviet Union? Might there be significant, unknown anomalies in the Earth's gravitational or magnetic fields? Is it significant that the weather over the Soviet Union is less predictable and typically poorer than the weather over Kwajalein?

Collins's 'experimenters' regress' appears in this dispute, though again it appears in a form inverse to that normal in science. The issue is not whether the Vandenberg – Kwajalein experiments are performed incompetently but whether they are performed too well. It is asserted by some (and denied by others) that missiles to be tested receive 'special treatment' to ensure test success. It is claimed that the learning which is possible in repeated firings over the same range means that results cannot be extrapolated to 'once-off' firings towards the Soviet Union. Thus one person involved in missile testing told me of a surveying error at Vandenberg Air Force Base that was discovered because of its small but persistent effects on accuracy, but which might well not have been discovered in an operational silo from which no missile was ever fired.

29. Harvey (1981). I owe this interesting speculation to Harry Collins.
30. Aside from the arguments put forward to me in interviews, I am drawing here on the following published statements of the critics' case: Anderson (1970, 1981, 1982); Schlesinger (1974); Cockburn & Cockburn (1980); Fallows (1981, pp. 139–70); Marshall (1981); Metcalf (1981); Tsipis (1981); Kaplan (1983, pp. 374–6); Powers (1983, pp. 88–97). Published rebuttals by defenders of the facticity of missile accuracy claims are fewer, but see Draper (1981) and Marsh (1982). Arguments from both sides can be found in the testimony to the Townes Panel on MX basing reported in Mann (1981).

Missile accuracy figures have also been shown to depend on 'auxiliary hypotheses'. As described above, extrapolation from test to operational circular errors is a theoretical as well as an empirical process. How much error is attributed to guidance system gyroscope drift, to uncertainties in knowledge of the Earth's gravitational field, etc., will affect the final estimate of operational circular error probable. Defenders of the facticity of missile accuracy claims regard the auxiliary hypotheses involved as well established. Our knowledge of the gravitational field is quite good enough, writes General Marsh, and the argument that one cannot predict accuracy over an untried trajectory 'defies all the demonstrated rigor of physical laws from Newton to Einstein'.[31]

Note, though, that only a very limited sub-set of auxiliary hypotheses has actually been challenged. Abstractly, it would, for example, be possible for critics to challenge 'physical laws' themselves. But they have not done so. Thus Anderson suggested a possible effect from 'magnetic drift due to unknown electronic charge on the re-entry vehicle'. Subsequent use of electromagnetic theory led to the conclusion that for such an effect to be significant the charge on each re-entry vehicle would have to be implausibly large.[32] Critics *could* have responded to this by challenging the validity of electromagnetic theory, but did not do so.

Thus much of the dispute has turned not on physics but on engineering, in precisely a form that emphasises judgments of similarity. As described above, the process of operational testing involves modifying the missile before it is test-fired. Defenders claim the modifications to be inconsequential, while critics disagree. One retired general told the *Washington Star*: 'About the only thing that's the same [between the tested missile and the Minuteman in the silo] is the tail number; and that's been polished'.[33] What is involved here is not merely disagreement about what modifications are in fact made (though such disagreement does exist) but about their significance – an issue where experience and tacit knowledge seem to play a key role. It can, for example, be argued that 'special' maintenance of the missile guidance system prior to test launch (if it occurs – which is itself a point of dispute) is necessary to make up for probable deterioration caused by the process of transporting the missile from its operational silo to Vandenberg.

There is at least some differentiation of the 'community structure of technological practice' evident in judgments concerning missile testing.

31. Marsh (1982), p. 37.
32. Anderson (1981), p. 10; Bunn & Tsipis (1983), p. 47.
33. Hadley (1979), p. 2.

Though the Air Force conducts a small number of tests in directions different from that from Vandenberg to Kwajalein, defenders of the facticity of ICBM accuracies dismiss the need to test over trajectories paralleling war-time firing.[34] The feeling in the Navy, on the other hand, appears to be different. Submarine-launched ballistic missile (SLBMs) are tested over a much wider variety of trajectories:

SLBM modernization engineering development can confidently proceed to provide ICBM accuracies with the mobile and survivable SLBM platform. Future tests of such an SLBM system would continue to benefit from the fact that the submarine's broad based ocean area launches enjoy the ability to test over diverse launch latitudes, trajectory ranges and flight azimuths. This provides what is the best proof for national confidence that the submarine launched ballistic missile operational test performance truly represents performance under wartime conditions.[35]

Defining 'accuracy'

One aspect of the challenge to missile testing has been a deconstruction of the meaning of 'accuracy'. The measure of accuracy, the circular error probable, is as stated above conventionally defined as the radius of the circle round the target within which 50% of warheads will fall in repeated firing. What the critics noted was that many of the potential error sources they identified would not simply cause an increase in the radius of the circle but would cause the mean point of impact to be offset from the target. Systematic *bias* would be present as well as random error (Fig. 13.1).

Defenders of the facticity of accuracies responded by arguing that they were aware of the possibility of biases, that biases were small and well understood, and that the circular error probable figures they produced were indeed errors around the target, and not around the mean point of impact. Nevertheless, the prominence given by the critics to 'bias' highlighted their claim that the accuracy of a missile was not a simple 'naturalistic' fact about it but an artifact of the process of testing. Thus the key *New York Review of Books* article by Cockburn & Cockburn introduced the matter

The propositions and conclusions that follow . . . deal with a problem that has led a semi-secret existence for a decade. The shorthand phrase often used to describe this problem is 'the bias factor'. Bias is a term used to describe the distance between the center of a 'scatter' of missiles and the center of the intended target. This distance, the most important component in the computa-

34. See Marsh (1982), p. 4.   35. Topping (1981), p. 7.

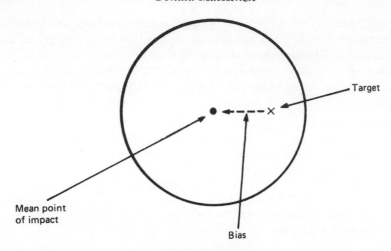

Figure 13.1. Bias

tion of missile accuracy, accounts for the fact that the predictions of missile accuracy cited above ['that a missile fired 6,000 miles can land within 600 feet of a target no more than fifty yards in diameter'] are impossible to achieve with any certainty, hence the premises behind 'vulnerability', the MX, and Presidential Directive 59 [Carter's 'hawkish' reorientation of American nuclear strategy] are expensively and dangerously misleading.[36]

Bias, their argument then went, could be corrected in repeated firings over a given test-range; but

the laborious adjustments needed to reduce bias . . . will be difficult to achieve over the novel northern trajectory, since no previous test data are available. The first shot is presumptively the only shot.[37]

Whether the circular error probable measured likely distance from the target or merely likely distance from the mean point of impact was, on one level, an arbitrary issue of definition. But in the context of the controversy this statistical issue took on deep symbolic meaning. No wonder then that an author such as Fallows could find no consensus:

Several men who have devoted large portions of their working lives to the details of missile technology patiently explained to me, over the course of many months, the concept of 'bias' I set out in the next few pages. Several other people, of comparable intelligence and integrity, who have been equally attentive to the details of nuclear policy, assured me that the first set of experts was wrong, and that a missile's 'accuracy' . . . really does measure how close its warheads would come to its intended target.[38]

36. Cockburn & Cockburn (1980), p. 40.   37. Ibid., p. 42.   38. Fallows (1981), p. 150.

Related issues of bias obviously also entered into American estimates of Soviet missile accuracies. To the extent that it was assumed that those estimates were derived simply from observation of where Soviet re-entry vehicles impacted (not a correct assumption), then possible Soviet deception was not the only issue. The best an American analyst could hope to do was to measure dispersion about the mean point of impact, since the target would be unknown. But 'bias', then, would effectively be artificially set to zero by this procedure.

Constructing implications

The early, basic challenge to the results of testing – that it did not prove that missiles might not turn into unexploding duds in operational use – had a fairly simple structure to it. The dominant strand of the challenge was represented by proponents of the manned bomber. Even here, however, things were complicated by adoption of the challenge by Senator Barry Goldwater when, as part of his 1964 campaign against incumbent President Lyndon Johnson, he charged the Democratic Administration with building America's defence upon a 'Maginot line' of untestable missiles.[39] The more recent challenge – that missile accuracy figures are not facts – is far more complex. No single social interest appears to underly the challenge, and indeed it has found a variety of different audiences. Nor has there been consensus as to what the implications of the challenge would be, were it to be accepted. It is thus an interesting demonstration of the plasticity, or constructed nature, of implication.[40]

At least four threads can be detected in the more recent challenge. The first is a continuation into the 1980s of the above 'hawkish' criticism of the bomber lobby and Goldwater. Here Arthur Metcalf, the iconoclastic publisher of *Strategic Review,* has played a central role, criticising over-reliance on missiles and what he sees as exaggerated worries about missile vulnerability.[41] This thread of the criticism has found a sympathetic ear in some leading Air Force circles,[42] and Senator Goldwater has also entered the fray again, telling a 1985 interviewer that he had 'never been convinced of the accuracy' of either American or Soviet intercontinental ballistic missiles: 'I have seen nothing except computer language that will testify to the accuracy'.[43]

But at the high-point of the public prominence of the debate in the early 1980s this 'hawkish' thread was less apparent than a 'liberal' thread. For this audience the implication was not Metcalf's conclusion

39. See MacKenzie (1987).   40. See Barnes (1982b).   41. Metcalf (1981).
42. See Marshall (1981).   43. *Aviation Week* (1985).

that MX could safely be put into Minuteman silos and the money saved spent on advanced bombers. Here the implication was that the MX programme was unnecessary and should be cancelled. J. Edward Anderson, one of the most prominent critics, made clear his overall position when, in what is to my knowledge the first clear statement of the issue in the public domain, he wrote: 'I agree fully that it is of the utmost urgency that the arms race be stopped and have made my own contribution by leaving the field of missile-guidance engineering for enginering work in non-military fields.'[44] The numerous press reports of the issue in the early 1980s typically identified the challenge to the facticity of missile accuracies with criticism of MX, to such an extent that the different position adopted by Metcalf became hard to distinguish.

The third and fourth threads predate the public controversy of the early 1980s. The third thread has emerged from the RAND Corporation, centrally from a RAND analyst with considerable guidance system experience, Hyman Shulman.[45] Bias was the issue that most exercised Shulman, who had much experience of data from Minuteman test-firings.[46] He was also concerned at the small sample sizes on which many conclusions about accuracy were drawn. Shulman's influence was important in leading former RAND Director of Strategic Studies, James Schlesinger, to assert, while Secretary of Defense, that 'we can never know what degrees of accuracy would be achieved in the real world [as against test firing]'.[47]

However to the extent that this version of the challenge was seen by its proponents as leading to definitive conclusions, these were quite different from those of the first two threads. Despite the radicalism of Schlesinger's 'we can never know' statement, considerable effort was

---

44. Anderson (1970), p. 6.

45. Shulman's role is noted by Kaplan (1983, pp. 374–6), although his name is misspelt and his position not described wholly accurately.

46. A. Ernest Fitzgerald, who as Deputy for Management Systems in the Air Force was a vigorous critic of what he saw as the mismanagement of the Minuteman programme, writes:

CEP was demonstrated graphically by plotting test warhead impact points on a target of concentric circles about the central aim point. The graphic displays looked just like rifle targets used in shooting matches, and a tight grouping of shots near the bullseye was the objective. In one of the briefings I received in the Minuteman control room at Norton Air Force Base, I noticed that the Minuteman II test shots showed a tight grouping all right, but that missed the bullseye by a startling margin. (Fitzgerald 1972, p.123).

47. Schlesinger (1974), p. 17.

put into improving the Minuteman guidance software. More generally, though, this line of argument suggested a shift away from reliance on accuracy in counterforce attacks to reliance on large-yield weapons, 'big bombs'. True, increasing accuracy was a more 'efficient' way of increasing counterforce capability than increasing yield; but the converse of this was that unexpected operational degradations in accuracy were that much more serious than degradations in yield. Shulman, too, felt that the argument indicated the wisdom of supplementing inertial guidance with radio inputs, a proposal that cut against the dominant trend of thinking that emphasised the virtues of the self-contained nature of inertial guidance.

An earlier, much less sophisticated version of the aspect of this third thread that emphasised the virtues of large yields was also to be found in Air Force circles in the 1960s. Those circles, 'including the top leadership in the Air Force', were deeply suspicious of reliance on small-yield weapons, even though the emerging technology of MIRVs (multiple independently-targetable re-entry vehicles) meant that a given missile force could carry many more warheads if they were small-yield MIRVs:

They and other non-technical officers were reluctant to rely on a complicated gadget to guarantee target kill. Technically-oriented personnel with more faith in the capability of inertial guidance systems were willing to accept the existence of a yield/accuracy trade-off and were not opposed to small yields in principle.[48]

The fourth thread concerns the analysis of the accuracy of Soviet missiles. That analysis is one of the most important and highly charged aspects of the formation by the American intelligence community of 'the Soviet estimates'.[49] In the mid-1970s the Central Intelligence Agency came under increasing criticism for alleged 'liberalism' and understating of the Soviet threat. The right-wing 'B team' formed in 1976 to contest the CIA's analyses naturally turned to questions of missile accuracy. Uncertainty in missile accuracy figures became in the hands of the B team grounds for assuming the worst about Soviet capabilities:

B-team members argued from the theoretical point that the accuracy of the Soviet missiles was essentially unknowable and that there was no evidence to support CIA's position that Soviet ICBMs were less accurate than U.S. ones. Indeed the B-team held it was possible that Russian missiles could be even more accurate than our own, which, if true, suggested that the U.S. land-based Minuteman missiles would become vulnerable in the near future.[50]

48. Greenwood (1975), p. 38.   49. See Freedman (1977) and Prados (1986).
50. Prados (1986), pp. 251–2.

CONCLUSION

Testing is crucial. But the knowledge that testing generates, like that generated by experiment, is always potentially problematic. The auxiliary hypotheses involved in testing, the competence of the testers, the judgments of similiarity involved in making inferences from testing to use, the removal of modalities involved in the construction of test-based facts – in principle all these are always open to challenge. Matters of social interests, expectations and convention – issues of *credibility* – enter into what challenges are actually made and what their outcome is, and thus enter into the construction of knowledge from testing. Again, the analogy with experiment is strong.

To this extent my argument is wholly general. At the most abstract level, categorical distinctions between 'testable' and 'untestable' technologies are naive. With sufficient ingenuity, *any* technology can be made out to have been inadequately tested. Conversely, any set of tests can be defended as adequate. Emphatically, though, this does not mean that testing is irrelevant to our knowledge of technology, any more than experiment is irrelevant to our knowledge of the natural world. *Because* the construction of knowledge from testing is a social matter, because interests, accepted practice and the like are involved in it, our abstract freedom to fashion knowledge as we will does not translate into the practical possibility of always being able to do so.

Nowhere does this matter more than in nuclear weaponry. There are two sides of a tension to grasp here. Certainly we cannot expect any wholly compelling resolution to a dispute as to whether a particular nuclear weapons technology has or has not been adequately tested. Nevertheless, the extent of testing that is possible is a vitally important determinant of the extent of confidence in that technology. Beliefs about what constitutes adequate testing are conventional, but that does not make them any less significant. If the relevant decision-makers cannot be persuaded that by their standards of adequate testing a novel technology has been properly tested, then the likelihood of their adopting the technology, or using it, is diminished.

The political significance of this has to do with attempts to restrain the development of military technology by negotiated constraints upon testing. Currently, the Partial and Threshold Test-Ban Treaties constrain the superpowers to underground nuclear tests only, and to tests with a yield no greater than 150 kilotons. The SALT II Treaty also constrained ballistic missile flight testing by each superpower, for example prohibiting flight-tests involving more 'procedures for releasing or for dispersing' re-entry vehicles than the number of re-entry vehicles the missile of that type was permitted to carry.[51] Considerable interest still exists in extending these constraints, for example in a Comprehensive Test-Ban Treaty banning all nuclear tests beyond a certain small threshold of reliable detection.

Such bans *could*, I would argue, have important effects on the development of nuclear weapons technology. Indeed, one aspect of the general significance of testing in this area is that testing is an avenue through which social and political interests and constraints can have a direct effect on the development of technology.

An example is precision-guided re-entry vehicles. One way of increasing missile accuracy, and of circumventing many of the issues raised in the controversy discussed above, would be to equip the missile's re-entry vehicle(s) with a capacity to manoeuvre and a form of guidance system. Instead of coming back through the atmosphere on an unguided trajectory, as all currently deployed strategic ballistic missile warheads do, such a system might take a 'fix' on some aspect of the terrain around its target and use that fix to correct errors that had accumulated in its previous flight.

Though technically demanding, such a system does not seem to those involved to be impossible. A major perceived constraint on its development, however, in the United States (which, because of its advantages in electronic miniaturisation, would likely lead in any such development) is the political and legal difficulty of testing such a system in a way that those involved would count as adequate. Ballistic missile flight-testing is widely seen as involving some level of risk to those living below the flight-path, as, for example, when burnt-out rocket stages crash to Earth. In large part because of this, nearly all American ballistic flight-testing is conducted over the ocean.[52] One exception is the fairly short-range flight-tests conducted from Green River, Utah, to the White Sands Missile Range in New Mexico, but there, I was told, local residents are sent a cheque through the post before each flight-test which they can use to move into a motel for the day.

There is a patent difficulty, however, in testing a 'homing' re-entry vehicle over the ocean, because of the absence of terrain features for

51. The texts of these treaties will be found in Blacker & Duffy (1984).
52. Even for cruise missile flight-testing, where there are no discarded rocket stages, the United States has been forced to move to Canada's sparsely populated north in order to conduct 'adequate' flight-tests, a move that has provoked considerable political protest.

its guidance system to recognise; and it could also be argued that the short-range flight-testing possible at White Sands is not adequate either. The difficulty of testing such a technology, which is a 'political' as much as a 'technical' difficulty, seems to have counted against it. A homing re-entry vehicle was considered by the US Navy's Improved Accuracy Program in the late 1970s. Admiral Wertheim, Director of the Strategic Systems Program Office, told the Senate Armed Services Committee that 'though it has a number of advantages potential in it . . . its disadvantages, as we see, lie in the area of questions about the testability, our ability to conduct flight tests over land'. The previous year Captain Clark of the same office had testified that 'there are severe problems. One must test over land . . .'[53]

So an (informal) political constraint upon testing was here counting as an argument against the development of an accuracy-enhancing technology. Consciously constructed political constraints, in the form of carefully negotiated test ban treaties, seem likely to be all the more efficacious.

Nevertheless, the above analysis does suggest that we should not see in them a simple 'technical fix' for the arms race. That standards of adequate testing are conventional does not mean, I have argued, that they would simply be discarded and the military would gaily go on to deploy 'untested' systems. But there would certainly be pressure, under circumstances where testing of a particular type had been prohibited but where there were still eager proponents of the novel technology, to revise those conventions, for example to argue that computer simulation was an adequate form of testing.

It would therefore be naive to assume that conventions of adequate testing would remain unchanged in a situation where the political constraints on testing increased. Pressure for and conflict over nuclear weapons development would certainly not come to an end with a Comprehensive Test-Ban Treaty. Such a Treaty remains a worthwile goal, but it has to be realised that such a Treaty would simply inject one new element – a powerful one, to be sure – into the complex ferment of interests, expectations, conventions and credibility from which our knowledge of nuclear weaponry emerges. A sociology of technological testing has something to contribute to understanding that ferment. The analogy between technological testing and scientific experiment is an important resource for that sociology.

53. Clark (1977), p. 6544; Wertheim (1978), p. 6684.

ACKNOWLEDGEMENTS

I would like to thank the Nuffield Foundation, whose two grants to me on 'The Development of Strategic Missile Guidance Technology' made possible the research reported here. I would also like to thank the various participants in the controversy over missile accuracy, and other guidance engineers and intelligence analysts, who permitted me to interview them during research trips to the United States in 1984–87. Since in this paper I do not cite interviewees by name, I have tried wherever possible to support the claims made by reference to published articles, rather than to anonymous interview material. Grateful thanks too to Matt Bunn and to the interviewees who provided me with much essential documentation.

References

Anderson, J.E. (1970). [Letter to editor]. *Scientific American*, **222**, 6.
Anderson, J.E. (1981). First strike: myth or reality. *Bulletin of the Atomic Scientists*, **37**, 6–11.
Anderson, J.E. (1982). Strategic missiles debated: missile vulnerability–what you can't know! *Strategic Review*, **10**, 38–42.
Anon. (1985). MX missile launch. *Aviation Week and Space Technology*, **18** November, 25.
Arkin, W.M. & Fieldhouse, Richard W. (1985). *Nuclear Battlefields: Global Links in the Arms Race*. Cambridge, Mass.: Ballinger.
Aviation Week (1985). Interview with Goldwater. *Aviation Week and Space Technology*, **25** February, 111–12.
Barnes, B. (1982a). *T.S. Kuhn and Social Science*, London: Macmillan.
Barnes, B. (1982b). On the implications of a body of knowledge. *Knowledge: Creation, Diffusion, Utilization*, **4**, 95–110.
Blacker, C.D. & Duffy, G. ed. (1984). *International Arms Control: Issues and Agreements*. 2nd edn. Stanford: Stanford University Press.
Bloor, D. (1976). *Knowledge and Social Imagery*. London: Routledge.
Bunn, M. & Tsipis, K. (1983). *Ballistic Missile Guidance and Technical Uncertainties of Countersilo Attacks*. Report no. 9, MIT Program in Science and Technology for International Security. Cambridge, Mass.: MIT.
Clark, G. (1977). Testimony to the Armed Services Committee, United States Senate. *Department of Defense Authorization, Fiscal Year 1978*. pp. 6542–7. Washington, DC: US Government Printing Office.
Cockburn, A. & Cockburn, A. (1980). The myth of missile accuracy. *The New York Review of Books*, **20** Nov., 40–2.

Collins, H. (ed.) (1981). Knowledge and controversy. *Social Studies of Science*, 11, 158p.

Collins, H. (1985). *Changing Order: Replication and Induction in Scientific Practice*. London: Sage.

Constant, E.W. (1980). *The Origins of the Turbojet Revolution*. Baltimore: Johns Hopkins University Press.

Constant, E.W. (1983). Scientific theory and technological testability: science, dynamometers, and water turbines in the 19th Century. *Technology and Culture*, 24, 183–98.

Draper, C.S. (1981). Imaginary problems of ICBM. *New York Times*, 20 Sept., section 4, 20.

Drucker, A.N. (1962). Performance analysis of rocket vehicle guidance systems, In *Inertial Guidance*, ed. G.R. Pitman, pp. 329–91. New York: Wiley.

Fallows, J. (1981). *National Defense*. New York: Random House.

Fieldhouse, R.W. (1986). Chinese nuclear weapons: an overview. In *World Armaments and Disarmament: SIPRI Yearbook 1986*, Stockholm International Peace Research Institute, pp. 97–113. Oxford: Oxford University Press.

Fitzgerald, A.E. (1972). *The High Priests of Waste*. New York: Norton.

Freedman, L. (1977). *U.S. Intelligence and the Soviet Strategic Threat*. London: Macmillan.

Greenwood, T. (1975). *Making the MIRV: A Study of Defense Decision-Making*. Cambridge, Mass.: Ballinger.

Hadley, A.T. (1979). Our ever-ready strategic forces: don't look closely if you want to believe. *Washington Star*, 1 Jul., section B, 1–3.

Harvey, B. (1981). Plausibility and the evaluation of knowledge: a case-study of experimental quantum mechanics. *Social Studies of Science*, 11, 95–130.

House of Representatives Armed Services Committee (1979). *Hearings on Department of Defense Authorization FY 1980*. Washington DC: US Government Printing Office.

Hussain, F. (1981). *The Future of Arms Control: Part IV. The Impact of Weapons Test Restrictions* (Adelphi Papers, no. 165). London: International Institute for Strategic Studies.

IISS (1985). *The Military Balance, 1985–86*. London: International Institute for Strategic Studies.

Jackson, H. (1984). Victims of the nuclear colonists. *The Guardian*, 30 Jun., 17.

Kaplan, F. (1983). *The Wizards of Armageddon*. New York: Simon and Schuster.

Keller, B. (1985). Imperfect science, important conclusions. *New York Times*, 28 Jul., 4E.

Latour, B. (1987). *Science in Action*. Milton Keynes: Open University Press.

Latour, B. & Woolgar, S. (1979). *Laboratory Life: the Social Construction of Scientific Facts*, Beverley Hills: Sage.

MacKenzie, D. (1988). The problem with 'the facts': nuclear weapons policy and the social negotiation of data. In *Information and Government*, ed. R. Davidson & P. White, pp. 232–51. Edinburgh: Edinburgh University Press.

Mann, P.S. (1981), Panel reexamines ICBM vulnerability. *Aviation Week and Space Technology*, 13 Jul., 141–5.

Marsh, R.T. (1982). A rebuttal by General Marsh. *Strategic Review*, 10, 42–3.

Marshall, E. (1981). A question of accuracy. *Science*, 213, 1230–1.

Metcalf, A.G.B. (1981). Missile accuracy–the need to know. *Strategic Review*, 9, 5–8.

Meyers, A.L. (1986). Nuclear testing. *Bulletin of the Atomic Scientists*, 43, 66–7.

Moncrief, F.J. (1979). SALT verification: how we monitor the Soviet arsenal. *Microwaves*, Sept., 41–51.

Moore, R.A. (1960). The evaluation of missile accuracy. In *An Introduction to Ballistic Missiles, Volume IV: Guidance Technologies*, ed. R.F. Kiddle *et al.*, pp. 111–27. Los Angeles: Air Force Ballistic Missile Division and Space Technology Laboratories.

Pickering, A. (1981). Constraints on controversy: the case of the magnetic monopole. *Social Studies of Science*, 11, 63–93.

Pickering, A. (1985). Pragmatic realism and the macrosociology of experiment. Paper presented to the *BSHS/BSA Conference on The Uses of Experiment*, Bath, 30 Aug.–2 Sept.

Powers, T. (1983). *Thinking About the Next War*. New York: New American Library.

Prados, J. (1986). *The Soviet Estimate: U.S. Intelligence Analysis and Soviet Strategic Forces*. Princeton: Princeton University Press.

Richelson, J.T. (1986). Old surveillance, new interpretations. *Bulletin of the Atomic Scientists*, 42, 18–23.

Schlesinger, J.R. (1974). Testimony to the Subcommittee on Arms Control, International Law and Organization of the Committee on Foreign Relations, United States Senate. *US and Soviet Strategic Doctrine and Military Policies*. Washington, DC: US Government Printing Office.

Slay, A. (1978). Testimony to the Armed Services Committee, United States House of Representatives. *Hearings on Military Posture, Fiscal Year 1979*, part 3, book 1, pp. 299–358. Washington, DC: US Government Printing Office.

Smith, R.J. (1982). An Upheaval in U.S. Strategic Thought. *Science*, 216, 30–4.

Topping, R.L., (1981). Submarine launched ballistic missile improved accuracy. Paper presented to *Annual Meeting and Technical Display of the American Institute of Aeronautics and Astronautics*, 12–14 May, Long Beach, Calif. (AIAA-81-0935).

Tsipis, K. (1981). Precision and accuracy. *Arms Control Today*, May, 3–4.

Tsipis, K. (1984). The Operational Characteristics of Ballistic Missiles. In *World Armaments and Disarmament: SIPRI Yearbook 1984*, Stockholm International Peace Research Institute, pp. 379–419. London: Taylor and Francis.

Vincenti, W.G. (1979). The air-propeller tests of W.F. Durand and E.P. Lesley: a case study in technological methodology, *Technology and Culture*, 20, 712–51.

Wertheim, R. (1978). Testimony to the Armed Services Committee, United States Senate. *Department of Defense Authorization, Fiscal Year 1979*, pp. 6674–90. Washington, DC: US Government Printing Office.

14

THE EPISTEMOLOGY OF EXPERIMENT[1]

ALLAN FRANKLIN

Although all scientists and philosophers of science are agreed that science is based on observation and experiment, very little attention has been paid to the question of how we come to believe rationally in an experimental result,[2] or, in other words, the problem of the epistemology of experiment.[3] How do we distinguish between a genuine result and a result that is an artifact created by the apparatus? In this essay I suggest that there are various strategies that both provide justification for rational belief in an experimental result and which are also used by practising scientists. Although a devout sceptic might doubt the validity of pointer readings or chart recordings, I do not believe this is the important question about the validity of experimental results. I shall be concerned here primarily, although not exclusively, with observations or results that are interpreted within an existing theory.

Ian Hacking has made an excellent start on discussing these issues.[4] He points out that most modern experiments involve complex

1. The material is based on work supported by the National Science Foundation under Grant No. SES-8308260. Any opinions, findings, conclusions or recommendations expressed in this publication are those of the author and do not necessarily reflect the views of the National Science Foundation. Part of this paper has appeared as a chapter in my book, *The Neglect of Experiment* (Franklin, 1986) and is published by permission of Cambridge University Press. Part of the work on this essay was done while I was a visiting fellow at the Center for Philosophy of Science, University of Pittsburgh. I am grateful to the Center and its director, Professor Nicholas Rescher, for their support and hospitality.
2. Some writers on science have questioned this. They are, to use the statement by the contemporary British philosopher, Gordon Sumner, 'like blind men looking for a shadow of doubt.' Mr Sumner is better known as Sting, the lead singer of the rock group, The Police. The quotation is from the song 'King of Pain' from the album *Synchronicity.*
3. A preliminary discussion appears in Franklin (1984).
4. See Hacking (1983).

apparatus and so, at the very least, the results are loaded with the theory of that apparatus. Dudley Shapere has extended the idea of 'direct observation' to include theoretical beliefs explicitly. In his discussion of the solar neutrino experiments he states, '*X* is directly observed if (1) information is received by an appropriate receptor and (2) that information is transmitted directly, i.e., without interference, to the receptor from the entity *X* (which is the source of the information).'[5] The dependence on theory is clear. Theory tells us what an appropriate receptor is and that the information is transmitted without interference. I will argue below that for the purposes of validating experimental results this dependence on theory is not a serious problem.

EPISTEMOLOGICAL STRATEGIES

By epistemological strategies I mean arguments designed to establish, or to help establish, the validity of an experimental result or observation. These may be verbal, but may also include additional experimental work, as will be made clear in the examples discussed below.

Different experimental apparatus

Hacking suggests, in his discussion of the microscope, that if something can be observed using 'different' microscopes (his example is dense bodies in cells) then it is real. He argues that it would be a preposterous coincidence if the same patterns were produced by two totally different kinds of physical systems, such as ordinary, polarising, phase-contrast, fluorescence, interference, electron, or acoustic microscopes. 'Different' here is clearly a theory-laden term. It is our theory of light and of microscopes which tells us that these are different. This does not, however, invalidate Hacking's argument. In recent work, Colin Howson and I have shown that a hypothesis *h*, receives more confirmation from 'different' experiments than from repetitions of the 'same' experiment.[6] Here the hypothesis *h* is of the form 'the value of *X* is *a*', or '*A* has been observed'. We also argued that at least one reason for deciding that experiments are 'different' is that the theories of the apparatus are different. If the theory is the same then one can point to differences of size, geometry, or even the experimenters, etc. Theoretical context can also be important in determining whether two experiments are the 'same' or 'different'. Thus, experiments to determine the velocity addition law at velocities small compared to the speed of light or at

5. Shapere (1982), p. 492.   6. Franklin & Howson (1984).

speeds near the speed of light would be considered to be close to being the 'same' before 1905 when Newtonian mechanics, the only theory available, made no distinction between those conditions. After 1905, when Einstein's theory of special relativity, which made the velocity at which the experiment was performed an important parameter, became a serious competitor these experiments would be quite 'different'.[7] It seems clear then, that in this case theory-ladenness[8] is a virtue rather than a defect because it leads to more confirmation of the hypothesis.

Indirect validation

A variant of the use of different experimental apparatuses as an epistemological strategy is the strategy of indirect validation. Suppose we have an observation that can be made using only one kind of apparatus. Let us also suppose that the apparatus can produce other similar observations which can be corroborated by different techniques. Agreement between these different techniques gives confidence not only in the observations but also in the ability of the first apparatus to produce valid observations. This, then, provides an argument in support of the observation made only with that apparatus. An example of this is the observation of the microtrabecular lattice using electron microscopy. It is argued that other objects of similar size (e.g., microtubules) have been seen by both electron microscopy and with an ordinary light microscope. This supports the idea that electron microscopy can detect objects of this size, and thus helps to validate the observation of the lattice. Here the similarity of the observations, in this case in size, is of importance. The ability of the electron microscope to detect objects of very different size, particularly of larger size, would not be of much assistance in arguing for the lattice, because the well-corroborated theory of the apparatus indicates that size is an important parameter.

Intervention and prediction

A more difficult and interesting problem, discussed only briefly by Hacking, is how one deals with phenomena that can be observed with

7. Franklin & Howson (1984), pp. 54–5.
8. One should be careful here to distinguish between the theory of the apparatus and the theory of the phenomena or the theory under test. If the two theories are distinct then no obvious problems arise for the testing of the theory of the phenomena. Even if they are not distinct one can use the apparatus to test the theory of the phenomena (see A. Franklin, *et al.*, 'Can a Theory-Laden Observation Test the Theory', forthcoming).

only one technique (i.e., electron microscopy or radioastronomy), or how one validates a single experiment. Hacking points out that the theory of the microscope changed drastically in 1873 when Ernst Abbe showed the importance of diffraction in its operation.[9] The use of microscope images remained robust despite such a theory change. Hacking attributes this to the fact that experimenters 'intervened', that is stained, injected fluid, and in other ways manipulated the objects under observation. This type of intervention is, I believe, a special case of a more general strategy in which one predicts what will be observed after the intervention if the apparatus is working properly or as expected. When the predicted observation is made we increase our belief in both the proper operation of the apparatus and in its results.[10]

This latter strategy is not available for radioastronomy and can be used only with difficulty for electron microscopy. In the case of electron microscopy scientists can intervene by, for example, changing the temperature of the observed cells, or by inducing changes in the activity of the cell by chemical means. One may argue that where an electron microscope reveals the same things as an ordinary microscope, or a radiotelescope reveals the same things as an ordinary telescope, this might give us some confidence both in the observations and in the technique. But it does not seem totally satisfactory. It is precisely the phenomena visible *only* with an electron microscope or with a radiotelescope that are of greatest interest. How, then, can such observations be validated? Extrapolation from unique observations to other observations is notoriously dangerous. One strategy is the use of a well-corroborated theory of the apparatus. If, as is indeed the case for both the electron microscope and the radiotelescope, the proper operation of the apparatus depends on such a theory, then it can be argued that the evidence supporting the theory also gives reasons to believe the observations.

Properties of the phenomena as validations

Even without such a theory, one can attempt to validate the observations on other grounds. Consider the Galilean telescope. For centuries, Cremonini and other Aristotelians of the early seventeenth century have been ridiculed because they refused to look through Galileo's

9. Hacking (1983), pp. 194–7.
10: A Bayesian analysis shows this quite clearly. Let h be the statement, 'the apparatus is working properly' and let o be some observation such that $h|$ -o. Then if we observe o, $P(h|o) > P(h)$. It then follows that if the apparatus is working properly we have more reason to believe in its results.

telescope to observe the moons of Jupiter. Even some of those who did denied their existence. Their scepticism was not without some merit. Galileo's telescope was not very good, and it was only his expertise in using it that enabled him to observe the moons. Even granting the observation of specks of light, how might we assert their real existence as moons of Jupiter and not as artifacts created by the telescope? One cannot here resort to the theory of the telescope and its independence of the distance of the object, because no such theory existed at the time of Galileo's *Starry Messenger*. One could argue that on Earth, where one could check by direct observation, the telescope seemed to give valid images. But the extrapolation to astronomical distances involves many orders of magnitude. I suggest here that it is the observed phenomena themselves that argue for their validity. (Note that this argument is not historical, but rather one that might have been offered.) Although one might imagine that the telescope could create specks of light, it hardly seems possible that it would create them in such a way that they would appear to be a small planetary system with eclipses and other consistent motions. It is even more preposterous to believe that they would satisfy Kepler's Third Law (R^3/T^2 = constant), although that argument would not have been available until publication of Kepler's *Harmonices Mundi* in 1619, and perhaps not until later in the seventeenth century, when it was generally accepted as a law.

A similar argument was offered by Robert Millikan to support his published observation of the quantisation of electric charge. He 'found in every case the original charge on the drop an exact multiple of the smallest charge which we found that the drop caught from the air. The total number of changes which we have observed would be between one and two thousand, *and in not one instance has there been any change which did not represent the advent upon the drop of one definite invariable quantity of electricity or a very small multiple of that quantity*' (emphasis in original).[11] The consistency of the data argued for their validity and against their interpretation as artifacts. No remotely plausible malfunction of the apparatus could produce such a consistent result. Millikan gave similar arguments in his other papers on electric charge.[12]

Interestingly, the consistency of the data provided support for Fairbank's recent reports of observations of fractional charge of $1/3e$. 'Out of 26 repeat measurements, we have observed 11 residual-charge changes, *in every case* [emphasis added] of $+1/3e$.'[13] The experimenters also stated that the residual charges on their spheres 'fall into three

11. Millikan (1911), p. 360.   12. See Millikan (1910, 1913).
13. LaRue, Phillips & Fairbank (1981).

groups which have weighted averages of $(-0.343 \pm 0.011)e$, and $(+0.001 \pm 0.003)e$, and $(+0.328 \pm 0.007)e'$.[14] The fact that both total charges and changes in charge had only the values 0, or $\pm 1/3e$ supports their conclusion that fractional charges exist. Were other values to be observed, then one might doubt both their observations and conclusions.

It is ironic that recent work by Fairbank's group[15] gave results that were not 0 or $\pm 1/3e$. One problem in this experiment is that of possible experimenter bias. The results of the measurements of the residual charge are known to the experimenter when the final data selection is made. To guard against possible bias, Luis Alvarez (personal communication) suggested that a random number unknown to the selector be added to each result and subtracted only after final event selection was made. That was done and the results of that blind test were $(+0.189 \pm 0.02)e$ and $(+0.253 \pm 0.02)e$. However, Phillips reported that this procedure cannot be used for all the niobium spheres, and a similar charge was observed in such an event. The observation of charges that were not 0 or $\pm 1/3e$ has cast further doubt on Fairbank's results. Fairbank has told me that after these discordant results were obtained from both the blind test and normal runs, the experiment was carefully re-examined and an instrumental effect has been found that might account for the discrepancy. Tests are currently in progress and the question of fractional charges is still unresolved. The point here, however, is that the consistency of the data was used to argue that the results were not artifacts of the apparatus and the appearance of inconsistency led to a re-examination of the experiment.[16] It should be emphasised also that although these strategies support rational belief in a result, rational belief may still be wrong.[17]

Theory of the phenomena

Difficulties also attend to validating observations with a radiotelescope. The radiotelescope can detect some, but not all, of the sources visible with an optical telescope. The radiotelescope is also based on a well-cor-

---

14. LaRue *et al* (1981), p. 967.   15. Phillips (1983).
16. Consistency by itself, without other arguments such as one that no plausible malfunction of the apparatus could give such consistent results, does not guarantee the validity of a result. An apparatus can behave in a consistently artifactual manner.
17. There may be, and in fact have been, cases in which two experimental results that appear to be valid are contradictory. Further investigation is then obviously required to decide the issue. I am concerned here only with strategies for rational belief in a single experimental result.

roborated theory. I suggest that, in addition, part of the argument for the validity of the observations is that they can be explained using the existing, accepted theory of the phenomena.

An interesting example of how theory is used to validate phenomena is the recent discovery of the $W^{\pm}$,[18] the charged intermediate vector boson, required by the independently well-corroborated Weinberg – Salam unified theory of electromagnetic and weak interactions. Although the two experiments in which the $W^{\pm}$ was discovered comprised very complex apparatuses and the validity of the discovery also rested upon other epistemological strategies (including independent checks on the proper operation of parts of the apparatus, independent methods of calculating the effect from the data, and the elimination of possible sources of background which might simulate W decay), I argue that the theoretical explanation of the observations and the agreement with theoretical predictions helped, in part, to validate the observations.

The importance of theory is shown even in the design of the particle accelerator used in the experiment. 'The CERN Super Proton Synchrotron (SPS) Collider, in which proton and antiproton collisions at $\sqrt{s} =$ 540 GeV (centre of mass energy) provide a rich sample of quark – antiquark events, has been designed with the search [for $W^{\pm}$ bosons] as the primary goal.[19] The theoretical predictions are also clearly evident in the design of the experiments, the selection of events, the analysis of the data, and the validation of the observations. Both experiments were designed to detect electrons with high momentum transverse to the proton – antiproton directions, as predicted by theory:

We report here the results of a search for single electrons of high transverse momentum (P_T) which are expected to originate from reaction (1). [Reaction (1) was $\bar{p} + p \rightarrow W^{\pm}$ + anything followed by $W^{\pm} \rightarrow e^{\pm} + \nu(\bar{\nu})$]. Because the neutrino from W decay is not detected, the events from reaction (1) are expected to show a large missing transverse energy (of the order of the electron, P_T) along a direction opposite in azimuth to that of the electron.[20]

It should be emphasised that no other reaction known was expected to produce such high-P_T electrons. Events with high-P_r which had no jet structure, as required by theory, were selected as the final event sample. These events also showed evidence of missing transverse energy of the right magnitude for W decay. From these events the mass of the W could be calculated, giving (81 ± 5) GeV/c^2 and $(80 \pm^{10}_6)$ GeV/c^2, for the two experiments, respectively. The Weinberg – Salam theory pre-

18. Arnison *et al.* (1983); Banner *et al.* (1983).   19. Arnison *et al.* (1983), p. 104.
20. Banner *et al.* (1983), p. 477.

dicted the mass of the W to be (82 ± 2.4) GeV/c^2. The experimenters noted that their measurements were 'in excellent agreement with the expectation of the Weinberg – Salam model'.[21] In addition, both the number of observed events and the transverse momentum distribution of the W's were in good agreement with theoretical predictions.

It seems clear that the observations confirmed the theory. It also seems clear, as confirmed by discussions with colleagues working in high energy physics and with a member of the experimental group, that the agreement with theory helped to validate the observations. I have not done justice here to the full strength of the arguments offered by the scientists in support of their observations, which included examples of the other strategies suggested here. It is interesting, though, to speculate what would have happened had the measurements not agreed so well with the theoretical predictions. It is not clear whether this would have been regarded as an anomaly for the theory or as a doubtful measurement.

The fact that the consistent values of fractional charge observed by Fairbank and his collaborators were also those predicted by theory added to their believability. Had these consistent values been $\pm 1/2e$, a value not predicted by theory, one suspects they would have been treated more sceptically.

A somewhat different illustration of the complex interaction of observation and theoretical explanation is provided by the discovery of the synchrotron radiation from Jupiter, or, to put it more accurately, the discovery that the microwave radiation emitted from Jupiter was due to synchrotron radiation. In 1959, Sloanaker reported measurements of the intensity of radiation from Jupiter at a wavelength of about 10 cm, that gave temperatures, as computed from a black-body model, ranging from 300 °K to 1010 °K with a mean of 640 °K $\pm$ 85 °K.[22] These temperatures were inconsistent with temperatures obtained from earlier measurements at 3 cm and in the infrared region. An additional problem was that the considerable scatter in the data allowed a variable component in the radiation. At the same time, Drake and Hvatum reported measurements at 22 cm and 68 cm that required temperatures of 3000 °K and 70 000 °K, respectively, and which they regarded as too high to be plausible.[23] They combined their results with other measurements and concluded that Jupiter was emitting nonthermal radiation. They proposed the argument:

21. Arnison *et al.* (1983), p. 115.   22. Sloanaker (1959).   23. Drake & Hvatum (1959).

that the radiation originates as synchrotron radiation from relativistic particles trapped in the Jovian magnetic field, a situation similar to the terrestrial Van Allen belts. A Jovian field of 5 gauss and a total number of particles 10^6 times greater than the terrestrial system will sufffice to explain the observations.[24]

There is at least a hint that the failure of the data to be consistent with the black-body model, combined with the variation and scatter in the data themselves, cast some doubt on the measurements.[25]

Field then attempted to provide a consistent theoretical explanation of the observations. He noted the failure of the black-body model and considered four other possible explanations: thermal emission from deep in Jupiter's atmosphere, emission in the ionosphere, synchrotron radiation from relativistic electrons, and cyclotron emission by non-relativistic electrons, trapped in Jupiter's magnetic field.[26] The first two were eliminated because they disagreed with the observations. The third was rejected because it required a density of high energy electrons much greater than that seen near the Earth, which Field regarded as implausible. 'It seems therefore, that, on the basis of energy consider-ations, a belt of electrons trapped in a 1000 gauss field and replenished by the solar corpuscular emission could account for the decimeter radi-ation.'[27] A second paper refined the model and gave reasonably good agreement with the observations.[28]

The situation was resolved after further measurements on decametric radiation, wavelengths of the order of 10 m, and a new calculation by Warwick, based on the synchrotron radiation model.[29] One problem for Field's model had been the rather large magnetic field required, approximately 1000 gauss, which was regarded as unlikely. New mea-surements had suggested that the field had to be even larger, about 10 000 gauss. In addition, new polarisation measurements seemed to favour the synchrotron radiation model, although the evidence was not conclusive. Warwick's model explained both the decimetric and decametric radiation on the basis of the interaction of external particles with the ionosphere of Jupiter. The electron densities required were approximately the same for both ranges of radiation although a spec-trum of energies was needed. The Jovian magnetic field required was only about 10 gauss. In addition, this model which depended on exter-nal particles from the Sun explained the apparent correlation of the

24. Drake & Hvatum (1959), p. 330.
25. This view was confirmed in conversation with James Warwick, one of the participants in this episode.
26. Field (1959). 27. Field (1959), p. 1175. 28. Field (1960).
29. Warwick (1961).

radiation and the observed solar activity. The currently accepted model agrees with Warwick's original proposal, except that it is believed that the electrons are locally generated and do not come directly from the Sun.

This is not simply a case where a theoretical explanation provided additional confidence in a set of measurements. The observations helped to decide between the competing explanations, and the chosen explanation helped to validate what had been a confusing, and perhaps somewhat doubtful, set of observations. This complex relationship between theory and observation happens frequently in physics. This is not to say that the observations were made in such a way as to agree with theory, but only that the theoretical explanations rationally strengthen our belief both in these observations and also in other observations made using the same technique for which we have no such theoretical explanation.

The elimination of alternative explanations

Yet another strategy is illustrated in the observation of electric discharges in the rings of Saturn (SED), that were recorded in the fly-bys of Voyagers 1 and 2.[30] This strategy entails the elimination of all plausible sources of error and all alternative explanations; then the observation is valid.

One possible explanation of the observations was poor data quality in the telemetry link between the spacecraft and Earth. This was ruled out by the measured quality of the link (a form of calibration, to be discussed later) and by the fact that the observations were independent of the telemetry mode, a highly unlikely occurrence if it were a defect in the telemetry. Alternative explanations that were considered included discharges generated near the spacecraft through deleterious environmental phenomena – such effects having been seen when spacecraft are in the Earth's atmosphere. No such discharges were seen when the spacecraft was near Jupiter, arguing that this explanation was not correct. The possibility of dust-like particles interacting with the spacecraft and causing the events was eliminated by the fact that the observed time scale of the discharges was far longer than that expected from dust interactions. The fact that the observations occurred when Voyager 1 was outside the magnetosphere of Saturn also argued against any interaction with the plasma or particulate environment of Saturn. The SED were also observed below a frequency of 100 kHz.

30. Warwick *et al.* (1981, 1982).

This indicated that there was no ionosphere between Voyager 1 and the SED source, because the cut-off frequency for Saturn, below which signals are reflected, is 1.370 MHz. The observed peak in electron density (ionosphere) was far above Saturn's clouds indicating that they could not be the source of the SED. After studies giving these results were published, it was suggested that the rings of Saturn eclipsed part of the atmosphere and that for that region reflections would not occur, and thus the argument that all plausible alternative explanations had been eliminated was suspect. For Voyager 2 different parts of the atmosphere were eclipsed, and the discharges were again observed, arguing against the eclipse explanation. Having eliminated sources of error and alternative explanations the conclusion was that 'We believe, therefore, that the most probable source of SED is in Saturn's rings'.[31]

The observation of Saturn's electrical discharges was confirmed by Voyager 2. Events with the same duration and frequency distribution were seen. Thus, the character of the events allowed the same arguments to be made as those which had been given previously for Voyager 1. In addition, 'Voyager 2 detected SED more than 48 hours before the first inbound bow shock crossing, so they clearly are not due to phenomena occurring at the spacecraft inside Saturn's magnetosphere'.[32]

There were, however, some significant differences between the two sets of observations. Voyager 2 detected SED at a rate only one-third of that of Voyager 1. 'In addition, the episodes themselves were distributed much more symmetrically about the time of closest approach (most of the Voyager 1 episodes occurred after the Voyager 1 encounter)'.[33] Significant differences in the polarisation of the events were also observed. These led to the conclusion that 'although Voyager 2 confirmed the SED phenomenon, the striking differences in polarization, episode distribution, and number of events strongly suggest a source that changes with time'.[34] The character of the events allowed the use of the same arguments as to the existence of SED. All plausible sources of error and all alternative explanations could again be eliminated. Thus, the differences were then reasonably attributed to a changing source.

Calibration and experimental checks

Perhaps the most widely used strategy for validation of results is that of calibration and experimental checks. Here the argument is that the

31. Warwick *et al.* (1981), p. 243.   32. Warwick *et al.* (1982), p. 585.
33. Warwick *et al.* (1982), p. 586   34. Warwick *et al.* (1982), p. 586

ability of the apparatus to reproduce already known phenomena argues both for its proper operation and in favour of the results obtained. Calibration not only provides a check on the operation of the apparatus but also provides a numerical scale for the measurement of the quantity involved. An example of this is the calibration of the Princeton experiment detecting $K_2^0 \rightarrow \pi^+ \pi^-$ and thus demonstrating the existence of CP violation.[35] In order to demonstrate the ability of their experimental apparatus to detect the decays, $K_2^0 \rightarrow \pi^+ \pi^-$ they detected the already known phenomenon of regenerating K_1^0 mesons from K_2^0 mesons. (K_1^0 also decay to $\pi^+ \pi^-$). Their procedure was described as follows:

An important calibration of the apparatus and data reduction system was afforded by observing the decays of K_1^0 mesons produced by coherent regeneration in 43 gm/cm^2 of tungsten. Because the K_1^0 mesons produced by coherent regeneration have the same momentum and directions as the K_2^0 beam, the K_1^0 decay simulates the direct decay of the K_1^0 into two pions. The regenerator was successively placed at intervals of 11 in[ches] along the region of the beam sensed by the detector to approximate the spatial distribution of the K_2^0's. The K_1^0 vector momenta peaked about the forward direction with a standard deviation of 3.4 ± 0.3 milliradians. The mass distribution of these events was fitted to a Gaussian with an average mass 498.1 ± 0.4 MeV and standard deviation of 3.6 ± 0.2 MeV.[36]

The agreement with the known K^0 mass of 497.7 MeV and with the expected angle of 0° argued for the ability of the apparatus to detect two pion decays of the K_2^0. They then compared their sample of suggested K_2^0 decays, those for which $\cos\Theta > 0.99999$, with the regenerated K_2^0 decays:

The average of the distribution of masses of those events with $\cos\Theta > 0.99999$ is found to be 499.1 ± 0.8 MeV. A corresponding calculation has been made for the tungsten data resulting in a mean mass of 498.1 ± 0.4. The difference is 1.0 ± 0.9 MeV. Alternately we may take the mass of the K^0 to be known and compute the mass of the secondaries for two-body decay. Again restricting our attention to those events with $\cos\Theta > 0.99999$ and assuming one of the secondaries to be a pion, the mass of the other particle is determined to be 137.4 ± 1.8. Fitted to a Gaussian shape the forward peak has a standard deviation of 4.0 ± 0.7 milliradians to be compared with 3.4 ± 0.3 milliradians for the tungsten. *The events from the He gas appear identical with those from the coherent regeneration in tungsten in both mass and angular spread.*[37]

The agreement of the two sets of data gave confidence in their result.[38]

35. Christenson *et al.* (1964).   36. Christenson *et al.* (1964), p. 138.
37. Christenson *et al.* (1964), p. 139.
38. It is interesting to note that various theoretical explanations to preserve CP symmetry suggested that either the decaying particle was not a K_2^0 meson or that the decaying particles were not pions.

In any well designed experiment there will be checks to ensure uniform, if not proper, operation of the apparatus.[39] An example of this is the careful monitoring of event rate as a function of K^0 intensity, in the Princeton experiment discussed earlier. Failure to perform these checks, or to perform them incorrectly, can result in invalid or erroneous results. An interesting case in point is one of the tests of time-reversal invariance. Time-reversal invariance requires that the angular distributions (at the same energy) in the reactions $\gamma + d \rightarrow n + p$ and $n + p \rightarrow \gamma + d$ be identical. The angular distribution for the first reaction had already been well measured, and the test involved looking at the angular distribution of the second reaction. The efficiency as a function of angle was needed so that the correct distribution could be calculated from the raw data. The experimenters had planned to use the γ rays resulting from $n + p \rightarrow d + \pi^0$, in which the π^0 decays into 2 γ-rays, to measure that efficiency. This was done, and on the basis of the measured efficiency the two distributions differed significantly. (The probability that the two distributions were the same was reported as one in a million). The conclusion that time-reversal invariance was violated, a dramatic result, was announced at a meeting of the American Physical Society. Unfortunately, as a sophisticated Monte Carlo calculation showed later, the measured efficiency using the π^0 reaction was not correct. The calculation thus served as a check on the efficiency. The proper efficiency was used in the published version and the final result showed no difference between the angular distributions, and time-reversal symmetry was still preserved.[40]

An experimental apparatus can also be used to detect artifacts that are known in advance to be present. This is a somewhat different kind of experimental check or calibration. An example of this comes from the infrared spectroscopy of organic molecules[41] where it was not always possible to prepare a pure sample. Sometimes one had to put the substance in an oil paste or in solution. In such cases, one expects to observe, superimposed on the spectrum of the substance, the spectrum of the oil or the solvent, which one can compare with the spectrum of oil or solvent alone. This is illustrated in Figures 14.1 and 14.2 where the absorption lines due to oil and $CHCl_3$ (a solvent) are shown. Observation of this artifact gives confidence in the other observations. If the spectrum of oil or solvent has been measured independently, it also provides a calibration of the apparatus.

39. Unfortunately the current trend toward the publication of important results in letters journals sometimes precludes detailed discussion of these checks in the published paper, as Sir Brian Pippard has noted (address given to the British Society for the Philosophy of Science, London, Sept. 1980).
40. Bartlett *et al.* (1969).   41. Randall *et al.* (1949).

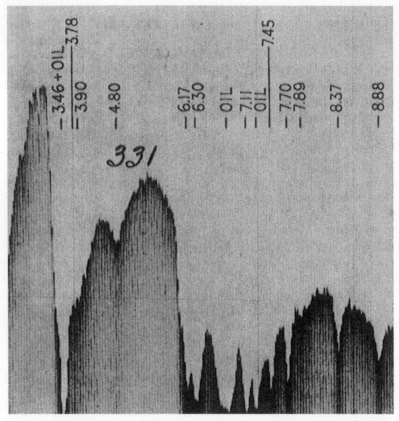

PLATE 58. Assignments: 6.17 μ   δNH₃⁺

6.30 μ   Carboxylate ion

Preparation: Oil paste

Figure 14.1. Infrared spectrum of an organic molecule prepared in an oil paste. The oil spectrum is clearly indicated. (Randall *et al.*, 1949)

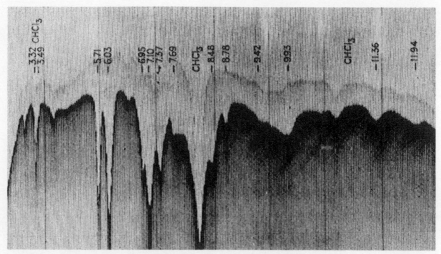

PLATE 21a.   Assignments: 5.71 μ   Ester C=O   Preparation: 10% solution in CHCl₃, 0.015 mm.
                          6.03 μ   N-acyl C=O

Figure 14.2. Infrared spectrum of an organic molecule in CHCl₃solution. The CHCl₃ spectrum is shown. (Randall *et al.*, 1949)

An interesting sidelight occurred in the report of these observations. An unexpected background was observed (see Figure 14.3, note absorption labelled 'background'):

Owing to what appears to have been a thin film of unknown nature deposited on the optical surfaces of the instrument as a result of an accident during its operation, there are three weak absorptions in these records which have no significance at any time since they were always present, even when no sample was placed in the beam.[42]

42. Randall *et al.* (1949), p. 97. A similar case occurred in a recent local trial. The defendant was convicted of heroin possession on the basis of infrared spectroscopy of a substance found in his possession. The spectrum of the substance matched that of heroin. An expert witness for the defense raised a legitimate objection that the spectroscope had not been run without a sample present to guard against contamination of the instrument. My own reaction on reading this in the newspaper was to note that if the instrument had been contaminated and the substance was something other than heroin, then what would have been seen was the spectrum of the other substance superimposed on the heroin spectrum. After the trial was over I spoke with a member of the jury and asked if my argument had been considered. He told me that although the jury considered the defense objection seriously they decided on a conviction because the defense had had ample opportunity to have the substance independently tested and had not done so. They inferred that the defense knew that the substance was heroin and avoided the test.

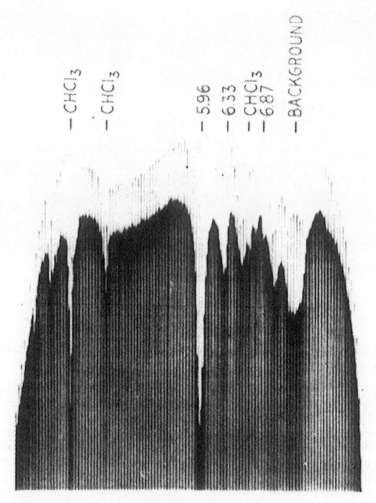

—CHCl3
—CHCl3
—5.96
—6.33
—CHCl3
—6.87
—BACKGROUND
PLATE 75a. Assignments:
5.96 μ Amide C=O
Preparation: 2% solution in
CHCl3, 0.04 mm.

Figure 14.3. Infrared spectrum of an organic molecule in $CHCl_3$ solution. The $CHCl_3$ spectrum and the unexpected background absorption are shown. (Randall *et al.* 1949)

The artifactual nature of the absorptions was clear, because they appeared when no sample was present. One might, however, be tempted to scepticism and use this background to cast doubt on the apparatus and on the other observations. That would not be correct. 'The correctness of the explanation of the source of these bands was seen after the instrument was taken apart and cleaned, shortly following the making of these records. After the cleaning, the spurious absorptions disappeared completely.'[43]

Prediction of lack of phenomenon

This suggests another strategy. If an effect disappears when you predict it will, then it is valid. This is, in fact, a particular instance of the strategy discussed earlier, in which observation of predicted behavior helps to validate a result, but it is so dramatic that it deserves discussion.

The 'crucial' experiment of Wu and associates[44] showing nonconservation of parity illustrates this very well.[45] In that experiment, Co^{60} nuclei were polarised, and the decay electrons detected. The asymmetry observed when the nuclei were polarised in two opposite directions demonstrated the violation of parity conservation. The polarisation could occur only at extremely low temperatures. One expected that as the temperature Increased the asymmetry would disappear. This is shown quite dramatically in Figure 14.4 and helps to validate the observation. The warm counting rates, with no polarisation, were independent of the magnetic field direction, arguing against any instrumental asymmetry.

If however, an effect does not disappear when it is predicted to, then one doubts both the proper operation of the apparatus and its results. This is illustrated in an experiment to measure the spin, or magnetic moment, of He^6 nuclei, using a Stern–Gerlach apparatus.[46] The He^6 nuclei passed through a long, inhomogeneous magnetic field. If the spin was one then a decrease in the counting rate by a factor of three should have been observed when the magnetic field was on, as compared with when it was off. This was indeed observed in an early data run. It was a startling result. Both accepted theory and all previous experimental work had shown that all even-even nuclei, of which He^6 is an example, have spin zero, although it had never been directly tested for He^6. In order to check on proper operation of the apparatus,

43. Randall *et al*, (1949), p. 97.   44. Wu *et al*. (1957).
45. For a detailed history of this episode see Franklin (1979).
46. Comins & Kusch (1958).

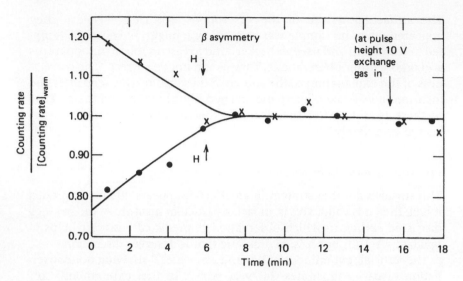

Figure 14.4. The decay electron counting rate as a function of time for the decay of polarised Co^{60} nuclei. The asymmetry between the two polarisations is shown. (Wu *et al.*, 1957)

the collimation of the beam was destroyed. Under those circumstances, no difference should have been observed between the field-off and field-on counting rates. The factor of three difference persisted and indicated that the magnetic field was reducing the efficiency of the detector. It was only an unfortunate coincidence that the reduction was the predicted factor of three. Subsequently the detector was more adequately shielded from the magnetic field. The final result showed no difference in counting rates, or spin zero. The importance of experimental checks is clear.

Another example of this is provided by the electron microscopic observations of the structure of myokinase and protamine by Ottensmeyer.[47] The electron micrographs he obtained are shown in Figures 14.5 and 14.6 (*a* and *b*) and do seem to indicate a structure. Dubochet[48] repeated these observations and after some difficulty obtained similar results for myokinase shown in Figure 14.7 (14.7 *1–3*). However, he also obtained similar micrographs even when no protamine or myokinase was present, casting doubt on both sets of observations [Figures 14.6 (*7* and *8*) and Figure 14.7 (*4–6*)]. As he remarked, 'These results demonstrate that a trained observer is astonishingly good

47. Ottensmeyer *et al.* (1975, 1978).   48. Dubochet (1976); Klug (1978).

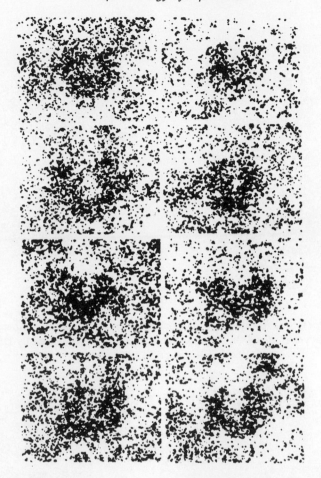

Figure 14.5. Electron micrograph of myokinase. (Ottensmeyer *et al.*,1975)

in selecting structures in noise'.[49] He suggested that other methods must be used to determine the structures.

Statistical validation

Statistical arguments can also be an important part of validating a result, or in establishing that a particular effect was seen. Thus, the difference between the π^+ and π^- lifetimes divided by their average lifetime, a difference of 0.053 ± 0.068 percent, is consistent with zero difference,

49. Dubochet (1978), p. 293.

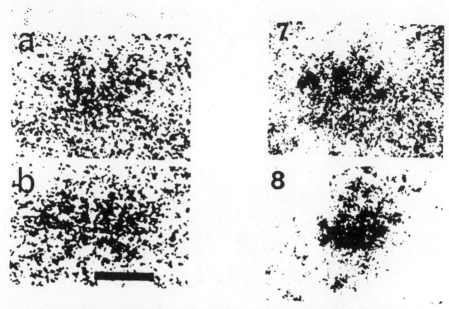

Figure 14.6. (a,b). Electron micrographs of salmon sperm protomine. (Ottensmeyer *et al.*,1978; (7,8). Electron micrographs obtained by Dubochet, (1978) with no protamine present.

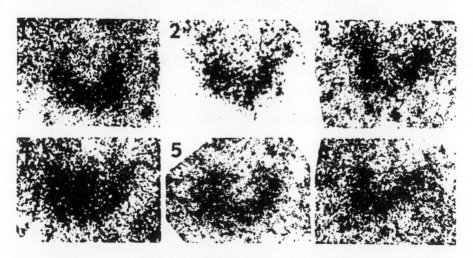

Figure 14.7. (1–3). Electron micrographs of myokinase obtained by Dubochet, (1978); (4–6). Electron micrographs obtained by Dubochet with no myokinase present.

and the CPT theorem remains valid. Consider, too, the editor of *Physical Review*, Samuel Goudsmit's comments concerning Telegdi's results on parity violation. He argued that the original result, 0.062 ± 0.027, which was only a little more than two standard deviations, was not sufficient to establish the effect. He compared it to the originally reported asymmetry in η decay of 0.072 ± 0.028 in which later experiments showed that there was no asymmetry. He also contrasted it to 'the overwhelming and compelling evidence' presented by the other two experiments, which were 13 and 22 standard deviations, respectively.

An interesting point arose in the 1960s when the search for new particles and resonances using bubble-chamber techniques occupied a substantial fraction of the time of those working in experimental high-energy physics. The usual technique was to plot the number of events as a function of invariant mass or the square of invariant mass and look for bumps above a smooth background. The usual informal criterion for a new particle was that it gave a three standard deviation effect above background, which had a probability of 0.27 percent. That was extremely unlikely in any single experiment, but Arthur Rosenfeld of the University of California is supposed to have pointed out that if one considered the number of such experiments done each year and the number of graphs drawn, then one would expect to observe a significant number of such three standard deviation effects. The informal criterion was then changed to four standard deviations, which has a probability of 0.0064 percent. The story may be apocryphal, but it did have wide circulation among physicists. Examination of the literature for the 1960s reveals several reported particles which were not confirmed by later observations and are no longer mentioned.

Statistical validation combined with theoretical prediction

An interesting strategy using both statistical argument and the theoretical explanation of the phenomenon is illustrated in the following. Let us assume that a mass spectrum has 1000 bins. The probability of observing three standard deviations from the expected distribution, assuming the deviations in the number of events in each bin is random, is 0.27 percent. The probability of observing a single three standard deviation effect in the entire spectrum (1000 bins) is quite high; in fact it is 93 percent. Thus, the observation of such an effect would not be strong evidence in favour of the existence of a particle. If, however, a theory (T) predicted the existence of a particle at a mass (m_0) and the

3 standard deviation effect was seen in that particular bin, then not only would the observation provide evidence in support of the theory, but the supported theory would support the validity of the observation, as opposed to it being a statistical fluctuation.[50]

In this essay I have presented a set of strategies that I believe provide grounds for rational belief in experimental results. Several of these strategies use the argument that the observation of predicted behavior, or of known phenomena, argues in favour of the proper operation of the apparatus, and this validates the observations or measurements. These include the following: intervention; experimental checks and calibration, as shown in the measurement of regenerated K_1^0 mesons as a check that the apparatus can measure K_2^0 decays, and the observation of the spectra of oil or solvent, as expected, in infrared spectroscopy of organic molecules; disappearance of an effect when predicted as in the disappearance of asymmetry in the decay of polarised nuclei as the sample warms up. As discussed earlier, failure to observe predicted behavior casts doubt on the observations or measurements, illustrated by the failure of the observed effect in the Stern–Gerlach experiment on He^6 to disappear when the collimation of the beam was destroyed.

Other strategies include the use of different experiments to validate an observation. It may also be very implausible for the data to be artifacts because of the consistent properties of the data as illustrated by the argument concerning Jupiter's moons and by Millikan's results. If one can eliminate all plausible sources of error and all alternative explanations, then an observation such as that of the electric discharges in the rings of Saturn is valid. One also has reason to believe in observations made with an apparatus based upon a well-corroborated theory, such as the radiotelescope or the electron microscope. Support for the theory also supports the observation. Similar support occurs when observations are explained or predicted by a well-corroborated theory, as shown in the discovery of the W bosons and the discovery of synchrotron radiation from Jupiter.

50. This can be seen quite easily. Let us suppose, for simplicity, that there are only two competing explanations of the effect at m_0: 1) C, that the 3 standard deviation effect is due to chance and 2) T, a theory that predicts a particle at m_0: $P(C) + P(T) = 1 = P(C/e) + P(T/e)$, where e is the observation of the 3 standard deviation effect at m_0. $P(T/e) - P(T) = P(C) - P(C/e)$ and because $P(T/e) > P(T)$, $P(C/e)$ is less than $P(C)$ and so the probability that the observation is due to chance goes down, increasing the validity of the observation.

I do not suggest that the strategies documented here are either exclusive or exhaustive, nor do I believe that any of them or any subset are necessary or sufficient conditions for rational belief. I do not believe such a general method exists. The fact that all of these strategies are illustrated by examples from the work of practising scientists argues against those who believe that rational argument plays little, if any, role in the validation of such results. Consider, for example, the work of Collins or Pickering.[51] To be fair, Collins and Pickering do discuss issues of validation, but they would regard the strategies as culturally accepted practices. I believe that there is a considerable distinction between a culturally accepted *practice* and a *reason* for rational belief. In particular, the case of the Saturn electric discharges demonstrates clearly that scientists do provide such arguments. I am not suggesting that because scientists behave this way, their behaviour has methodological significance. I believe that independent and reasonable justification has been given for these strategies, and I suggest that rational argument plays the major role in this issue.

51. Collins (1976); Pickering (1984).

References

Arnison G. *et al.* (1983). Experimental observation of isolated large transverse energy electrons with associated missing energy at $\sqrt{s} = 540$ Gev. *Physics Letters*, **122B**, 103–16.

Banner, M. *et al.* (1983). Observation of single isolated electrons of high transverse momentum at the CERN $\bar{p}$ collider. *Physics Letters*, **122B**, 476–85.

Bartlett, D.F. *et al.* (1969). $n + p \rightarrow d + \gamma$ and time reversal invariance. *Physical Review Letters*, **23**, 893–7.

Christenson, J.H. *et al.* (1964). Evidence for the 2π decay of the K_2^0 meson. *Physical Review Letters*, **13**, 138–40.

Collins, H. (1975). The seven sexes: a study in the sociology of a phenomenon, or the replication of experiments in physics. *Sociology*, **9**, 205–24.

Commins, E.D., & Kusch, P. (1958). Upper limit to the magnetic moment of He^6. *Physical Review Letters*, **1**, 208–9.

Drake, F.D. & Hvatum, S. (1959). Non-thermal microwave radiation from Jupiter. *The Astronomical Journal*, **64**, 329–30.

Dubochet, J. (1978). Comment on Prof. Ottensmeyer. *Chemica Scripta*, **14**, 293.

Field, G. (1959). The source of radiation from Jupiter at decimeter wavelengths. *Journal of Geophysical Research*, **64**, 1169–75.

Field, G. (1960). The source of radiation from Jupiter at decimeter wavelengths 2. Cyclotron radiation by trapped electrons. *Journal of Geophysical Research*, **65**, 1661–71.

Franklin, A. (1979). The discovery and nondiscovery of parity nonconservation. *Studies in History and Philosophy of Science*, 10(3), 201–57.

Franklin, A. (1984). The epistemology of experiment. *British Journal for the Philosophy of Science*, 35, 381–90.

Franklin, A. (1986). *The Neglect of Experiment*. Cambridge: Cambridge University Press.

Franklin, A. & Howson, C. (1984). Why do scientists prefer to vary their experiments? *Studies in History and Philosophy of Science*, 15, 51–62.

Hacking, I. (1983). *Representing and Intervening*. Cambridge: Cambridge University Press.

Klug, A. (1978). Direct imaging of atoms in crystal and molecules. *Chemica Scripta*, 14, 291–3.

LaRue, G., Phillips, J., & Fairbank, W.M. (1981). Observation of fractional charge of $(1/3)e$ on matter. *Physical Review Letters*, 46, 967–70.

Millikan, R.A. (1910) A new modification of the cloud method of determining the elementary electrical charge and the most probable value of that charge. *Philosophical Magazine*, 19, 209–28.

Millikan, R.A. (1911). The isolation of an ion, a precision measurement of its charge, and the correction of Stokes's Law. *Physical Review*, 32, 349–97.

Millikan, R.A. (1913). On the elementary electrical charge and the Avogadro constant. *Physical Review*, 2, 109–43.

Ottensmeyer, F.R. *et al.*, (1975). Electron microtephroscopy of proteins. *Journal of Ultrastructural Research*. 52, 193–201.

Ottensmeyer, F.R. *et al.*, (1978). The imaging of atoms: its application to the structure determination of biological macromolecules. *Chemica Scripta*, 14, 257–62.

Phillips, J. (1983). 'Residual charge on niobium spheres'. Unpublished Ph.D. Thesis, Stanford University.

Pickering, A. (1984). Against putting the phenomena first: the discovery of the weak neutral current. *Studies in History and Philosophy of Science*, 15(2), 85–117.

Randall, H.M. *et al.* (1949). *Infrared Determination of Organic Structures*. New York: Van Nostrand.

Shapere, D. (1982). The concept of observation in science and philosophy. *Philosophy of Science*, 49, 485–525.

Sloanaker, R. (1959). Apparent temperature of Jupiter at a wavelength of 10 cm. *The Astronomical Journal*, 64, 346.

Warwick, J. (1961). Theory of Jupiter's decametric radio emission. *Annals of the New York Academy of Sciences*, 95, 39–60.

Warwick, J. *et al.* (1981). Planetary radio astronomy observations from Voyager 1 near Saturn. *Science*, 212, 239–43.

Warwick, J. *et al.* (1982). Planetary radio astronomy observations from Voyager 2 near Saturn. *Science*, 215, 582–7.

Wu, C.S. *et al.* (1957). Experimental test of parity conservation in beta decay. *Physcial Review*, 105, 1413–15.

SELECT BIBLIOGRAPHY

These items have been selected from chapter bibliographies to give a representative but not comprehensive sampling of work discussed in this book.

Achinstein, P. & Hannaway, O. (ed.) (1985) *Observation, Experiment and Hypothesis in Modern Physical Science*. Cambridge, Mass. and London: MIT Press.

Allison, G.T. (1971). *Essence of Decision: Explaining the Cuban Missile Crisis*. Boston: Little, Brown & Co.

Bachelard, G. (1951). *L'Activité Rationaliste de la Physique Contemporaine*. Paris: Presses Universitaires de France.

Barnes, B. (1982a). *T.S. Kuhn and Social Science*. London: Macmillan.

Barnes, B. (1982b). On the implications of a body of knowledge, *Knowledge: Creation, Diffusion, Utilization*, **4**, 95–110.

Bechler, Z. (1974). Newton's 1672 optical controversies: a study in the grammar of scientific dissent. In *The Interaction between Science and Philosophy*, ed. Y. Elkana. Atlantic Highlands: Humanities Press.

Bennett, J.A. (1986). The mechanics' philosophy and the mechanical philosophy. *History of Science*, **24**, 1–28.

Brannigan, A. (1981). *The Social Basis of Scientific Discoveries*, Cambridge: Cambridge University Press.

Caneva, K.L. (1978). From galvanism to electrodyamics. *Historical Studies in the Physical Sciences*, **9**, 63–159.

Collins, H.M. (1975). The seven sexes: a study in the sociology of a phenomenon or on the replication of experiments in physics. *Sociology*, **9**, 205–24.

Collins, H.M. (1981). The place of the 'core set' in modern science. *History of Science*, **19**, 6–19.

Collins, H.M. (ed) (1981). Knowledge and Controversy: Studies of Modern Natural Sciences, *Social Studies of Science*, **11**, 1–158.

461

Collins, H.M. (1985). *Changing Order: Replication and Induction in Scientific Practice.* London and Beverley Hills: Sage.

Collins, H.M. & Pinch, T.J. (1979). The construction of the paranormal: nothing unscientific is happening. In ed. R. Wallis, pp., 237–70.

Collins, H.M. & Pinch, T.J. (1982). *Frames of Meaning: The Social Construction of Extraordinary Science.* London: Routledge and Kegan Paul.

Collins, H.M. & Shapin, S. (1983). The historical role of the experiment, In *Using History of Physics in Innovatory Physics Education,* ed. F. Bevilacqua & P.J. Kennedy, pp. 282–92. Pavia: Università di Pavia.

Conant, J.B. (ed.) (1948). *Harvard Case Histories in Experimental Science.* 2 vols. Cambridge, Mass.: Harvard University Press.

Constant, E.W. (1983). Scientific theory and technological testability: science, dynamometers, and water turbines in the 19th century. *Technology and Culture,* 24, 183–98.

Crombie, A.C. (1953). *Robert Grosseteste and the Origins of Experimental Science, 1100–1700.* Oxford: Clarendon Press.

Dear, P. (1985). *Totius in verba*: rhetoric and authority in the early Royal Society. *Isis,* 76, 145–61.

Dear, P. (1987). Jesuit mathematical science and the reconstruction of experience in the early 17th century. *Studies in History and Philosophy of Science,* 18, 133–75.

Dorling, J. (1973). Demonstrative inference. *Philosophy of Science,* 40, 360–72.

Duhem, P. (1954). *The Aim and Structure of Physical Theory.* Princeton: Princeton University Press.

Earman, J. & Glymour, C. (1980). Relativity and eclipses: the British eclipse expeditions of 1919 and their predecessors. *Historical Studies in the Physical Sciences,* 11, 49–85.

Edge, D. (ed.) (1964). *Experiment: a Series of Scientific Case Histories.* London: BBC.

Falconer, I.J. (1987). Corpuscles, electrons and cathode rays: J.J. Thomson and the 'Discovery of the Electron'. *British Journal for the History of Science,* 20, 241–76.

Farley, J. & Geison, G.L. (1974). Science, politics and spontaneous generation in nineteenth-century France: The Pasteur – Pouchet debate. *Bulletin of the History of Medicine,* 48, 161–98.

Feyerabend, P.K. (1975). *Against Method.* London: New Left Books.

Fleck, L. (1979). *Genesis and Development of a Scientific Fact.* Chicago and London: University of Chicago Press.

Frankel, E. (1976). Corpuscular optics and the wave theory of light: the science and politics of a revolution in physics. *Social Studies of Science,* 6, 141–84.

Franklin, A. (1986). *The Neglect of Experiment.* Cambridge: Cambridge University Press.

Franklin, A. & Howson, C. (1984). Why do scientists prefer to vary their experiments?. *Studies in History and Philosophy of Science,* 15, 51–62.

Galison, P. (1983). How the first neutral current experiments ended. *Reviews of Modern Physics,* 55, 477–509.

Galison, P. (1985). Bubble chambers and the experimental workplace. In ed. P. Achinstein & O. Hannaway, pp. 309–73.

Galison, P. (1987). *How Experiments End*, Chicago: University of Chicago Press.

Galison, P. (1988). *Image and Logic*. Chicago: University of Chicago Press. (in press).

Geison, G.L. & Secord, J.A. (1987). Pasteur and the process of discovery – the case of optical isomerism. *Isis*, **78**.

Gilbert, N. & Mulkay, M. (1984). Experiments are the key: participants' histories and historians' history of science. *Isis*, **75**, 105–25.

Gooding, D.C. (1980). Metaphysics versus measurement: the conversion and conservation of force in Faraday's physics. *Annals of Science*, **37**, 1–29.

Gooding, D.C. (1981). Final steps to the field theory: Faraday's study of magnetic phenomena, 1845–1850. *Historical Studies in the Physical Sciences*, **11**, 231–75.

Gooding, D.C. (1982a). Empiricism in practice: teleology, economy and observation in Faraday's physics. *Isis*, **73**, 46–67.

Gooding, D.C. (1982b). A convergence of opinion on the divergence of lines. *Notes and Records of the Royal Society*, **36**, 243–59.

Gooding, D.C. (1985a). In nature's school: Faraday as an experimentalist. In ed. D. Gooding and F. James pp. 105–35.

Gooding, D.C. (1985b) Experiment and concept formation in electromagnetic science and technology in England in the 1820s. *History and Technology*, **2**, 151–76

Gooding, D.C. (1986). How do scientists reach agreement about novel observations?. *Studies in the History and Philosophy of Science*, **17**, 205–30.

Gooding, D.C. (1989). *The Making of Meaning*. Dordrecht: Martinus Nijhoff. (in press).

Gooding, D.C. & James, F.A.J.L. (ed). (1985). *Faraday Rediscovered. Essays on the Life and Work of Michael Faraday. 1791–1867.* London: Macmillan; New York: Stockton Press.

Hacking, I (1983). *Representing and Intervening*. Cambridge: Cambridge University Press.

Hackmann, W.D. (1978). *Electricity from Glass: the History of the Frictional Electrical Machine 1600–1850*. Alphen aan den Rijn: Sijthoff and Noordhoff.

Hackmann, W.D. (1979). The relationship between concept and instrument design in eighteenth-century experimental science. *Annals of Science*, **36**, 205–24.

Hackmann, W.D. (1985). Instrumentation in the theory and practice of science: scientific instruments as evidence and as an aid to discovery. *Annali dell'Istituto e Museo di Storia della Scienza di Firenza*, **10(2)**, 87–115.

Harré, R. (1980). Knowledge. In *The Ferment of Knowledge: Studies in the Historiography of Eighteenth-century Science*, (ed.) G.S. Rousseau & R.S. Porter, pp. 11–54. Cambridge: Cambridge University Press.

Harré, R. (1981). *Great Scientific Experiments*. Oxford: Phaidon.

Heilbron, J.L. (1979). *Electricity in the Seventeenth and Eighteenth Centuries*. Berkeley: University of California Press.

Heilbron, J.L. (1980). Experimental naural philosophy. In *The Ferment of Knowledge: Studies in the Historiography of Eighteenth-century Science*, (ed) G.S. Rousseau & R.S. Porter, pp. 357–87. Cambridge: Cambridge University Press.

Heilbron, J.L. (1981). The electrical field before Faraday. In *Conceptions of Ether. Studies in the History of Ether, 1740–1900*, (ed.) G.N. Cantor & M.J.S. Hodge, pp. 187–213. Cambridge: Cambridge University Press.

Hofmann, J.R. (1987a). Ampère's invention of equilibrium apparatus: a response to experimental anomaly. *British Journal for the History of Science*, **20**, 309–41.

Hofmann, J.R. (1987b). Ampère, electrodynamics and experimental evidence. *Osiris*, **3**, 45–76.

Holton, G,. (1978). Subelectrons, presuppositions and the Millikan – Ehrenhaft dispute. In *The Scientific Imagination: Case Studies*, ed. G. Holton, pp. 25–83. Cambridge: Cambridge University Press.

Home, R.W. (1985). The notion of experimental physics in early eighteenth-century France. In *Charge and Progress in Modern Science*, ed. J.C. Pitt, pp. 107–32. Dordrecht and Boston: Reidel.

James, F.A.J.L. (1985). 'The optical mode of investigation': light and matter in Faraday's natural philosophy. In ed. D.C. Gooding & F. James, pp. 137–62.

James, W. (1978). *Pragmatism and the Meaning of Truth*, Cambridge, Mass. and London: Harvard University Press.

Knorr-Cetina, K. (1981). *The Manufacture of Knowledge: An Essay on the Constructivist and Contextual Nature of Science*. Oxford and New York: Pergamon.

Knorr-Cetina, K. & Mulkay, M. (ed). (1983). *Science Observed: Perspectives on the Social Study of Science*. London and Beverley Hills: Sage.

Krige, J. & Pestre, D. (1986). The choice of CERN's first large bubble chambers for the proton synchrotron (1957–1958). *Historical Studies in the Physical and Biological Sciences*, **16**, 255–78.

Kuhn, T.S. (1961). The function of measurement in modern physical science. *Isis*, **52**, 161–90; reprinted in T.S. Kuhn (1977).

Kuhn, T.S. (1962). A function for thought experiments. In *L'Aventure de la Science: Mélanges Alexandre Koyré*, vol. 2, pp. 307–34. Paris: Hermann; reprinted in T.S. Kuhn (1977).

Kuhn, T.S. (1976). Mathematical versus experimental traditions in the development of physical science. *Journal of Interdisciplinary History*, **7**, 1–31; reprinted in T.S. Kuhn (1977).

Kuhn, T.S. (1977). *The Essential Tension*. Chicago and London: University of Chicago Press.

Latour, B. (1983). Give me a laboratory and I will raise the world. In ed. K. Knorr-Cetina & M. Mulkay, pp. 141–70.

Latour, B. (1986a). Visualisation and cognition: thinking with eyes and hands. *Knowledge and Society*, **6**, 1–40.

Latour, B. (1986b). *Science in Action*, Milton Keynes: Open University Press; Cambridge, Mass.: Harvard University Press.

Latour, B. & Woolgar, S. (1979). *Laboratory Life*, 1st edn. Beverly Hills and London: Sage; revised edn. (1986). Princeton University Press.

Laymon, R. (1978). Newton's *experimentum crucis* and the logic of idealization and theory refutation. *Archive for History of Exact Sciences*, 1, 389–405.

Law, J. (ed) (1986). *Power, Action and Belief: a New Sociology of Knowledge?* Sociology Review Monograph 32. London: Routledge and Kegan Paul.

Lenoir, T. (1986). Models and instruments in the development of electrophysiology. *Historical Studies in the Physical and Biological Sciences*, 17, 1–54.

Lohne, J.A. (1968). Experimentum crucis. *Notes and Records of the Royal Society*, 23, 169–99.

Lynch, M. (1985a). *Art and Artifact in Laboratory Science: a Study of Shop Work in a Research Laboratory*. London: Routledge.

Lynch, M. (1985b). Discipline and the material form of scientific images: an analysis of scientific visibility. *Social Studies of Science*, 15, 37–66.

Medawar, P. (1964). Is the scientific paper a fraud?. *Saturday Review*, (August 1), 43–4.

Musgrave, A. (1974). Logical versus historical theories of confirmation. *British Journal for the Philosophy of Science*, 25, 1–23.

Naylor, R.H. (1974a). Galileo and the problem of free fall. *British Journal for the History of Science*, 7, 105–34.

Naylor, R.H. (1974b). The evolution of an experiment: Guidobaldo del Monte and Galileo's *Discorsi* demonstration of the parabolic trajectory. *Physis*, 16, 323–46.

Naylor, R.H. (1976). Galileo: the search for the parabolic trajectory. *Annals of Science*, 33, 153–72.

Naylor, R.H. (1980). The role of experiment in Galileo's work on the law of fall. *Annals of Science*, 37, 363–78.

Nickles, T. (1985). Beyond divorce: current states of the discovery debate. *Philosophy of Science*, 52, 177–207.

Nickles, T. (1987). The reconstruction of scientific knowledge. *Philosophy and Social Action*, 13, 91–104.

Nickles, T. (1988). Reconstructing science: discovery and experiment. In *Theory and Experiment*, ed. D. Batens & J.P. van Bendegem, pp. 33–53. Dordrecht: Reidel.

North, J.D. (1981). Science and analogy. In *On Scientific Discoveries*, (ed.) M.D. Grmek, R.S. Cohen & G. Cimino, pp. 115–40. Dordrecht and Boston: Reidel.

Nye, M.J. (1980). N-Rays: an episode in the history and psychology of science. *Historical Studies in the Physical Sciences*, 11, 125–56.

Pestre, D. (1988). Comment se prennent les décisions de très gros équipements dans les laboratoires de 'science lourde' contemporains: un récit suivi de commentaire. *Revue de synthèse*. (in press).

Pickering, A. (1981a). The hunting of the quark. *Isis*, 72, 216–36.

Pickering, A. (1981b). Constraints on controversy: the case of the magnetic monopole. in ed. H.M. Collins, pp. 63–93.

Pickering, A. (1984a). Against putting the phenomena first: the discovery of the weak neutral current. *Studies in History and Philosophy of Science*, 15, 85–117.

Pickering, A. (1984b). *Constructing Quarks: a Sociological History of Particle Physics.* Edinburgh: Edinburgh University Press; Chicago: University of Chicago Press.

Pickering, A. (1987). Forms of life: science, contingency and Harry Collins. *British Journal for the History of Science*, 20, 213–21.

Pinch, T.J. (1977). What does a proof do if it does not prove?. In *The Social Production of Scientific Knowledge*, ed. E. Mendelsohn, P. Weingart & R. Whitley, pp. 171–215. Dordrecht: Reidel.

Pinch, T.J. (1981). The sun-set: the presentation of certainty in scientific life. In ed. H.M. Collins, pp. 131–56.

Pinch, T.J. (1985a). Towards an analysis of scientific observation: the externality and evidential significance of observation reports in physics. *Social Studies of Science*, 15, 3–35.

Pinch, T.J. (1985b). Theory testing in science – the case of solar neutrinos: do crucial experiments test theories or theorists?. *Philosophy of the Social Sciences*, 15, 167–87.

Pinch, T.J. (1986). *Confronting Nature: the Sociology of Solar Neutrino Detection.* Dordrecht: Reidel.

Price, D.J. de Solla (1980). Philosophical mechanism and mechanical philosophy: some notes towards a philosophy of scientific instruments. *Annali dell'Istituto e Museo di Storia della Scienza di Firenze*, 5, 75–85.

Price, D.J. de Solla. (1984). Of sealing wax and string. *Natural History*, 93, 48–56.

Rudwick, M.J.S. (1985). *The Great Devonian Controversy: the Shaping of Scientific Knowledge among Gentlemanly Specialists.* Chicago: University of Chicago Press.

Schaffer, S. (1980). Natural philosophy. In *The Ferment of Knowledge: Studies in the Historiography of Eighteenth-century Science*, ed. G.S. Rousseau & R.S. Porter, pp. 55–80. Cambridge: Cambridge University Press.

Schuster, J.A. & Yeo, R.R. (ed.) (1986). *The Politics and Rhetoric of Scientific Method: Historical Studies.* Dordrecht: Reidel.

Settle, T. (1961). An experiment in the history of science. *Science*, 133, 19–23.

Shapere, D. (1982). The concept of observation in science and philosophy. *Philosophy of Science*, 49, 485–525.

Shapin, S. (1982). History of science and its sociological reconstructions. *History of Science*, 20, 157–211.

Shapin, S. (1984). Pump and circumstance: Robert Boyle's literary technology. *Social Studies of Science*, 14, 481–520.

Shapin, S. (1988). The house of experiment in seventeenth-century England. *Isis*, **79**, in press.

Shapin, S. & Schaffer, S. (1985). *Leviathan and the Air Pump: Hobbes, Boyle and the Experimental Life.* Princeton: Princeton University Press.

Shapiro, A.. (1979). Newton's 'achromatic dispersion law': heoretical background and experimental evidence. *Archive for History of Exact Sciences*, 21, 91–128.

Shelley, M. (1831). *Frankenstein, or, the Modern Prometheus*, (ed.) M. Hindle. Harmondsworth: Penguin Books (1985).

Stuewer, R. (1975). *The Compton Effect.* New York: Science History Publications.

Swenson, L. (1972). *The Ethereal Aether: A History of the Michelson – Morley –*

Tweney, R. (1985). Faraday's discovery of induction: a cognitive approach. In ed. D.C. Gooding & F. James, pp. 189–209.

Van Helden, A. (1983). The birth of the modern scientific instrument, 1550–1700. In *The Uses of Science in the Age of Newton* ed. J.G. Burke, pp. 49–84. Berkeley: University of California Press.

Vincenti, W.G. (1979). The air propeller tests of W.F. Durand and E.P. Lesley: a case study in technological methodology. *Technology and Culture*, 20, 712–51.

Wallis, R. (ed.). (1979). *On the Margins of Science: the Social Construction of Rejected Knowledge*. Sociological Review Monograph 27. Keele: University of Keele.

Williams, L.P. (1983). What were Ampère's earliest discoveries in electrodynamics?. *Isis*, 74, 292–308.

Williams, L.P. (1985). Faraday and Ampère: a critical dialogue. In ed D.C. Gooding & F. James, pp. 83–104.

Williams, L.P. (1986). Why Ampère did not discover electromagnetic induction. *American Journal of Physics*, 54, 306–11.

Wilson, D.B. (1982). Experimentalists among the mathematicians: physics in the Cambridge Natural Sciences Tripos, 1851–1900. *Historical Studies in the Physical Sciences*, 12, 325–71.

Wise, N. & Smith, C. (1988). *Energy and Empire*. Cambridge: Cambridge University Press. (in press).

Worrall, J. (1976). Thomas Young and the 'refutation' of Newtonian optics: a case study in the interaction of philosophy of science and history of science. In *Method and Appraisal in the Physical Sciences*, ed. C. Howson, pp. 107–80. Cambridge: Cambridge University Press.

Worrall, J. (1982). The pressure of light: the strange case of the vacillating 'crucial experiment'. *Studies in the History and Philosophy of Science*, 12, 133–71.

1993 ADDENDA

Dear P. (ed.) 1991. *The Literary Structure of Scientific Arguments: Historical Studies*. Philadelphia: University of Pennsylvania Press.

Gooding, D. 1990. *Experiment and the Making of Meaning*. Dordrecht: Kluwer.

Holmes, F.L. 1992. *Hans Krebs: the Formation of a Scientific Life*. Oxford: Oxford University Press.

James, F.A.J.L. (ed.) 1989. *The Development of the Laboratory: Essays on the Place of Experiment in Industrial Civilization*. London: Macmillan.

Le Grand, H. (ed.) 1990. *Experimental Inquiries: Historical Philosophical and Social Studies of Experimentation in Science*. Dordrecht: Kluwer.

Lenoir, T. and Elkana, Y. (eds.) 1988. 'Practice, Context and the Dialogue between Theory and Experiment', *Science in Context*, 2 (1).

Lynch, M. and Woolgar, S. (eds.) 1990. *Representation in Scientific Practice*. Cambridge, MA.: M.I.T. Press.

Pickering, A. (ed.) 1992. *Science as Practice and Culture*. Chicago and London: University of Chicago Press.

NAME INDEX

474

SUBJECT INDEX

Academy of Sciences, Paris 92n53, 135–46,
178, 339, 356
accelerator 16–17, 387ff, 393–8, 443; size
of 395–402
acceptance of theories 135ff, 187ff
accounts of experiment xiv, 12;
by scientists 309–10; dialogue form
120ff; of experiments 6–13, 159–61,
166ff, 283ff, 301ff, 309–13;
see also experimental discourse
accuracy;
definition of 410, 425–27; of
experiment: see hallmarks
Acta eruditorum (Leipzig) 94, 97
action;
at a distance 209; in experiment 191ff,
215ff, 275
Adelaide Gallery, London 362
Admiralty 207
ad hoc hypothesis 155, 255;
and knowledge 208
agency xvi, 217
agreement see consensus
Air Force, US 425, 427–9
air pump 32, 43, 363
American Physical Society 449
American War of Independence 55
analogy 45ff, 206, 210
Annales de chimie et de physique 138–9, 145
Annual Register 351, 363
apparatus:
heroic 32, 110ff; operation of 449ff;
theory of 439ff; see also instruments
appraisal of theories 146ff
argument 5ff;
and discovery 306ff; audience of 314ff;

dialogue form of 167ff; in experiment
59–60, 70–1, 124ff, 154, 299ff, 438ff,
455ff; observational 129–31;
rationality of 458–9
art of seeing 109–11, 261
artefacts, human 411ff;
see also experiment, instrument,
phenomena, testing
articulation 9, 165, 188, 192–3, 214–16
artisans 34ff
Ashmolean Society, Oxford 339, 361
Athanaeum 339, 346, 353
Atlas 363

barometer 105ff
Bayesians 304n–11
belief;
in experimenters 18; in observers
110–12; rationality of 22, 437ff
Ben Nevis Observatory 226, 233, 238–40,
243
Berkeley Bevatron 387, 396
Berlin, Royal Academy of Sciences 51
big science 17ff, 399
Birmingham, accelerator at 389
Birr Castle 112–13
black boxing 5, 217, 310–11
black spot 145
Board of Trade, Meteorological Department
237n49
Bologna 96–8
Book of Nature 6, 173, 215
Bradford Grammar School 240
Bristol Advocate 346
Bristol Institution 346